Voice Over IPv6

Voice Over IPv6

Architectures for
Next Generation VoIP Networks

by Daniel Minoli

AMSTERDAM • BOSTON • HEIDELBERG • LONDON
NEW YORK • OXFORD • PARIS • SAN DIEGO
SAN FRANCISCO • SINGAPORE • SYDNEY • TOKYO

Newnes is an imprint of Elsevier

Newnes

ELSEVIER

Newnes is an imprint of Elsevier
30 Corporate Drive, Suite 400, Burlington, MA 01803, USA
Linacre House, Jordan Hill, Oxford OX2 8DP, UK

 Recognizing the importance of preserving what has been written,
Elsevier prints its books on acid-free paper whenever possible.

Library of Congress Cataloging-in-Publication Data

Minoli, Daniel, 1952-
  Voice over IPv6 : architectures for next generation VoIP networks / Daniel Minoli.
    p. cm.
  ISBN-13: 978-0-7506-8206-0 (pbk. : alk. paper)
  ISBN-10: 0-7506-8206-X (pbk. : alk. paper) 1. Internet telephony. I.
Title: Architectures for next generation VoIP networks. II. Title.
  TK5105.8865.M5715 2006
  621.382'12--dc22

                  2006005260

British Library Cataloguing-in-Publication Data
A catalogue record for this book is available from the British Library.

ISBN-10: 0-7506-8206-X
ISBN-13: 978-0-7506-8206-0

For information on all Newnes publications
visit our website at www.books.elsevier.com

06 07 08 09 10 11   10 9 8 7 6 5 4 3 2 1

Printed in the United States of America

For Anna

Contents

Preface xi

Acknowledgments xiii

Chapter 1: Introduction 1
    1.1   Overview 1
    1.2   Introductory Overview of IPv6 4
    1.3   Introductory Overview of VoIP.. 17
    1.4   Third-Generation 3G VoIP Networks 28
    1.5   Deployment/Penetration Issues 29
    1.6   Line of Investigation... 31
    Appendix A: Basic IPv6 Terminology..................................... 32
    Appendix B: Basic Bibliography 41

Chapter 2: Basic VoP/VoIP Concepts 49
    2.1   Introduction and Background... 49
    2.2   Voice Digitization and Encoding 54
    2.3   Signaling 78
    2.4   Numbering 91
    2.5   VoIP and Wireless Networks 94
    2.6   Conclusion 106

Chapter 3: Basic VoIP Signaling and SIP Concepts 107
    3.1   Introduction.. 107
    3.2   Overview 108
    3.3   Fundamental SIP Functionality... 108
    3.4   Overview of Operation.. 109
    3.5   Structure of the Protocol.. 114
    3.6   SIP Details.. 115
    Appendix A 115
    A.1   Definitions.. 115
    A.2   SIP Messages 119
    A.3   General User Agent Behavior 123
    A.4   Canceling a Request 134
    A.5   Registrations... 136

Contents

A.6   Querying for Capabilities 140
A.7   Dialogs 142
A.8   Initiating a Session 147
A.9   Modifying an Existing Session 152
A.10 Terminating a Session 153
A.11 Proxy Behavior 155
A.12 Transactions 160
A.13 Transport 162
A.14 Additional Details 165

Chapter 4: Basic "Presence" Concepts 167
4.1   Introduction 167
4.2   Abstract Model for a Presence and Instant Messaging 168
4.3   Instant Messaging/Presence Protocol Requirements 177
4.4   SIP Applications 181
4.5   Conclusion 191

Chapter 5: Issues with Current VoIP Technologies 193
5.1   General Enterprise Security Issues 193
5.2   What is NAT? 206
5.3   STUN—Simple Traversal of User Datagram Protocol (UDP) Through Network Address Translators (NATs) 215
5.4   Overview of MIDCOM Approaches 235
5.5   Pragmatic Approaches using SIP Border Gateways 253

Chapter 6: Basic IPv6 Concepts 259
6.1   Introduction 259
6.2   Terminology 259
6.3   IPv6 Header Format 260
6.4   IPv6 Extension Headers 261
6.5   Packet Size Issues 273
6.6   Flow Labels 273
6.7   Traffic Classes 274
6.8   Upper-Layer Protocol Issues 274
6.9   Semantics and Usage of the Flow Label Field 276
6.10 Formatting Guidelines for Options 277
6.11 Introduction to Addressing 279
6.12 IPv6 Addressing 279
6.13 IANA Considerations 289
6.14 Creating Modified EUI-64 format Interface Identifiers 290
6.15 64-Bit Global Identifier (EUI-64) Registration Authority 291
6.16 Additional Technical Details 292

Chapter 7: Using IPv6 to Support 3G VoIP **293**

   7.1   Overview of VoIPv6 Positioning 293

   7.2   IPv6 Infrastructure... 297

   7.3   IPv6 Addressing Mechanisms 301

   7.4   Configuration Methods 313

   7.5   Routing and Route Management... 315

   7.6   Deployment Status 317

Chapter 8: Issues Related to Transitioning to IPv6 .... **323**

   8.1   Introduction.. 323

   8.2   Dual-IP Layer Operation 325

   8.3   Common Tunneling Mechanisms 327

   8.4   Configured Tunneling 333

   8.5   Automatic Tunneling... 334

   8.6   Application Aspects of IPv6 Transition 337

References **343**

About the Author **349**

Index **351**

Preface

Voice over IP (VoIP) in particular and Voice over Packet (VoP) in general have been advocated and studied since the mid 1970s. It was the advent of DSP technology for voice compression in the late 1980s and early 1990s that gave these services the impetus they needed to enter the mainstream. Commercial-grade technologies and services started to appear in the 1996-7 timeframe and books on the topic started to appear in 1998, with Mr. Minoli's co-authored Delivering Voice over IP book (Wiley, April 1, 1998) being the first text on the market on this topic. A lot has transpired since then. Now, enterprise networks, cellular carriers, voice-over-cable carriers, "triple-play" carriers, "pure-play VoIP carriers," and even traditional voice carriers are all moving rather aggressively to a VoIP paradigm.

A fair degree of commercial success can be acknowledged as of 2006. Small-to-medium size enterprises are using the technology to save money on trunking costs. Large-size enterprises are using the technology for mobility support and related functional enhancements, including "presence-related functions" and unified messaging. Large-size companies are also using this technology for Contact Center support, particularly for hosted ACD-capabilities and for virtual Contact Centers (where agents are distributed throughout al large geographic area.) Carriers are deploying these services to generate new revenues, stem the movement away from traditional TDM services, and enter new markets (e.g., "triple-play" applications.)

However, there are two fundamental problems that currently pose a significant risk to the scalability of VoIP to a large-population base along with guaranteed "industrial-grade" service levels. The first problem is lack of de-facto intrinsic QoS in many of the IP networks deployed around the globe (both at the carrier level and at the enterprise level.) The second problem relates to end-to-end integrity of the signaling and bearer path for VoIP, specifically the fact that VoIP packets have a "difficult" time being carrier across firewalls, not only because of protocol considerations, but, at the practical level, because of the Network Address Translation (NAT) issues. Additionally, one can also cite overall security concerns as another potentially problematic issue.

What were the "bragging rights" of the large national voice carriers across the world (particularly under the auspices of the ITU-T) were the assertions that, in true fact, "...(with TDM) any one in the world could call anyone else in the world, any time, any where, and be able to support a good-quality telephonic conversation..." That is an elusive, currently-unachievable goal for the VoIP industry...

This book is a vade mecum on IPv6 opportunities for carrier-class VoIP. Specifically, it looks at the use of IPv6 to support a next-generation, carrier-class VoIP environments, which we call VoIPv6. IPv6 offers the potential of achieving the scalability, reacheability, end-to-end interworking, QoS, and commercial-grade robustness for VoIP that are mandatory mileposts of the technology if indeed it is to replace the TDM infrastructure around the world, in the true sense of the word sens the marketing hype... Specifically, IPv6 can be employed to addresses the QoS and NAT issues. While some technologists may argue that "there is nothing one can do about QoS with IPv6 that one cannot do with IPv4," or "NATs will not disappear just because

IPv6 has a larger native address space because there are a number of reasons for isolating a private address domain behind a firewall that are unrelated with the scarcity of IPv4 addresses," such arguments exhibit some naiveté: one can cite literally dozens of examples where ostensibly the claim that "there is nothing that technology $n + 1$ can do that technology n can't do" has been made in the past 40 years, yet, the technology marches on inexorably to new approaches and new modalities. Specifically, in this context, the question we believe is operative is not if "IPv4 will be replaced," but simply "when it will be replaced". The position of this textbook is not that there are certain things that cannot be done one way or another with IPv4, just that, in our opinion, the most optimal way to build next-generation VoIP systems that have the scope, reach, ubiquity, reliability, and robustness of the current commercial PSTN is with IPv6 – other approaches may be possible, but may not be as optimal.

This book is the first book of its kind to address this issue as a macro-level scalability requirement. The book basically is comprised of two sections: an opening section (Chapters 1-5) that looks at VoIP applications and motivations, and the second part (Chapters 6-8) focuses on the IPv6 itself.

After an introduction in Chapter 1, we provide a quick tutorial in Chapter 2 on VoIP and in Chapter 3 on SIP and signaling. Chapter 4 continues that basic discussion by examining the area of "presence," which represents a value-added set of VoIP-based capabilities. Chapter 5 discusses the issues associated with current VoIP implementations, as highlighted above.

Chapter 6 provides a basic introduction to IPv6. Chapter 7 discusses the delivery of voice-over-packet in an IPv6 environment. Chapter 8 provides some basic discussion of the transition issues including interworking between IPv6 and IPv4. Europe and Asia are currently leading in the planning of IPv6 networks.

In general, telecom areas of (industry) growth in the short-to-medium term (say four-to-seven years) include VoIP, wireless, security, IPv6, Grid Computing, Ubiquitous Computing, and nanotechnology (nanoelectronics, nanophotonics, and quantum computing). This book covers several of these themes.

This book should prove useful to VoIP equipment vendors, VoIP service providers (e.g., cellular carriers, voice-over-cable carriers, "triple-play" carriers, "pure-play VoIP carriers," and even traditional voice carriers), enterprise customers, researchers, planners, and educators. Portions of this book are based on particularly-selected RFC material: given the emerging nature of the topic and the status of the 3G service proposals, this contextual assembly of the information should be useful and should read as logically well organized. We explicitly credit the developers of these RFCs as the "intellectual" trust behind the material. This derivative assembly of material comments on or otherwise explains selected RFC material, and intend to assist in their implementation/understanding/dissemination. The synthesis provided in this text is much more than an anthology of relevant specs or documents – but recognizing that the IETF has created an extremely valuable and very lucid exposition on these exact issues, we find that said materials provide an excellent foundation for a study of VoIPv6; hence, their synthesis herewith. The book in not principally designed to be a pure pedagogical introduction to VoIP itself, however, a fair amount of basic material in included to make the book relatively self-contained; for basic introductions to VoIP there are several textbooks, including, but not limited to, this author's market-first books listed in the Reference section; the principal focus here is on the IPv6 perspective.

There is an extensive body of research literature on the topic of SIP, IPv6, NAT, and related issues (as noted in the appendix of Chapter 1). This text is intended only as a first-read on the topic. The bibliography in Chapter 1 is an excellent source of useful additional information.

Acknowledgments

The following researchers are amply acknowledged for the contributions made to the topics discussed in this book, which were sources for the present treatment, analysis, and discussion:

S. Deering, R. Hinden; R. Gilligan, E. Nordmark, M. Day, J. Rosenberg, H. Sugano; J. Weinberger, C. Huitema, R. Mahy; H. Schulzrinne, G. Camarillo, A. Peterson, R. Sparks, M. Handley, E. Schooler; P. Srisuresh, J. Kuthan, A. Molitor, and A. Rayhan.

Introduction

1.1 Overview

Voice over Internet Protocol (VoIP), in particular, and Voice over Packet (VoP), in general, have been advocated and studied since the late-1970s. The reality is, however, that until the late 1990s, voice- and video-over-packet networks have been mostly used for pre-stored kinds of media solutions, namely, for *one-way download* of sound files or video files that were played in nonreal time at the user's personal computer, or at most, for a simplex transmission path with a relatively large intrinsic delay, such as Internet radio. At this juncture, however, there is a discernible movement afoot in the industry to affect the transition to a converged, fully multimedia-enabled, real-time packet-based communication infrastructure for both enterprise networks and for carriers' network environments in support of commercial-grade real-time voice, commercial-grade video, and commercial-grade Video-On-Demand (VOD) services. These converged networks will allow voice, video, data, and images to be delivered anywhere in the world, at any time, and with any kind of user's communication device and network access service [MIN199801], [MIN199802], [MIN200201], [MIN200202].

It was the advent of Digital Signal Processing (DSP) technology for voice and video compression in the late 1980s and early 1990s that gave multimedia-over-packet services the impetus they needed to enter the mainstream. Commercial-grade VoIP technologies and services started to appear in the early 2000s in First-Generation (1G) networks. A lot has transpired since then; enterprise networks, cellular carriers, Voice-over-Cable (VoCable) carriers, "triple-play" carriers, "pure-play VoIP carriers," and even traditional voice carriers are all moving rather aggressively to a VoIP paradigm. This is a Second-Generation (2G) technology. Next generation third-wave (3G) technology is just 2–3 years away.

A fair degree of commercial success can be acknowledged with regard to 2G VoIP networks as of press time. Small-to-medium sized enterprises are using the technology to save money on trunking costs. Large-size enterprises are using the technology for mobility support and related functional enhancements, including "presence-related functions" and unified messaging. Roaming from Ethernet-based phones to cellular service is beginning to be supported. Large-sized companies are also using this technology for contact center support, particularly for hosted Automatic Call Distribution (ACD)-capabilities and for virtual contact centers (where agents are distributed throughout a large geographic area). Carriers are deploying these services to generate new revenues, replace (substitute) the revenue movement away from traditional Time Division Multiplexing (TDM) services and, enter new markets (for example, "triple-play" applications.)

Despite recent successes, there are two fundamental problems that currently pose a significant risk to the unconstrained scalability of VoIP to a large-population base along with guaranteed "industrial-grade" service levels. The first problem is lack of de facto intrinsic Quality of Service (QoS) in many of the IP networks deployed around the globe (both at the carrier level and at the enterprise level). The second problem relates to end-to-end integrity of the signaling and bearer path for VoIP, specifically the fact that VoIP packets have a "difficult" time being carried across firewalls, not only because of protocol considerations, but, at the practical level, because of Network Address Translation (NAT) issues. Additionally, one can also cite overall security concerns, such as eavesdropping and hacking as another potentially problematic issue. Next-generation 3G VoIP networks are now on the drawing board to address these issues, specifically, scalability and commercial-grade reliability; these networks are based on IPv6.

The "bragging rights" of the large international voice carriers across the world (particularly under the auspices of the International Telecommunication Union [ITU]) were the assertions that, in true fact, "...(with TDM) anyone in the world could call anyone else in the world, any time, any where, and be able to support a good-quality telephonic conversation ..." Frankly, this is an elusive, currently-unachievable goal for the VoIP industry. IPv6 offers the potential of achieving the scalability, reachability, end-to-end interworking, QoS, and commercial-grade robustness levels for VoIP that are mandatory mileposts of the technology if it is indeed to eventually replace the TDM infrastructure around the world, in the true sense of the word sans the marketing hype. Specifically, IPv6 deals with the QoS and NAT issues.

This is the first book of its kind to address the issue of macro-level scalability requirements and IPv6's opportunity in this context.

IPv6 is considered to be the next-generation Internet protocol [HUI199701], [HAG200201], [MUR200501], [SOL200401], [ITO200401], [MIL199701], [MIL200001], [GRA200001], [DAV200201], [LOS200301], [LEE200501], [GON199801], [DEM200301], [GOS200301], and [WEG199901]. The current version of IPv4 has been in use for almost thirty years and exhibits (some) challenges in supporting emerging demands for address space cardinality, high-density mobility, and strong security—this is particularly true in developing domestic and defense department applications utilizing peer-to-peer networking. IPv6 is an improved version of the Internet protocol that is designed to coexist with IPv4 and eventually provide better overall internetworking capabilities than IPv4 [IPV200401]. Proponents see applications in a whole gamut of environments, including VoIP, third-generation Wi-Fi, and security [ISL200501].

IPv6 was initially developed in the early 1990s because of the anticipated need for more addresses based on forecasted Internet growth; that is, cell phone deployment, PDA introduction, smart appliances, and billions of new users in developing countries (e.g., in China, India, and so on.). New technologies such as VoIP, always-on Internet access (for example, DSL and cable), Ethernet to the home, and evolving ubiquitous computing applications are driving and/or will be driving this need even more in the next few years.

NAT-based accommodation is but a short-term solution against this anticipated growth phenomenon and a better solution is needed. Basic Network Address Translation (Basic NAT) is a method in which IP addresses are mapped from one group to another, transparent to end users. Specifically, private "nonregistered" addresses are mapped to a small set (as small as 1) of legal addresses. This, however, impacts the general addressability, accessibility, even "individuality" of the device. Network Address Port Translation (NAPT) is a method by which many network addresses and their Transmission Control Protocol/User Datagram Protocol (TCP/UDP) ports are translated into a single network address and its TCP/UDP ports. Together, these two operations, referred to as traditional NAT, provide a mechanism utilized in IPv4 to connect a realm with private addresses to an external realm with globally-unique registered addresses [SRI200101].

There is a recognition that NAT techniques makes the Internet, the applications, and even the devices more complex, which in turn implies a cost overhead [IPV200501]. The expectation is that IPv6 can make every IP device less expensive, more powerful, and even consume less power; the power issue is not only important for environmental reasons, but also improves operability (e.g., longer battery in portable devices, such as cell phones) [IPV200501]. The market has associated IPv6 with "lots of addressing," but planners do not yet realize that it does a lot of other things too [MAL200501]. IPv6 can improve the Internet or a firm's intranet, with benefits such as:

- expanded addressing capabilities;
- serverless autoconfiguration (plug-and-play) and reconfiguration;
- more efficient and robust mobility mechanisms;
- end-to-end security, with built-in, strong IP-layer encryption and authentication;
- streamlined header format and flow identification;
- enhanced support for multicast and QoS and,
- extensibility: improved support for options/extensions.

Corporations and government agencies will be able to achieve a number of improvements with IPv6. While the basic functions of Internet protocols are to move information across networks, IPv6 has more capabilities built into its foundation than IPv4. A key capability—and the reason research began to replace IPv4—is the significant increase in address space. Consumers look for "plug-and-play" simplicity, collaboration (e.g., distributed games), and mobility. IPv6 is a natural convergence protocol for tomorrow's IP-centric world, as depicted in Figure 1.1 [ISL200501].

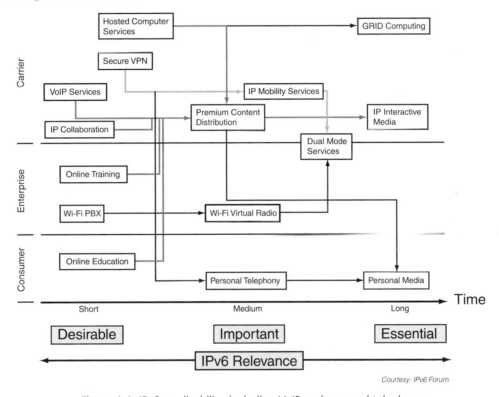

Courtesy: IPv6 Forum

Figure 1.1: IPv6 applicability, including VoIP and personal telephony.

For example, in an IPv6 environment, all equipment can have a "legal"/globally-unique IP address, so that equipment can be uniquely tracked; while today, all Ethernet devices have a unique Media Access Control (MAC) address. Not all devices are Ethernet, hence, the challenge in this respect. Today's inventory management cannot be achieved with IP; it follows that someone has to manually walk around at various times during the inventory cycle to ensure each desktop computer is where it is supposed to be. With IPv6, one can use the network to verify that such equipment is there. Even non-IT equipment in the field can also be tracked, by having an IP address permanently assigned to it.

More network security is embedded in IPv6 than in IPv4. Also, it has extensive automatic configuration (autoconfiguration) mechanisms; with IPv4 one needs to add mechanisms such as Dynamic Host Configuration Protocol (DHCP) and others to make it happen, while IPv6 reduces the IT burden making configuration essentially plug-and-play.

In general, telecom areas of (industry) growth in the short-to-medium term (say four to seven years) include VoIP, wireless, security, IPv6, grid computing, ubiquitous computing, and nanotechnology (nanoelectronics, nanophotonics, and quantum computing). This book covers several of these themes.

IPv6 is now gaining momentum globally, with a lot of interest and activity in Europe and Asia. There is also incipient traction in the U.S., and it's only a matter of time before a transition will need to occur worldwide. For example, the U.S. Dept. of Defense (DoD) announced in 2003 that from October 1, 2003, all new developments and procurements had to be IPv6-capable. The DoD's goal is to complete the transition to IPv6 for all intra- and internetworking across the agency by 2008 [IPV200501]. In 2005, the U.S. Government Accountability Office (GAO) recommended that all agencies become proactive in planning a coherent transition to IPv6 [MAL200501].

VoIP systems, and even more so mobile devices, require end-to-end interworking between entities such as VoIP network elements/gatekeepers/gateways, Session Initiation Protocol (SIP) servers/proxies/registrars, IEEE 802.11 wired and wireless LANs, protocol firewalls, stateful firewalls, application gateways/session border controllers, (IP security protocol) IPSec appliances, Virtual Private Network (VPN) nodes, DHCP servers, DNS servers/proxies, desktop systems and desktop (as well as back-end) applications. Some of these elements do not work as well as one would like in an IPv4 environment. Naturally, these observations apply to pure IP VoIP networks, with end-to-end IP delivery across the enterprise network or across a carrier's (or carriers') network(s). A number of institutions are currently deploying a hybrid approach (particularly where the Private Branch Exchanges (PBX) has been fully depreciated but is still working), which consists of an IP upgrade to serve the "station side," but a traditional TDM switching fabric and TDM trunking on the "network side." Only a fully IP-based system can deliver all of the mobility, productivity, and enhancements that are promised by VoIP. Hence, a transition to the pure IP environment is only a matter of time; in turn, this will raise the NAT, firewall, and scalability issues we are describing.

Some proponents see aggressive inroads for VoIP (while others see a more modest path to penetration.) For example, U.K. market research firm Analysys claimed in 2005 that traditional residential telephone service "will whither away in Western Europe in the next four years"; specifically, mobile service and VoIP will account for 60% of all residential voice service expenditures by 2010 according to the firm [MOB200501]. "The mass market for voice services in Western Europe is being transformed by the substitution of mobile and new VoIP services for traditional fixed-voice services, and we expect that in five years 45% of voice minutes will be made from a mobile or VoIP connection, compared to 28% in 2004." According to Analysys, overall, a quarter of households in Western Europe will switch away from so-called Plain Old Telephony Services (POTS) by 2010 to a combination of mobile and VoIP service, the report concluded. About 45% of all voice minutes by residential customers will be made from one of those two newer types of connections, according to the research firm. By 2010, VoIP over broadband connections could account for 9.6% of all voice minutes. Other observers are more conservative in their estimates. Because VoIP over broadband is inexpensive, the percentage of users' incomes spent on voice service will decrease in coming years. All of this points to the need for VoIP scalability and IPv6 as a subtending technology to support such scalability.

1.2   Introductory Overview of IPv6

IP was designed in the 1970s with the purpose of connecting computers that were in separate geographic locations. Computers in a campus were connected to each other by means of local networks, but these local networks were separated into stand-alone islands. Internet, as a name to designate the protocol, and later the name of the worldwide information network, simply means "internetwork"—that is, a connection between networks. In the beginning, the protocol had an only military use, but computers from universities, users, and enterprises were soon added. Internet as a worldwide information network is the result of the practical application of the IP protocol, that is, the result of the interconnection of a large set of information networks existing in the world [IPV200501]. The current generation of IP protocol is IPv4. Starting in the early 1990s, developers realized that the communication needs of the twenty-first century needed a protocol with some new features and capabilities, while at the same time retaining the useful features of the existing protocol.

While link-level communication does not generally require a node identifier (address) because the remote device is intrinsically identified with the link's remote sink, communication over a group of links (a network) does require unique node identifiers (addresses). The IP address is an identifier that is assigned to each device connected to an IP network. In this setup, different elements taking part in the network (servers, routers, user computers, gateways, etc.) communicate among them using their IP address as the identifier. The issue relates to the question if a device can have its own dedicated, permanent, globally-unique, "registered" IP address, or if it needs to operate only via the use of a temporary nonglobally-unique IP address. Version 4 of the IP protocol addresses consist of four octets. For ease of human conversation, IP protocol addresses are represented as numbers separated by periods, for example: 166.74.110.83, where the decimal number is a shorthand (and corresponds to) the binary code described by the byte in question (an 8-bit number takes a value from 0–255 range). Since the IPv4 address has 32 bits there are nominally 2^{32} different IP addresses (approximately 4 billions nodes, if all combinations are used).

IPv6 is the Internet's next-generation protocol, which was at first called *IPng, Internet Next Generation*. IPv6 is the upgrade of the data networking protocol in which the Internet at large, and nearly all enterprise networks are based on. The Internet Engineering Task Force (IETF) developed the basic specifications during the 1990s to eventually support a migration to a new environment. IPv6 is defined in RFC 2460, "Internet Protocol, Version 6 (IPv6) Specification," S. Deering, R. Hinden (December 1998), which obsoletes RFC 1883 (the "version 5" reference was employed for another use: an experimental real-time streaming protocol. To avoid any confusion, it was decided not to make use of this nomenclature). The IPv6 protocol apparatus is described by the 100+ IETF RFCs identified in Appendix B (some have been obsoleted and/or replaced.)

1.2.1 IPv6 Benefits

IPv4 has demonstrated, by means of its long life, to be a flexible and powerful mechanism. However, IPv4 is starting to exhibit limitations, not only with reference to the need for raw increase of the IP address space, driven, for example, by new populations of users in large countries like China and India, along with new technologies with "always connected devices" (xDSL, cable, PDAs, UMTS mobile telephones, etc), but also in reference of VoIP both in terms of the NAT issue as well as the QoS issue.

IPv6 tackles these problems by creating a new IP address format, so that the number of IP addresses will not exhaust for several decades or longer, even though an entire new crop of devices are expected to connect to the Internet, under the thrust of the coming wave of Ubiquitous Computing (also known as *Pervasive Computing*). IPv6 also adds improvements in areas such as routing and network autoconfiguration. New devices that connect to the Internet will be "plug-and-play" devices. With IPv6, one is not required to configure dynamic nonpublished local IP addresses, the gateway, the subnetwork mask, or any other parameters. The equipment is plugged into the network and it obtains all requisite configuration data [IPV200501].

The advantages of IPv6 can be summarized as follows:

> *Scalability*: IPv6 has 128-bit addresses versus 32-bit IPv4 addresses. With IPv4, the theoretical number of available IP addresses is $2^{32} \sim 10^{10}$. IPv6 offers a 2^{128} space. Hence, the number of available unique node addresses are $2^{128} \sim 10^{40}$.

Security: IPv6 includes security in its specifications such as information encryption and authentication of the source of the communication.

Real-time applications: To provide better support for real-time traffic (e.g., VoIP, IPTV), IPv6 includes in its specifications "labeled flows." By means of this mechanism, routers can recognize the end-to-end flow to which transmitted packets belong to. This is similar to the service offered by Multiprotocol Label Switching (MPLS), but it is intrinsic with the IP mechanism rather than an add-on. Also, it preceded this MPLS feature by a number of yeas.

Plug-and-play: IPv6 includes in its standard a "plug-and-play" mechanism that facilitates the connection of equipment to the network. The requisite configuration is automatically made (as is the case today with Ethernet).

Mobility: IPv6 includes more efficient and enhanced mobility mechanisms.

Optimized protocol: IPv6 embodies IPv4 best practices, but removes unused or obsolete IPv4 characteristics; this results in a better-optimized Internet protocol.

Addressing and routing: IPv6 improves the addressing and routing hierarchy.

Extensibility: IPv6 has been designed to be extensible and offers optimized support for new options and extensions.

1.2.2 Network Address Translation Issues in IPv4

As noted, IPv4 theoretically allows up to 2^{32} addresses, based on a four-octet address space. Legal, globally-unique addresses are assigned by the Internet Assigned Numbers Authority (IANA). IP addresses are addresses of computer nodes at layer 3. Each device on a network (whether the Internet or an intranet) must have a unique address; in IPv4 the layer 3 address is a 32-bit (4-byte) binary address used to identify a host's network ID as well as the host's own ID. As noted, it is represented by the nomenclature a.b.c.d (each of these being from 1 to 255–0 has a special meaning); for example, 167.168.169.170, or 232.233.229.209, or 200.100.200.100, etc. The network portion can contain either a network ID or a network ID and a subnet ID.

The IP address can be from an officially-assigned range, or from an internal (but not globally-unique) block. Internal intranet addresses may be in other ranges, for example, in the 10.0.0.0 or 192.0.0.0 range. In the latter case, a NAT function is employed to map the internal addresses to an external legally-assigned number when the private-to-public network boundary is crossed by a packet. This, however, imposes a number of limitations, particularly since the number of registered external addresses available to a company is almost invariably much smaller (as small as 1) than the number of internal devices requiring an address.

The 32-bit address can be represented as AdrType|netID|hosted. Every network and every host or device has a unique network address, by definition, although such address may not be globally-unique. Figure 1.2 depicts the traditional address classes.

Address Class A. Class A uses the first bit of the 32-bit space (bit 0) to identify it as a Class A address; this bit is set to 0. Bits 1 to 7 represent the network ID, and bits 8 through 31 identify the PC, terminal device, VoIP handset, or host/server on the network. This address space supports $2^7 - 2 = 126$ networks and approximately 16 million devices (2^{24}) on each network. By convention, the use of an "all 1s" or "all 0s" address for both the Network ID and the Host ID is prohibited (which is the reason for subtracting the 2 above.)

Address Class B. Class B uses the first two bits (bit 0 and bit 1) of the 32-bit space to identify it as a Class B address; these bits are set to 10. Bits 2 to 15 represent the network ID, and bits 16 through 31 identify the PC, terminal device, VoIP handset, or host/server on the network. This address space supports $2^{14} - 2 = 16,382$ networks and $2^{12} - 2 = 65,134$ devices on each network.

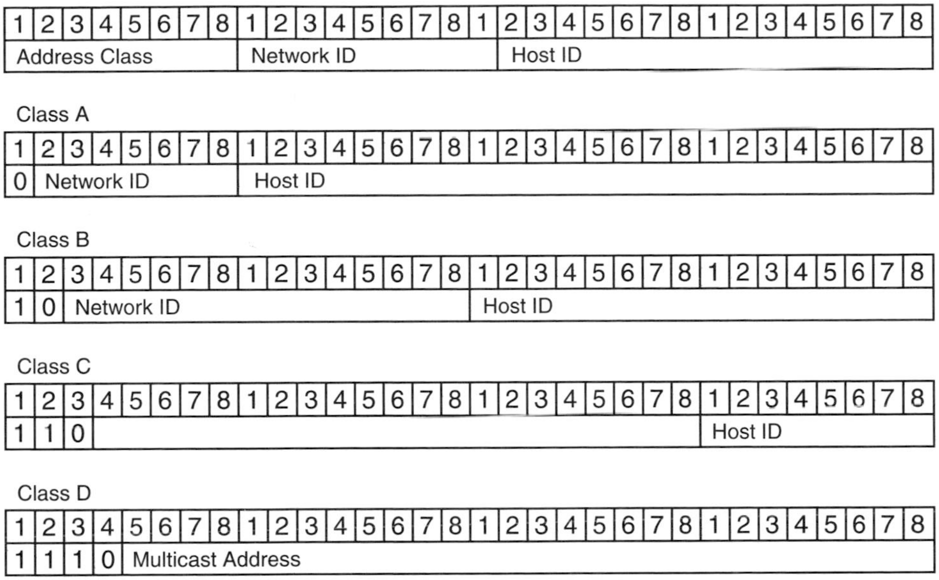

1	2	3	4	5	6	7	8	1	2	3	4	5	6	7	8	1	2	3	4	5	6	7	8	1	2	3	4	5	6	7	8
Address Class								Network ID								Host ID															

Class A

1	2	3	4	5	6	7	8	1	2	3	4	5	6	7	8	1	2	3	4	5	6	7	8	1	2	3	4	5	6	7	8
0	Network ID							Host ID																							

Class B

1	2	3	4	5	6	7	8	1	2	3	4	5	6	7	8	1	2	3	4	5	6	7	8	1	2	3	4	5	6	7	8
1	0	Network ID														Host ID															

Class C

1	2	3	4	5	6	7	8	1	2	3	4	5	6	7	8	1	2	3	4	5	6	7	8	1	2	3	4	5	6	7	8
1	1	0																						Host ID							

Class D

1	2	3	4	5	6	7	8	1	2	3	4	5	6	7	8	1	2	3	4	5	6	7	8	1	2	3	4	5	6	7	8
1	1	1	0	Multicast Address																											

Figure 1.2: Traditional address classes for IP address.

Address Class C. Class C uses the first 3 bits (bit 0, bit 1, and bit 2) of the 32-bit space to identify it as a Class C address; these bits are set to 110. Bits 3 to 23 represent the network ID, and bits 24 through 31 identify the PC, terminal device, VoIP handset, or host/server on the network. This address space supports about 2 million networks ($2^{21} - 2$) and $2^8 - 2 = 254$ devices on each network.

Address Class D. This class is used for broadcasting: multiple devices (all devices on the network) receive the same packet. Class D uses the first 4 bits (bit 0, bit 1, bit 2, and bit 3) of the 32-bit space to identify it as a Class D address; these bits are set to 1110.

Classless Interdomain Routing (CIDR) is yet another mechanism that was developed to help alleviate the problem of exhaustion of IP addresses and growth of routing tables. CIDR is described in RFC 1518, RFC 1519, and RFC 2050. The concept behind CIDR is that blocks of multiple addresses (for example, blocks of Class C addresses) can be combined, or aggregated, to create a larger classless set of IP addresses, with more hosts allowed. Blocks of Class C network numbers are allocated to each network service provider; organizations using the network service provider for Internet connectivity are allocated subsets of the service provider's address space as required. These multiple Class C addresses can then be summarized in routing tables, resulting in fewer route advertisements. CIDR mechanism can be applied to blocks of Class A, B, and C addresses [TEA200401]. All of this assumes, however, that the institution in question already has an assigned set of "legal" IP addresses; it does not address the issue of how to get additional "legal" (registered) globally-unique IP addresses.

During the 1980s, a great quantity of "legal" addresses were allocated to firms and organizations without stringent control; as a result, many organizations have more addresses that they actually need, giving rise to the present dearth of available "registerable" Layer 3 addresses. Furthermore, not all IP addresses can be used due to the fragmentation described above. One approach to the issue would be a renumbering and a reallocation of the IPv4 addressing space. However, this is not as simple as it seems since it requires worldwide coordination efforts. Moreover, it would still be limited for the human population and the quantity of devices that will be connected to Internet in the medium-term future [IPV200501].

At this juncture, and as a temporary and pragmatic approach to alleviating the dearth of addresses, NAT mechanisms are employed by organizations and even home users. This mechanism consists of using only a legal IPv4 address for an entire network to access to Internet. The many internal devices an organization may have are assigned IP addresses from a specifically-designated range of Class A or Class C addresses that are locally-unique, but are duplicatively used and reused within various organizations. Figure 1.3 depicts the NAT arrangement diagrammatically.

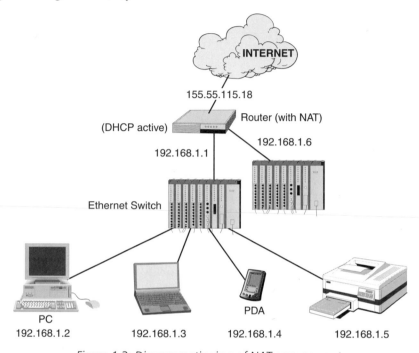

Figure 1.3: Diagrammatic view of NAT arrangement.

Many protocols cannot travel through a NAT device; hence, NAT implies that a number of applications (for example, with VoIP) cannot be used effectively in all instances. As a consequence, these applications can only be used within the scope of the intranet. Examples include the following [IPV200501]:

- Multimedia applications such as videoconference applications, VoIP through the Internet, or video-on-demand/IPTV do not work smoothly through NAT devices. Multimedia applications make use of Real-Time Transport Protocol (RTP) and Real-Time Control Protocol (RTCP); in turn, these use UDP with dynamic allocation of ports (NAT does not directly support this environment).
- Kerberos[1] authentication needs the source address; unfortunately, the source address in the IP header is often modified by NAT devices.
- IPSec is used extensively for data authentication, integrity, and confidentiality. However, when NAT is used, IPSec operation is impacted, since NAT changes the address in the IP header.
- Multicast, although is possible in theory, requires complex configuration in a NAT environment; hence, in practice it is not utilized as often as should be the case.

NAT disappears with IPv6.

---

[1] A network security package developed at the Massachusetts Institute of Technology that depends on passwords and symmetric cryptography (e.g., Data Encryption Standards (DES)) to implement ticket-based, peer entity authentication service and access control service distributed in a client-server network environment.

1.2.3 IPv6 Address Space

As we have already noted, IPv6 provides more IP addresses than does IPv4. The format of IPv6 addressing is described in RFC 2373. We have indicated that an IPv6 address consists of 128 bits, rather than 32 bits of the IPv4 addresses. The number of bits correlates to the address space, is as follows in Table 1.1:

Table 1.1: IPv6 and IPv4 address space.

IP Version	Size of Address Space
IPv6	128 bits, which allows for 2^{128} or 340,282,366,920,938,463,463,374,607, 431,768,211,456 (3.4×10^{38}) possible addresses.
IPv4	32 bits, which allows for 2^{32} or 4,294,967,296 possible addresses.

The relatively large size of the IPv6 address is designed to be subdivided into hierarchical routing domains that reflect the topology of the modern-day Internet. The use of 128 bits provides multiple levels of hierarchy and flexibility in designing hierarchical addressing and routing. The IPv4-based Internet currently lacks this flexibility [MSD200401].

The IPv6 address is represented as eight groups of 16 bits each, separated by the " : " character. Each 16-bit group is represented by four hexadecimal digits (also known as *hex digits*); that is, each digit has a value between 0 and 15 (0, 1, 2, ... A, B, C, D, E, F with A = 10, B = 11, and so on to F = 15). What follows is an IPv6 address (fictitious) example:

3223:0ba0:01e0:d001:0000:0000:d0f0:0010

An abbreviated format exists to designate IPv6 addresses when all endings are 0. For example:

3223:0ba0::

is the abbreviated form of the following address:

3223:0ba0:0000:0000:0000:0000:0000:0000

Similarly, only one 0 is written, removing 0's in the left side, and four 0's in the middle of the address. For example, the address:

3223:ba0:0:0:0:0::1234

is the abbreviated form of the following address:

3223:0ba0:0000:0000:0000:0000:0000:1234

There is also a method to designate groups of IP addresses or subnetworks that is based on specifying the number of bits that designate the subnetwork, beginning from left to right, using remaining bits to designate single devices inside the network. For example, the notation:

3223:0ba0:01a0::/48

indicates that the part of the IP address used to represent the subnetwork has 48 bits. Since each hexadecimal digit has 4 bits, this points out that the part used to represent the subnetwork is formed by 12 hex digits—that is, "3223:0ba0:01a0". The remaining digits of the IP address would be used to represent nodes inside the network.

There are some special IPv6 addresses, as follows:

- **Auto-return or loopback virtual address.** This address is specified in IPv4 with 127.0.0.1 address. In IPv6 this address is represented as ::1.

- **Not specified address (::).** It will never be allocated to any node since it is used to indicate absence of address.

- **IPv6 over IPv4 dynamic/automatic tunnels**. They are designated as IPv4-compatible IPv6 addresses, and allows the sending of IPv6 traffic over IPv4 networks in a transparent manner. They are represented as ::, for example, ::156.55.23.5.

- **IPv4 over IPv6 addresses automatic representation**. They allow for IPv4 only nodes to still work in IPv6 networks. They are designated as "mapped from IPv4 to IPv6 addresses." They are represented as ::FFFF:, for example, ::FFFF.156.55.43.3.

Appendix A provides a basic glossary of IPv6 terms and concepts, all of which will be discussed in detail throughout the text.

1.2.4 Basic Protocol Constructs

Table 1.2 shows the core protocols that comprise IPv6 apparatus [MSD200401] (see Appendix B for a more inclusive listing).

Table 1.2: Key IPv6 protocols.

Protocol	Description
IPv6: RFC 2460	IPv6 is a connectionless, "unreliable" datagram protocol used for routing packets between hosts.
Internet Control Message Protocol for IPv6 (ICMPv6): RFC 2463	Internet Control Message Protocol for IPv6 (ICMPv6) enables hosts and routers that use IPv6 communication to report errors and send simple status messages.
Multicast Listener Discovery (MLD): RFC 2710, RFC 3590, RFC 3810	Multicast listener discovery enables you to manage subnet multicast membership for IPv6. MLD is a series of three ICMPv6 messages that replace the Internet Group Management Protocol (IGMP) v3 that is used for IPv4.
Neighbor discovery (ND): RFC 2461	Neighbor discovery is a series of five ICMPv6 messages that manage node-to-node communication on a link. Neighbor discovery replaces Address Resolution Protocol (ARP), ICMPv4 router discovery, and the ICMPv4 redirect message and provides additional functions.

Like IPv4, IPv6 is a connectionless, unreliable datagram protocol used for routing packets between hosts. "Connectionless" means that a session is not established before exchanging data; "unreliable" means that delivery is not guaranteed. IPv6 always makes a best-effort attempt to deliver a packet. An IPv6 packet might be lost, delivered out of sequence, duplicated, or delayed. IPv6 does not attempt to recover from these types of errors. The acknowledgment of packet delivery and the recovery of lost packets are done by a higher-layer protocol, such as TCP [MSD200401]. From a packet-forwarding perspective, IPv6 operates just like IPv4. An IPv6 packet, also known as an *IPv6 datagram*, consists of an IPv6 header and an IPv6 payload, as shown Figure 1.4.

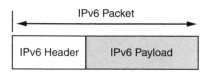

Figure 1.4: IPv6 packet.

The IPv6 header consists of two parts: the IPv6 base header, and optional extension headers. Functionally, the optional extension headers and upper-layer protocols, for example TCP, are considered part of the IPv6 payload. Table 1.3 shows the fields in the IPv6 base header [MSD200401], while Figure 1.5 depicts the header graphically, along with the extension header mechanism. IPv4 headers and IPv6 headers are not directly interoperable—hosts or routers must use an implementation of both IPv4 and IPv6 in order to recognize and process both header formats. This gives rise to a number of complexities in the migration process between the IPv4 and the IPv6 environments; however, techniques have been developed to handle these migrations.

Table 1.3: IPv6 base header.

IPv6	Length	Function
Version	8 bits	Identifies the version of the protocol. For IPv6, the version is 6.
Class	8 bits	Intended for originating nodes and forwarding routers to identify and distinguish between different classes or priorities of IPv6 packets.
Flow label	20 bits	Defines how traffic is handled and identified. A flow is a sequence of packets either sent to a unicast or a multicast destination. This field identifies packets that require special handling by the IPv6 node. The following list shows the ways the field is handled if a host or router does not support flow label field functions: • if the packet is being sent, the field is set to zero, • if the packet is being received, the field is ignored.
Payload length	16 bits	Identifies the length, in octets, of the payload This field is a 16-bit unsigned integer. The payload includes the optional extension headers, as well as the upper-layer protocols, for example, TCP.
Next header	8 bits	Identifies the header immediately following the IPv6 header. The following list shows examples of the next header: 0 = Hop-by-hop options 1 = ICMPv4 4 = IP in IP (encapsulation) 6 = TCP 17 = UDP 43 = Routing 44 = Fragment 50 = Encapsulating security payload 51 = Authentication 58 = ICMPv6 59 = None 60 = Destination options
Hop limit	8 bits	Identifies the number of network segments, also known as links or subnets, on which the packet is allowed to travel before being discarded by a router. The hop limit is set by the sending host and is used to prevent packets from endlessly circulating on an IPv6 internetwork. When forwarding an IPv6 packet, IPv6 routers must decrease the hop limit by 1, and must discard the IPv6 packet when the hop limit is 0.
Source address	128 bits	Identifies the IPv6 address of the original source of the IPv6 packet.
Destination address	128 bits	Identifies the IPv6 address of the intermediate or final destination of the IPv6 packet.

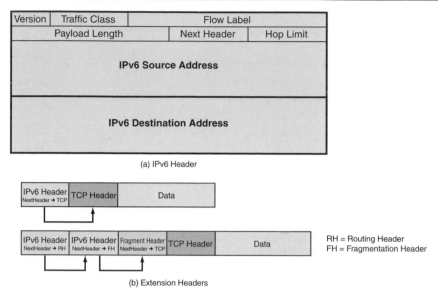

(a) IPv6 Header

(b) Extension Headers

Figure 1.5: IPv6 header mechanism (graphical representation).

1.2.5 IPv6 Autoconfiguration

IPv6 "autoconfiguration" is a new characteristic of the protocol that facilitates network management and systems setup tasks by users. This characteristic is often designated as "plug-and-play" or "connect-and-work." This facilitates initialization of user devices; after connecting a device to an IPv6 network, one or several IPv6 globally-unique addresses are automatically allocated.

The "autoconfiguration" process is flexible, but it is also complex; the complexity arises from the fact that various policies are or may be defined and implemented by the network administrator. The administrator determines the parameters that will be assigned automatically. At a minimum (and/or when there is no network administrator), the allocation of a "link-local" address is often included. The "link-local" address allows the communication with other nodes placed in the same physical domain. Note that "link" has somewhat of a special meaning in IPv6, as follows: A communication facility or medium over which nodes can communicate at the link layer—that is, the layer immediately below IPv6. Examples are: Ethernet (simple or bridged), PPP links, X.25, frame relay, or Asynchronous Transfer Mode (ATM), networks, Internet (or higher) layer "tunnels," such as tunnels over IPv4 or IPv6 itself [DEE199801].

Two autoconfiguration basic mechanisms exist: (1) Stateful and, (2) Stateless. Both mechanisms can be used in a complementary manner and/or simultaneously to define configuration parameters, as discussed next [IPV200501].

Stateless autoconfiguration is also described as *serverless*. There is no need for a configuration server to supply profile information. In this environment, manual configuration is required only at the host level; a minimal configuration at the router level is occasionally needed. The host generates its own address using a combination of the information that it possesses (in its interface or network card), and the information that is periodically supplied by the routers. Routers determine the prefix that identifies networks associated to the link under discussion. The "interface identifier" identifies an interface inside a subnetwork and is often, and by default, generated from the MAC address of the network card. The IPv6 address is built combining the 64 bits of the interface identifier with the prefixes that routers determine as belonging to the subnetwork.

If there is no router, the interface identifier is self-sufficient to allow the PC or VoIP handset to generate the "link-local" address. The link-local address is sufficient to allow the communication between several nodes connected to the same link (the same local network).

Stateful configuration requires a server to send the information and parameters of network connection to nodes and hosts. The configuration server maintains a database with all addresses allocated and a mapping of the hosts to which these addresses have been allocated, along with any information related with all requisite parameters. In general, this mechanism is based on the use of DHCPv6.

Stateful autoconfiguration is often employed when there is a need for rigorous control in reference to the address allocated to hosts; in stateless autoconfiguration, the only concern is that the address be unique. Depending on the network administrator policies, it may be required that some addresses be allocated to specific hosts and devices in a permanent manner; here, the stateful mechanism is employed on this subset of hosts, but the control of the remaining parameters and/or nodes could be less rigorous. In some environments/applications there are no policy requirements on the importance of the allocated addresses, but there may be rules on the parameters; for example, that they be allocated in a certain "static" manner, with information stored in a server. In this situation the "stateless" mechanism can be used.

IPv6 addresses are "rented" to an interface for a fixed-established time (including an infinite time). When this "lifetime" expires, the link between the interface and the address is invalidated, and the address can be reallocated to other interfaces. For the suitable management of addresses expiration time, an address goes through two states (stages) while is affiliated to an interface [IPV200501]:

1. First, an address is in a "preferred" state, so its use in any communication is not restricted.
2. Second, an address may become "deprecated," indicating that its affiliation with the current interface will (soon) be invalidated.

While it is in the "deprecated" state, the use of the address is discouraged, although it is not forbidden. However, when possible, any new communication (for example, the opening of a new TCP connection) must use a "preferred" address. A deprecated address should only be used by applications that already used it before and in cases where it is difficult to change this address to another address without causing a service interruption.

To ensure that allocated addresses (granted either by manual mechanisms or by autoconfiguration) are unique in a specific link. the "link duplicated addresses detection algorithm" is used. The address to which the duplicated address detection algorithm is being applied to is designated (until the end of this algorithmic session) as an "attempt address." In this case, it does not matter that such address has been allocated to an interface—received packets are discarded.

Next, we'll describe how an IPv6 address is formed. The lowest (rightmost) 64 bits of the address identify a specific interface, and are designated as *interface identifier*. The highest (leftmost) 64 bits of the address identify the "path" or the "prefix" of the network or router in one of the links in which such interface is connected. The IPv6 address is formed by combining the prefix with the interface identifier.

It is possible for a host or device to have IPv6 and IPv4 addresses simultaneously. Most of the systems that currently support IPv6 allow the simultaneous use of both protocols. This way, it is possible to support communication with IPv4 only networks, as well as IPv6 only networks, and the use of the applications developed for both protocols [IPV200501].

Is it possible to transmit IPv6 traffic over IPv4 networks; this is accomplished using tunneling methods. This approach consists of "wrapping" the IPv6 traffic as IPv4 payload data. IPv6 traffic is sent "encapsulated" into IPv4 traffic and at the receiving end this traffic is separated and then parsed as IPv6 traffic. Transition mechanisms are methods used for the coexistence of IPv4 and/or IPv6 devices and networks. For example, an "IPv6-in-IPv4 tunnel" is a transition mechanism that allows IPv6 devices to communicate among them through an IPv4 network. The mechanism consists of creating the IPv6 packets in a normal way and introducing them in an IPv4 packet. The reverse process is undertaken in the destination machine, which receives an IPv6 packet.

There is a significant difference between the procedures to allocate IPv4 addresses, which focus on the parsimonious use of addresses (since addresses are a scare resource and should be managed with caution), and the procedures to allocate IPv6 addresses, that focus on flexibility. Internet Service Providers (ISPs) deploying the IPv6 systems follow the *Regional Internet Registries (RIRs) policies relating how to assign IPv6 addressing space among their clients. RIRs are recommending ISPs and operators to allocate to each IPv6 client a /48 subnetwork; this allows clients to manage their own subnetworks without using NAT.* (The implication is that the need for NAT disappears in IPv6).

In order to allow its maximum scalability, the IPv6 protocol uses an approach based on a basic header, with minimum information (refer back to Figure 1.5). This differentiates it from IPv4 where different options are included in addition to the basic header. IPv6 uses a header "concatenation" mechanism to support supplementary capabilities. The advantages of this approach include the following:

- The size of the basic header is always the same, and is well known. The basic header has been simplified compared with IPv4, since only eight fields are used instead of twelve. The basic IPv6 header has a fixed size; hence, its processing by nodes and routers is more straightforward. Also, the header's structure aligns to 64 bits, so that new and future processors can process it in a more efficient way.
- Routers placed between a source point and a destination point (that is, the route that a specific packet has to pass through) do not need to process or understand any "following headers." In other words, in general, interior (core) points of the network (routers) only have to process the basic header, while in IPv4 all headers must be processed. This flow mechanism is similar to the operation in MPLS, yet precedes it by several years.
- There is no limit to the number of options that the headers can support (the IPv6 basic header is 40 octets in length, while IPv4 one varies from 20 to 60 octets, depending on used options).

In IPv6, interior/core routers do not perform packet fragmentation, rather the fragmentation is performed end-to-end. That is, source and destination nodes perform, by means of the IPv6 stack, the fragmentation of a packet and then the reassembly, respectively. The fragmentation process consists of dividing into smaller packets the "fragmentable" part of the source packet, and adding to each one the "unfragmentable" part [IPV200501].

Finally, a "jumbogram" is an option that allows an IPv6 packet to have a payload greater than 65,535 bytes. Jumbograms are identified with a 0 value in the payload length in the IPv6 header field, and include a jumbo payload option in the hop-to-hop options header. It is anticipated that these packets will be used specially for multimedia traffic.

1.2.6 Applications

IPv6 networks have a large number of possible applications that range, among others, from corporate intranets, institutional networks and extranets, mobility networks, hotspot networks, ubiquitous computing networks, 3G wireless networks, VoIP carrier networks, the global Internet and, government networks including the global grid. Figure 1.1 identified a gamut of such possible applications. *The emphasis in this book is the VoIP application.*

IPv6 enables new network models, specifically, integration and end-to-end security, as depicted graphically in Figure 1.6.

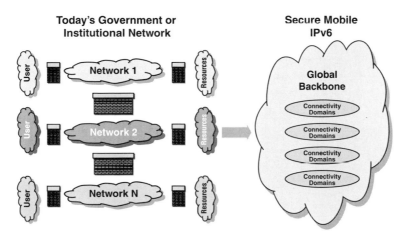

Figure 1.6: IPv6 integration and end-to-end security.

As noted, among other applications, some see applicability of IPv6 to carrier's 3G Wi-Fi® (IEEE 802.11a/b/g/n technologies) and to Voice over Wi-Fi (VoWi-Fi) applications. Carriers seek to add off-net mobility to existing voice services and extend existing branded services beyond coverage areas. Applications include in-building coverage (leveraging Wi-Fi indoor systems to improve coverage within shielded buildings), virtual calling card, call forwarding, off-net roaming, enabling of mobility in multimedia services (leverage the mobility and high speed of Wi-Fi to expand market, for example, personal video conferencing.) Issues for second-generation Wi-Fi solutions that appeared in the mid-2000s include the following [ISL200501]:

- Provisioning a large number of Access Points (APs).
- Radio planning and manual AP configuration not being cost effective.
- Desire for auto configuring, self-healing, meshed network topology.
- User authentication and security.
- Web login is fine for Enterprise but not Service Provider (SP).
- Security is a major issue for Enterprise usage of public Wi-Fi services.
- Limited mobility.
- In-building roaming works fine for enterprise, but not yet for SPs.
- Dropped connections are undesirable for multimedia services.

Third-generation Wi-Fi services are expected to provide carrier-class secure mobility. This is achieved by cost effective deployment of thousands of access points. In turn, this requires network discovery and auto configuration; over-the-air and network level security (IEEE 802.11i along with peer-to-peer connectivity). Finally, one also needs seamless roaming from home to a hotspot network (which requires Virtual Local Area Network [VLAN] switching along with Mobile IPv6.) Figure 1.7 depicts a view of the 3G Wi-Fi environment, while Figure 1.8 depicts graphically the improvements in VoIP mobility support in a 3G VoWi-Fi future state as enabled by IPv6.

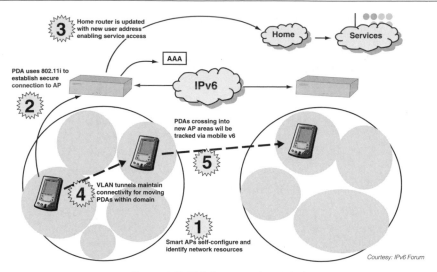

Figure 1.7: Third-generation Wi-Fi.

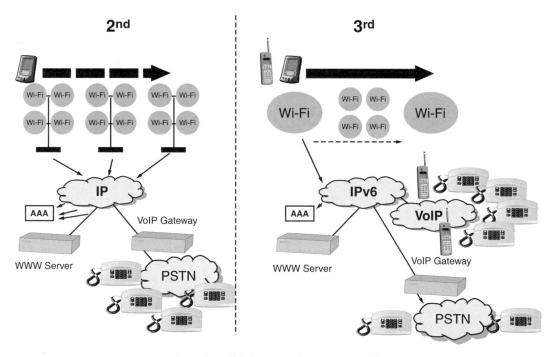

Figure 1.8: Third-generation: user mobility.

1.2.7 Transition Approaches

As discussed earlier, a number of transition mechanisms exist. The most common approach under consideration is the dual-stack approach illustrated in Figure 1.9 [CAB200501]. This topic will be re-examined in later chapters.

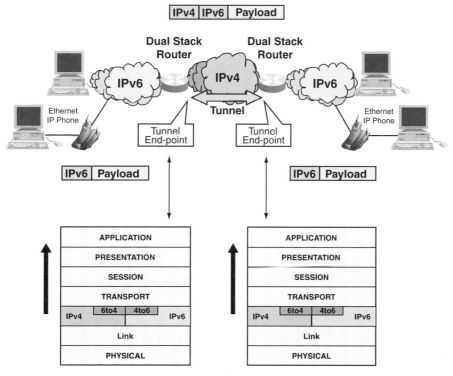

Figure 1.9: Dual-stack transition.

This preliminary overview of IPv6 highlights the advantages of the new protocol and its applicability to a whole range of applications, including VoIP.

1.3 Introductory Overview of VoIP

1.3.1 Overview

As implied in the previous sections, data networks based on packet technology in general, and IP in particular, have progressed to the point that it is now possible to support voice and multimedia applications over such networks. Enterprise networks, cellular carriers, voice-over-cable carriers, "triple-play" carriers, "pure-play VoIP carriers," and even traditional voice carriers are all moving rather aggressively to a VoIP paradigm. This is driven by the interest in convergence, which so far has been an elusive goal, in practical terms for the past 30 years. IP promises to be the "Holy Grail" of convergence.

Given this extensive deployment of data networking resources, the questions that have presented themselves in the past few years are, "Was it possible to use the investment already made in enterprise networks to carry real-time voice in addition to the data?" and, "Is it possible to deploy IP-based networks in the carrier's domain to carry real-time voice in place of circuit-based technology?" As we stated at the beginning of this chapter, the desire to build one integrated network goes back to the 1970s, if not earlier. The Advanced Research Projects Agency, with project DACH-15-75-C0135 (and many other projects with many other researchers), funded the author's work in 1975 to look at the feasibility of *integrated voice and data packet networking* [MIN199801], [MIN199802], [MIN197902], [MIN197903], [MIN197904], [MIN197906]. During the past decade or so, the answers to these questions have been given in the affirmative. At press time, a large equipment manufacturer was stating that they had placed 3 million IP phones with 15,000 customers worldwide; other leading manufacturers probably have a similar deployed base.

For enterprise environments, many companies have already deployed IP-based backbones that provide both broadband capabilities and QoS-enabled communications. Large companies have deployed broadband services such as ATM, MPLS, Packet over Synchronous Optical Network (POS), or metro-packet services such as metro-gigabit Ethernet, or Resilient Packet Ring (RPR) in their wide area cores. High-speed wide area technology provides increased bandwidth across the enterprise's regional, nationwide, and international networks. Link layer switching technology, particularly in terms of the switched Local Area Networks (LANs) and ancillary Layer 2 switches, has come a long way in the past ten years, providing higher-capacity, lower-contention services across the enterprise campus network. IEEE 802.1p/802.1Q protocols provide prioritization capabilities at the Ethernet level. Network-level QoS-supporting protocols, such as differentiated services (*diffserv*) in IPv4, IPv6, and MPLS, along with supporting router equipment and traffic management policies are now being deployed in many enterprises' networks. All of this opens the door for the possibility of carrying voice over the enterprise network.

For carrier environments, many providers have taken the opportunity of modernizing the existing Public Switched Telephone Network (PSTN) with an IP-based infrastructure that supports multiple services (described by some as Services over IP (SoIP) in general, and VoIP in particular. These carriers include both wireline and wireless providers. Other carriers with a "greenfield" voice environment; for example, cable TV operators or hotspot providers have chosen a VoIP approach from the get-go. Figure 1.10 depicts a typical carrier/enterprise VoIP environment.

1.3.2 First-Generation 1G VoIP Networks

A number of approaches have been advanced over the years to accomplish media/service integration. IP-based networks have emerged as the only viable mechanism to achieve this. Integrated Services Digital Network (ISDN) research started in Japan in the early 1970s (before the idea started to get some real attention in North America and in Europe in the late 1970s and early 1980s) with the explicit goal of developing and deploying integrated networks. However, a lot of the mainstream work has been in supporting voice and data over "circuit-switched" TDM networks. Only some early packet-over-data work (but not limited to [MIN197901], [MIN197902], [MIN197903], [MIN197904], [MIN197905], [MIN197906], [MIN197907], [MIN197908]), and then some Fiber Distributed Data Interface II (FDDI II) and integrated voice/data LAN (IEEE 802.9) work, looked at voice support in a noncircuit-mode network. Even for ATM, the emphasis has been mostly on data services, but some VoP usage has taken place [MIN199802]. The idea of carrying voice-over-data networks has received considerable commercial attention in the past ten years.

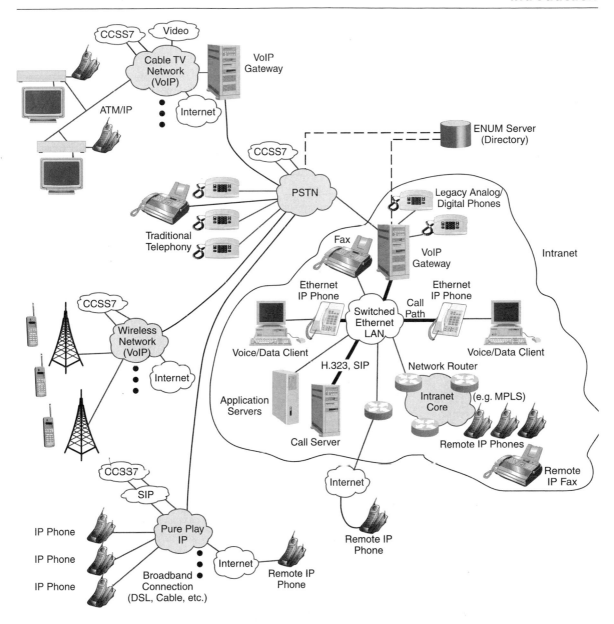

Figure 1.10: The VoIP environment.

The basic 1G commercial VoIP products started to appear in the mid- to late-1990s. These products were rudimentary and typically only supported basic telephony services; for example, fundamental intranet-level dial-tone and PSTN trunk bypass. In addition to failing to have a rich feature set, this generation of products lacked commercial-grade reliability and had very limited signaling capabilities (advanced features depend intrinsically on sophisticated signaling capabilities.) In general, the equipment also did not support QoS, inline power, and E911 services.

1.3.3 Second-Generation 2G VoIP Networks

Connectionless IP-based networks are ubiquitous, and so there is a desire to carry business-quality voice over them. 2G VoIP products were brought to the market in the early- to mid-2000s. These products started to support more advanced telephony services; for example, conferencing, least-cost-routing, QoS, and some wireless and security features. Reliability and manageability improved. Most importantly, the gamut of VoIP equipment started to support signaling capabilities in a more intrinsic manner and based on somewhat mature standards (for example, H.323v3 and/or SIP). The equipment also typically supported inline power and basic E911 services. However, integration with legacy (or other vendor's) e-mail systems remained problematic, as is support of unified messaging and centralized directories (LDAP-based directories, for example to support single sign-on features.)

Many standards bodies, industry forii, and other entities (e.g., IETF, IEEE for power over Ethernet) have published specifications in recent years, and a whole battery of voice-over-data network equipment has appeared and/or is appearing. For example, a plethora of IP phones for enterprise applications and IP-to-public network gateways has entered the market.

Noteworthy is the standardization work on Session Initiation Protocol, where over 60 RFCs had been developed by the Internet Engineering Task Force (IEFT) by 2005.

The major challenge in this regard is that public IP networks and many enterprise networks do not intrinsically support QoS features (even though the capabilities are now built into the end user VoIP equipment). Also, there are a host of security issues including firewalling, NAT-ing, and proxying. In addition to the NAT/security/end-to-end seamlessness problems, one has to deal with the lack of true internetworking between suppliers, forcing a firm or organization to buy all the equipment from a single vendor, thereby institutionalizing the long-existing enslavement of the telephone handset to the Private Branch Exchange (PBX) node. Additionally, one has do deal with the relatively high handset costs, typically ranging from $400–$1000 depending on features—even the softphone (which are enslaved to the VoIP elements because of vendor-proprietary signaling or signaling extensions) can be relatively expensive typically ranging from $200–$350.

Table 1.4 depicts some market observations and statistics to position the evolution of VoIP services in context, as it relates to 2G systems. In spite of the interest, impetus and hype that has been seen in the past ten years, the actual deployment so far has been relatively modest from a global perspective; however, the rollout of 2G systems is now reaching an inflection point, with a lot of market critical mass and momentum.

Table 1.4: Some market statistics to position the evolution of VoIP services.

Penetration Trends (Market Information)
Despite the growing popularity of VoIP calling, VoIP phones in the enterprise will not represent the majority of installed PBX base until 2009, says a new report from Insight Research. PBXs are the phone systems used by large companies to route calls to individual offices, and though shipments between 2004 until 2009 of the newer VoIP PBX phones are expected to grow at a compounded rate of more than 20% over the forecast period while the older TDM-based phone technology decline at roughly the same rate, the older TDM phone technology will continue to dominate the installed base until the end of the decade, says a new report from Insight Research. According to Insight Research, "By 2009, the installed base of IP equipment will dominate the enterprise landscape, but that's still five years away. The cost of going VoIP is certainly a factor here, since the price of newer IP phones will continue to be about 25% higher than the TDM alternative, even as volume shipment of VoIP phones takes off. VoIP never was and never will be the least expensive way to deliver voice to the enterprise—but the allure of VoIP's rich applications will slowly convert enterprise legacy customers." [INS200401]

According to a study Forrester Research conducted in 2004, about 15% of the 818 companies surveyed said they had completed or were in the process of rolling out VoIP systems [REA200401].
Frost & Sullivan, found in a 2003 survey that fewer than 4% of the new contact center systems installed in North America are VoIP systems. While that is a small number, she expects it to jump to 35% by 2006 [REN200301].
Infrastructure Trends (Market Information)
"... InformationWeek Research survey of 140 businesses shows that of those who combined their voice and data traffic onto one network, 16% went back to separate networks. Half of them say that administrative and management problems prompted the split and 44% cite insufficient IT infrastructure. Some 38% also cite consistency-of-service issues and difficulties in resolving service problems ... {There is an issue as to} how much needs to be upgraded or replaced. In the survey, nearly 60% of the respondents say they upgraded routers and switches, 58% upgraded servers, 51% upgraded bandwidth, 38% upgraded network-administration software...only 16% did not need any upgrade on any part of their network infrastructures..." [INF200401]
"... switching from your PBX to a voice-enabled data network is not as straightforward as calling up your nearest supplier and making an appointment. Chances are your router-based data network is counted among the estimated 85% of networks in use today that are not ready to support VoIP services without modification. And even if you are among the fortunate 15%, can you be sure that your network will reliably handle VoIP traffic over a long-term rollout, and what will the impact be on your other business-critical applications? These are the questions forward-thinking CIOs and system administrators are asking themselves before considering—let alone committing to—a wholesale VoIP switch..." [INT200401]

1.3.4 Pragmatic Enterprise 2G VoIP Deployment Approaches

The subsections that follow focus on enterprise-level applications of VoIP. A number of 2G converged communication architectures that can be utilized by enterprise organizations have emerged in recent years [MIN199801], [MIN199802], [MIN200201], [MIN200202]. Typically two issues are of interest to firms: (1) cost-reductions either in toll charges or in run-the-engine costs as vendor proprietary PBX hardware is replaced with standardized server based technology and, (2) functional enhancements, specifically mobility to support a nomadic (mobile) workforce. For enterprises, ultimately a pure-IP environment supporting end-to-end IP telephony, unified messaging, and VoWi-Fi to cellular roaming will be required to secure all of the advancements possible with VoIP, including presence and cost-effectiveness. For carriers, the motivation for a pure-IP environment is the support of user needs, in conjunction with enterprise migrations to this technology. This is where IPv6 will have the most impact.

However, stakeholders are approaching the migration with a degree of pragmatism, as it should in fact be the prudent case. In this section, we'll briefly review the possible approaches. All of them are IPv4-based.

The array of available VoIP platforms and architectural alternatives complicates the decision regarding the selection of a single target architecture (and, ultimately, of a vendor/product suite). Given this, the pragmatic approach followed by for-profit organizations is to identify a set of stage-based architectures that allow the organization to seamlessly migrate to a cohesive, all-inclusive, end-stage (final) architecture. In all instances, any proposed architecture must allow for ease of migration and consistent cost-effectiveness in order to be a viable approach.

Considering the installed base of PBXs and station equipment, it becomes imperative that any new IP-based system be able to integrate effectively with the existing solution and facilitate inflection toward more mature solutions. The interworkability of IP allows for next-generation telephony systems to be architected in more of a decentralized fashion than was possible before (specifically the voice-supporting switching gear need not be housed in the immediate proximity of the users served as was the case with traditional PBXs). In VoIP, the IP network per se becomes the switch, with call processing and applications residing in separate servers distributed across the network.

The major 2G architectural alternatives that have evolved at the commercial/deployment level are as follows:

- VoIP trunking only;
- Traditional PBXs with IP adjunct extensions (enhancements);
- Hybrid TDM and IP systems, and
- Pure IP server-based telephony systems.

At this instance in time, most enterprises appear to have a preference for traditional PBX with IP adjunct extensions, while the rest favors an IP server-based approach.[2]

1.3.5 2G VoIP Trunking Only

This first stage solution uses the intranet to handle PBX-to-PBX tie lines, for site-to-site communication. See Figure 1.11. Currently some companies are using ADPCM over-AAL1-over ATM. While ATM provides excellent QoS to the directly-attached (virtual) circuits (tie lines), these firms are not getting the most bandwidth efficiency possible, especially if constant bit rate services are used. A reduction in bandwidth of 75% is achievable using IP trunking and going from 32 kbps to 8 kbps voice.

The advantage of this approach is that it allows least cost routing, as shown in Figure 1.12. Notice that generally this requires an external gateway, unless the PBX has a gateway blade that can be utilized. One limitation of this approach is that calls cannot be transferred multiple times because of the repetitive encoding/decoding stages.

---

[2] From a supplier perspective, Avaya and Cisco have the largest market share in today's enterprise market; competitors such as Alcatel and Nortel are closing the market gap.

Before

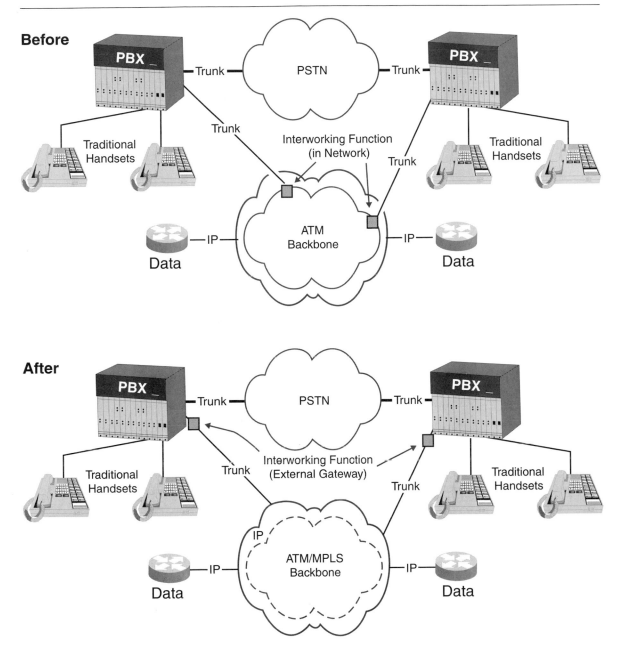

After

Figure 1.11: VoIP trunking.
Top: Before. Bottom: After.

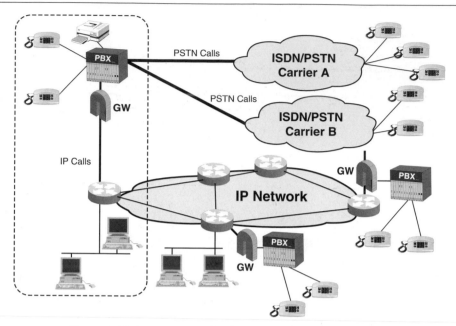

Figure 1.12: Least cost routing in VoIP trunking environments.

1.3.6 2G Traditional PBXs with IP Adjunct Extensions (Enhancements)

This second-stage solution uses an adjunct (either as add-on hardware/blade in a slot on the PBX or as an external server to the PBX), to handle a small set of IP phones typically limited to one department or site (say, ~10% of the total population). The major portion of the enterprise-wide signaling and trunking, however, is still handled by the traditional PBX. This approach allows initial deployment of a VoP island, and the coexistence and intercommunication of the conventional corporate telephony network (conventional phones connected to PBX) and the local(lized) IP telephony network. The scenario is suitable when the local IP telephony network is built out gradually in an institution that already has a conventional telephony network. In a later stage, the conventional telephony network and the PBX can be totally replaced by the IP telephony network. See Figure 1.13.

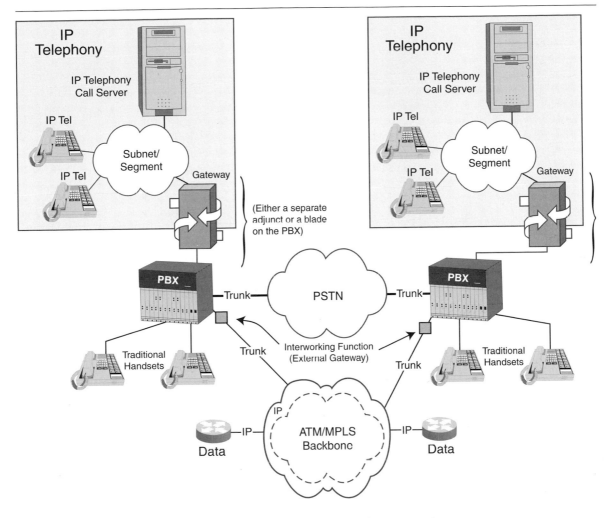

Figure 1.13: Traditional PBXs with IP adjunct extensions.

1.3.7 2G Hybrid TDM and IP Systems

This third-stage solution uses a *major* intrinsic PBX upgrade (but not replacement) to handle a larger set of IP phones (say, between 30% and 50% of the total population), while at the same time handling a nontrivial amount of traditional telephone handsets. Trunk-to-trunk signaling and transport is still handled by the traditional PBX, and it could entail both TDM trunking and IP trunking (at this stage the ratio of trunking technology between TDM and IP is probably around 50/50). This approach allows the coexistence and interconnection of major portions of the surviving conventional telephony network (conventional phones connected to PBX) with the local IP telephony network, while at the same time supporting a major and robust shift (migration) to an IP-based environment. See Figure 1.14. The scenario also makes optimal use of traditional handsets, because such handsets can continue to be employed until an end-of-life situation requires a migration and/or the inventory of traditional handsets is depleted due to normal wear-and-tear. When such stage is reached, the affected population of users are migrated to the IP side of the system.

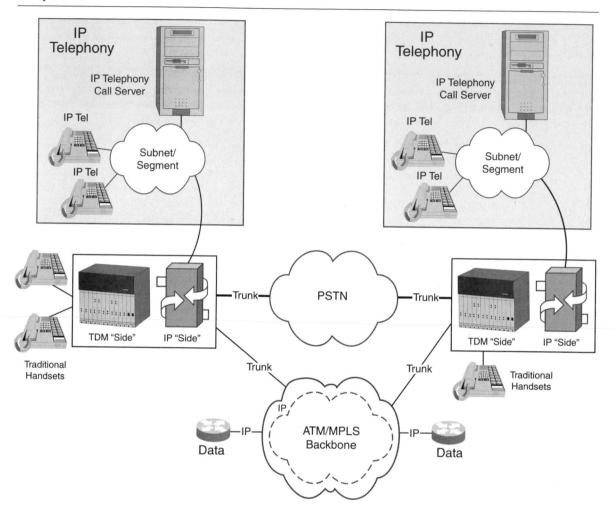

Figure 1.14: Hybrid environments

1.3.8 2G Pure IP Server-Based Telephony Systems

This final-stage solution applies either in a completely-greenfield situations, or when an existing PBX is fully depreciated. At this stage, IP telephony can be considered as a complete replacement alternative to a traditional PBX. See Figure 1.15. In fact, this can position a company for a possible next-move at a point in the future to 3G VoIP based on IPv6.

An issue of interest to many large firms, however, is if the call control equipment and the telephone handsets are truly open (such as would be the case if both classes of equipment fully implemented a set of specifications, e.g., SIP). For decades firms were forced to purchase the PBX hardware and the handsets from the same vendor; in fact, the money-making proposition for the vendor was the handset equipment, not the switching fabric (the PBX). At this juncture, firms would like to be able to obtain the equipment from a variety of suppliers, and large organizations may be taking a "wait-and-see" approach on VoIP until the interworking question is fully resolved. The motivation is the desire to be able to retain the (fairly expensive) handsets if/when the switching elements (e.g., SIP proxy) is replaced with equipment from another vendor.

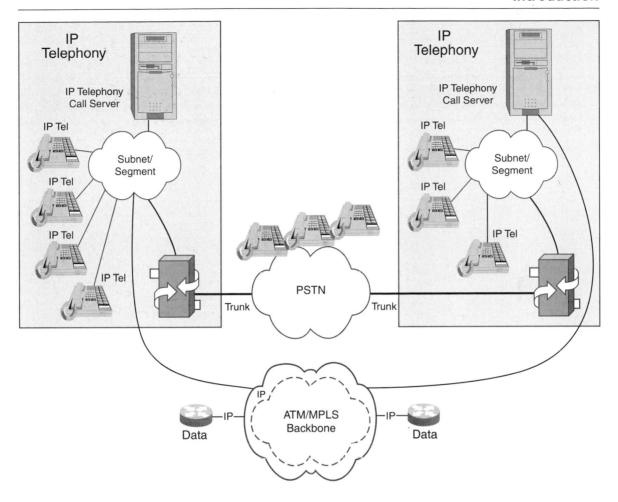

Figure 1.15. IP-telephony fully replacing PBX.

1.3.9 Possible Evolution Paths for 2G Deployments

A number of pragmatic VoIP deployment strategies for the pre-VoIPv6 environment are possible, as seen in Figure 1.16. Many enterprise companies follow "Approach Alpha" depicted in this figure, which is fairly conservative and allows the company to gradually enter the VoIP space, while at the same time taking into account (and taking advantage of) the embedded base.

An approach such as "Approach Delta" is relatively radical and not easily cost-justifiable in typical corporate settings, especially for large firms (say firms with 10,000+ employees). In fact, as of 2005 only about a couple dozen large firms (firms with >10,000 employees) has migrated to VoIP, as can be gleaned from the open press, and even these tend to be in the HQ-with-high-branch-population environments (for example, banks and brokerage firms), where the remote branch offices appear to be relatively stand-along small offices, rather than environments with a handful of massive campuses. Approach Delta is basically a "fork lift" approach; it may be applicable to greenfield environments, extreme end-of-life situations, or small environments (25–200 stations.)

All approaches listed in the preceding subsections have intrinsic "pros" and "cons," and may, given further analysis, be potential avenues of deployment for an organization.

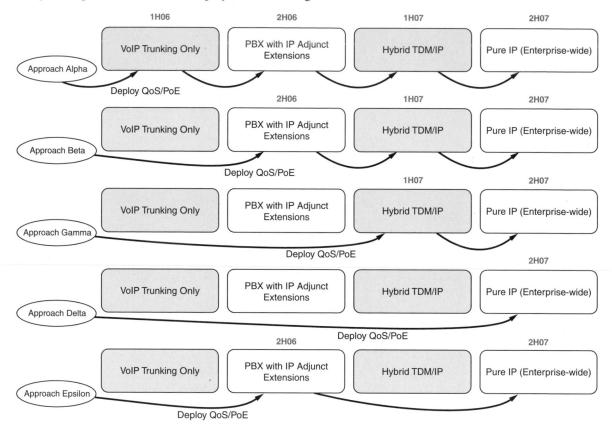

Figure 1.16: Possible deployment strategies (pre-VoIPv6) – enterprise environments.

1.4 Third-Generation 3G VoIP Networks

The next-generation VoIP network constructs are just around the corner. These networks will be characterized by the following features:

- Full end-to-end IP-based (specifically IPv6-based);
- Fully accessible to/from any user in the world;
- SIP-based for advanced signaling;
- Seamless integration with corporate enterprise networks from a protocol and security perspective;
- QoS-enabled in the Wireless LAN (WLAN) environment;
- Integrations with 3G cellular systems;
- Commercial-grade service levels/reliability/security;
- End-to-end QoS enabled (across multiple public networks);
- Support of low bit-rate video/videoconferencing;
- Media/network independent fully converged system, regardless of the (potential mixed) use of cable TV networks, cellular networks, hotspot networks, pure-play IP networks, traditional networks, and/or enterprise networks;

- Heavy support of "presence" features (presence/proximity, multimodal and collaborative communications); and,
- Full integration with other media to support a true unified messaging environment.

Each of the features listed previously could be discussed in a stand-alone text. This book focuses on the IPv6 technology and its implications with reference to other items on this desiderata list.

Figure 1.17 depicts some deployment strategies vis-à-vis VoIPv6. "Approach Gimmel" is a supercharged fork lift to 3G VoIP and probably will not happen within the next couple of years, although there may be limited-scope opportunities in some institutional and/or government applications. "Approach Aleph" is effectively a continuation of the approaches discussed above for the VoIPv4 introduction; this evolution takes a measured and conservative approach to the deployment of technology, and may be a reasonable approach to the migration is some cases; for example Early Adopters. "Approach Bet" is an evolution strategy that is more ambitious than Approach Aleph, but not as radical as Approach Gimmel.

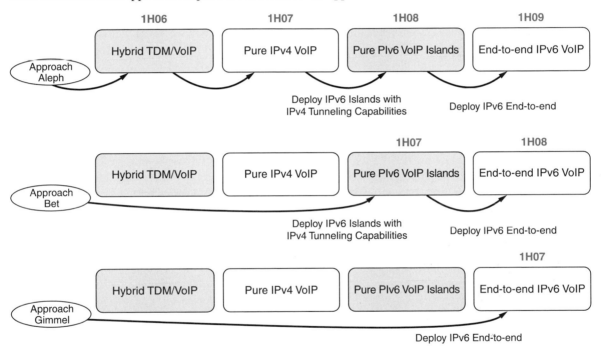

Figure 1.17: Possible enterprise deployment strategies vis-à-vis VoIPv6.

1.5 Deployment/Penetration Issues

The following observations from [DOM200501] provide a counterpoint to this book and are included herewith to provide balance.

"Despite the initial expectations raised, IPv6 is clearly far from being extensively deployed, and therefore it is too early to claim any complete success for it yet. There are several reasons for this. The first is that the dire warnings regarding IPv4 address exhaustion have not yet materialized. Recent studies analyzing past data forecast that IPv4 addresses will hold out beyond 2030 unless new conditions arise that bring about a change in the current trend in address consumption, such as a strong demand for addresses for mobile devices or the addition of a large number of users in China or India. There are several explanations for this change in expectations: tough political control from the RIRs over address assignment, address reuse in dial-up accesses, and so on, but the deployment of has probably played the most important part in slowing down address consumption. NATs allow the reuse of a few publicly registered addresses in the provision of connectivity to a much larger number of systems. NATs are in widespread use, and are now serving both large organizations and residential users. Although NATs' implementation has deferred the address scarcity problem, this has not been achieved without cost. Firstly, connectivity has become asymmetric because some nodes are more capable of receiving externally-initiated communications than others. Secondly, end-to-end functions depending on the preservation of the original IP address throughout the communication, such as IPsec security, are now precluded. Addressing these issues is now one of the goals of IPv6 supporters. There have also been technological obstacles to the success of IPv6. While basic IPv6 standards have been available for some time, the standardization process has not been smooth in some key areas, since several important issues, such as DHCPv6 or Mobile IPv6, have taken a considerable amount of time to resolve. There have also been some changes made to the core specification in recent years, such as the deprecation of site-local addresses, or updates to the programming interfaces. There are also some problems to which we are only now beginning to find solutions, such as multihoming in IPv6, or what security model to be deployed. But even once the technology is fully available, there will still be a great many challenges to be overcome. One of the biggest of these is the requirement for applications using the socket interface to be ported to a new programming interface to be able to use IPv6, due to dependencies imposed by the socket interface on the specific protocol to be used. Fortunately, most operating systems already provide support for IPv6. Communication hardware providers have been less enthusiastic, and have provided – barring some notable exceptions – inferior support for IPv6, compared to IPv4, in terms of both functionality and performance. Major service providers have also been understandably reluctant to change equipment in their operational networks to support a protocol with a relatively low number of users and applications. It is clear that the migration process will entail significant costs and complexities for networking organizations. Finally, there have been no new killer IPv6-based applications or services to attract users."

However, this reference continues with:

"Notwithstanding the points raised above, there is some good news for IPv6, and this could be a key moment in the migration process. The achievement of a critical mass of IPv6 users may become a reality with the strong political support coming from many Asian countries. Additionally, specifications for some 3G mobile networks require the deployment of IPv6, so some near-term growth in the number of IPv6 users can be expected. IPv6 is also seen as an opportunity for European and Asian communication hardware and software providers that have lagged behind North American providers in sales for IPv4 equipment to gain a new advantage. This, along with the enthusiastic work of organizations promoting IPv6, such as the IPv6 Forum, or the numerous IPv6 Task Forces all over the world, has generated political awareness in the European Union. An example outcome of this political interest is the growing trend for the requirement of IPv6 support in newly-issued public contracts. Some technologies that can only be deployed in their current form using IPv6 are also generating some expectations, such as the deployment of end-to-end security on the network layer, which requires public addressability that can only be provided by IPv6; or the possibility of providing full multihoming support for small networks or even residential users. Some of these technologies may evolve sufficiently to convince even the last remaining dyed in the wool IPv6-agnostics."

1.6 Line of Investigation

As already stated, this book looks at IPv6 opportunities; specifically it looks at the use of IPv6 to support a next-generation, carrier-class VoIP environments, which we call VoIPv6. As noted, IPv6 offers the potential of achieving the scalability, reacheability, end-to-end interworking, QoS, and commercial-grade robustness for VoIP that is required if the technology is indeed to replace the TDM infrastructure around the world. Specifically, IPv6 deals with the QoS and NAT issues.

This is the first book of its kind to address this issue as a macro-level scalability requirement. The book basically is comprised of two sections: an opening section (Chapters 1–5) that looks at applications and motivations, and the second part (Chapters 6–8) focuses on the IPv6 itself.

After an introduction in Chapter 1, we provide a quick tutorial in Chapter 2 on VoIP, and in Chapter 3 on SIP and signaling. Chapter 4 continues that basic discussion by examining the area of "presence," which represents a value-added set of VoIP-based capabilities. Chapter 5 discusses the issues associated with current VoIP implementations, as highlighted above.

Chapter 6 provides a basic introduction to IPv6. Chapter 7 discusses the delivery of voice-over-packet in an IPv6 environment. Chapter 8 provides some basic discussion of the transition issues including interworking between IPv6 and IPv4. Europe and Asia are currently leading the way in the planning of IPv6 networks.

This book should prove useful to VoIP equipment vendors, VoIP service providers (for example, cellular carriers, voice-over-cable carriers, "triple-play" carriers, "pure-play VoIP carriers," and even traditional voice carriers), enterprise customers, researchers, planners, and educators.

Appendix A: Basic IPv6 Terminology

This Appendix for Chapter 1 provides a basic glossary of IPv6 terms and concepts that are loosely based on references [IPV200501] and [MSD200401].

6over4	*An IPv6 mechanism designed to favor the coexistence with IPv4; it provides unicast and multicast connectivity through an IPv4 infrastructure with multicast support, using the IPv4 network as a logical multicast link.*
6over4 Address	*A |64-bit prefix|:0:0:WWXX:YYZZ address, where WWXX:YYZZ is the hexadecimal representation of w.x.y.z (a public or private IPv4 address), used to represent a device in 6over4 technology.*
6to4	*An IPv6 mechanism designed to favor the coexistence with IPv4, that provides unicast connectivity between IPv6 networks and devices through an IPv4 infrastructure. 6to4 uses a public IPv4 address to build a global IPv6 prefix.*
6to4 Address	*A |64-bit prefix|:0:0:WWXX:YYZZ address, where WWXX:YYZZ is the hexadecimal representation of w.x.y.z (a public or private IPv4 address), used to represent a device on 6over4 technology. A 2002:WWXX:YYZZ:|SLA ID|:|interface ID| address, where WWXX:YYZZ is the hexadecimal representation of w.x.y.z (a public or private IPv4 address), used to represent a device on 6to4 technology.*
6to4 Machine	*An IPv6 device that is configured, at least, with one 6to4 address (a global address with a 2002::/16 prefix). 6to4 devices do not require manual configuration and they create 6to4 addresses by means of classical autoconfiguration mechanisms.*
6to4 Router	*Router designed to favor the coexistence with IPv4; it provides unicast connectivity between IPv6 networks and devices through an IPv4 infrastructure. 6to4 uses a public IPv4 address to build a global IPv6 prefix.*
Address	*Network layer identifier assigned to an interface or set of interfaces that can be used as source or destination field in IP datagrams.*
	The IPv6 128-bit address is divided along 16-bit boundaries. Each 16-bit block is then converted to a 4-digit hexadecimal number, separated by colons. The resulting representation is called colon-hexadecimal. This is in contrast to the 32-bit IPv4 address represented in dotted-decimal format, divided along 8-bit boundaries, and then converted to its decimal equivalent, separated by periods [MSD200401]. The following example shows a 128-bit IPv6 address in binary form:
	0010000111011010000000000110100110000000000000000000000010111100111011 00000010101010100000000011111111111111110001010001001110001011010
	The following example shows this same address divided along 16-bit boundaries:
	0010000111011010 0000000011010011 0000000000000000 0010111100111011000000 1010101010 0000000011111111 1111111000101000 1001110001011010
	The following example shows each 16-bit block in the address converted to hexadecimal and delimited with colons.
	21DA:00D3:0000:2F3B:02AA:00FF:FE28:9C5A
	IPv6 representation can be further simplified by removing the leading zeros within each 16-bit block. However, each block must have at least a single digit. The following example shows the address without the leading zeros:
	21DA:D3:0:2F3B:2AA:FF:FE28:9C5A

This Appendix for Chapter 1 provides a basic glossary of IPv6 terms and concepts that are loosely based on references [IPV200501] and [MSD200401].

Address Autoconfiguration	The automatic configuration process for IPv6 addresses in an interface. The process for configuring IP addresses for interfaces in the absence of a stateful address configuration server, such as Dynamic Host Configuration Protocol version 6 (DHCPv6).
Address Maximum Valid Time	Time during a unicast address, obtained by means of stateless autoconfiguration mechanism, keeps in valid state.
Address Resolution	Procedure of link addresses resolution for the next-hop address in a link. In an IPv6 context, the process by which a node resolves a neighboring node's IPv6 address to its link-layer address. The resolved link-layer address becomes an entry in a neighbor cache in the node. The link layer address is equivalent to ARP in IPv4, and the neighbor cache is equivalent to the Address Resolution Protocol (ARP) cache. The neighbor cache displays the interface identifier for the neighbor cache entry, the neighboring node IPv6 address, the corresponding link-layer address, and the state of the neighbor cache entry.
Aggregatable Unicast Global Address	Also known as global addresses, these addresses are identified by means of the prefix format 001 (2000::/3). IPv6 global addresses are equivalent to IPv4 public addresses and they are whole routable and reachable in the IPv6 Internet fragment.
Anycast Address	A unicast address that identifies several interfaces and is used for the delivery from one to one between-several. With an appropriate route, datagrams addressed to an anycast address will be deliver to a single interface, the nearest one.
AS	See Autonomous System.
Attempt Address	Unicast address where uniqueness is no longer checked.
Automatic IPv6 Tunnel	Automatic creation of tunnels using IPv4 compatible addresses.
Automatic Tunnel	An IPv6 over IPv4 tunnel in which end points are specified by means of the use of tunnels logical interfaces, routes, and IPv6 source and destination addresses.
Autonomous System (AS)	A network domain that belongs to the same administrative authority.
Colon Hexadecimal Notation	The notation used to represent IPv6 addresses. The 128 bits address is divided in 8 blocks of 16 bits. Each block is represented as an hexadecimal number and moves apart from next block by means of colon orthographic sign (:). Inside each block, zeros left placed are removed. An example of an IPv6 unicast address represented in hexadecimal notation is 3FFE: FFFF:2A1D:48C:2AA:3CFF:FE21:81F9.
Compatibility Addresses	IPv6 addresses used when IPv6 traffic is sent through an IPv4 infrastructure. Some examples are: IPv4 compatible addresses, 6to4 addresses, and ISATAP addresses.

This Appendix for Chapter 1 provides a basic glossary of IPv6 terms and concepts that are loosely based on references [IPV200501] and [MSD200401].

Compressing Zeros	Some types of addresses contain long sequences of zeros. In IPv6 addresses, a contiguous sequence of 16-bit blocks set to 0 in the colon-hexadecimal format can be compressed to :: (known as double-colon). The following list shows examples of compressing zeros [MSD200401]:
	The link-local address of FE80:0:0:0:2AA:FF:FE9A:4CA2 can be compressed to FE80::2AA:FF: FE9A:4CA2.
	The multicast address of FF02:0:0:0:0:0:0:2 can be compressed to FF02::2.
	Zero compression can only be used to compress a single contiguous series of 16-bit blocks expressed in colon-hexadecimal notation. One cannot use zero compression to include part of a 16-bit block. For example, one cannot express FF02:30:0:0:0:0:0:5 as FF02:3::5.
	Zero compression can be used only once in an address, which enables you to determine the number of 0 bits represented by each instance of a double-colon (::). To determine how many 0 bits are represented by the ::, one can count the number of blocks in the compressed address, subtract this number from 8, and then multiply the result by 16. For example, in the address FF02::2, there are two blocks (the FF02 block and the 2 block). The number of bits expressed by the :: is 96 (96 = (8 - 2) × 16) [MSD200401].
Correspondent Node	A node that communicates with a mobile node that is out of its own network.
Default Path	The route with a ::/0 prefix. The default route, gathers all destinations and is the route used to obtain next destination address when there are no more matching routes.
Default Routers List	A list supported by each device where all routers, from which a no null router lifetime value advertisement has been received, appear.
Destination Cache	Table supported by each IPv6 node that maps each destination address (or addresses range) with the next router address to which the datagram has to be sent. Moreover it stores the associated path MTU.
Distance Vector	A routing protocol mechanism that propagates routing information as network identifier and its distance as hops numbers.
Domain Names System	An storage hierarchical system and its associated protocol to store and recover information about names and IP addresses.
Double Colon	Compressing continuous series of 0 blocks, into IPv6 addresses like "::". For example, FF02:0:0:0:0:0:0:2 multicast address is expressed as FF02::2.
Dual Stack Architecture	An IPv6/IPv4 node architecture in which two complete protocols stack implementations exist, one for IPv4 and another one for IPv6, each with its own implementations of the transport layer (TCP and UDP).
Dynamic Host Configuration Protocol (DHCP)	A configuration protocol with "stateful" state that provides IP addresses and other configuration parameters to connect to an IP network.
Encapsulating Security Payload	An IPv6 extension header and trailer that provides data source authentication, data integrity and confidentiality and a not-reply service for the loading of the datagram encapsulated by the header and trailer.
EUI	See Extended Unique Identifier.
EUI-64 Address	64 bits link layer address that is used as basis to generate interface identifiers in IPv6.
Extended Unique Identifier (EUI)	Link layer address defined by the Institute of Electrical and Electronic Engineers (IEEE).
Extension Headers	Headers placed between IPv6 header and higher level protocols headers, and are used to provide with additional functionalities to IPv6.

This Appendix for Chapter 1 provides a basic glossary of IPv6 terms and concepts that are loosely based on references [IPV200501] and [MSD200401].

Flow	*Datagram series exchanged between a source and a destination that require an special treatment at middle routers, and defined by a specific source and destination IP address, just as by a flow label with a non 0 value.*
Format Prefix	*High order bits with a fixed value that define an IPv6 type address.*
Fragment	*A portion from a message sent by a host into an IPv6 datagram. Fragments contain a fragmentation header.*
Fragmentation	*Process in which the source device divides the message of an IPv6 datagram in some number of fragments, so all fragments have a properly MTU to its destination.*
Fragmentation Header	*An IPv6 extension header that contains information needed for reassembly to be used in the receiving node.*
Global Address	*See global unicast aggregatable address.*
Group Identifier	*Leftmost 112 bits or leftmost 32 bits (according to RFC 2373 recommendation) of an IPv6 multicast address, that identifies a multicast group.*
Higher Level Checksum	*Calculation of the checksum, realized in ICMPv6, TCP and UDP, that uses the IPv6 pseudo-header.*
Higher Level Protocol	*Protocol that uses IPv6 as transport and it is placed in the upper layer than IPv6, such as ICMPv6, TCP and UDP.*
Hop-By-Hop Option Header	*An IPv6 extension header that contains options that must be processed by all intermediate routers as well as final router.*
Hosts File	*A text file used to contain name-IP address correspondences. In Windows XP or .NET server is located at \SystemRoot\System32\Drivers\Etc directory. In Unix devices is located at /etc directory.*
Host-To-Host Tunnel	*An IPv6 over IPv4 tunnel in which end points are devices.*
Host-To-Router Tunnel	*An IPv6 over IPv4 tunnel in which the tunnel begins in a host and ends in an IPv6/IPv4 router.*
ICMPV6	*See Internet Control Message Protocol For IPv6.*
Interface	*A representation of a physical or logical link of a node to a link. An example of a physical interface is a network interface. An example of a logical interface is a tunnel interface.*
Interface Identifier	*Last 64 bits of a unicast or anycast IPv6 address.*
Internet Control Message Protocol For IPv6 (ICMPV6)	*Protocol for Internet Control Messages for IPv6. A protocol that provides error messages for the routing and delivers IPv6 datagrams and information messages for diagnostics, neighbor discovery, multicast receiver discovery, and IPv6 mobility.*
Intra-Site Automatic Tunneling Addressing Protocol (ISATAP)	*A technology of coexistence that provides IPv6 unicast connectivity between devices placed in an IPv4 intranetwork. ISATAP obtains an interface identifier from the IPv4 address (public or private) assigned to the device. This identifier is used for the establishment of automatic tunnels through IPv4 infrastructure.*
IP6.Int	*The DNS domain created for the IPv6 reverse resolution. The reverse resolution has the purpose of fixing the name of a device by means of its address.*
IPv4 Compatible Address	*A 0:0:0:0:0:0:w.x.y.z or ::w.x.y.z address, where w.w.y.z is the decimal representation of a public IPv4 address. For example, ::131:107:89:42 is an IPv4 compatible address. These addresses are used in Automatic IPv6 tunnels.*
IPv4 Node	*A node that implements IPv4; it can send and receive IPv4 packets. It can be an only IPv4 node or a dual IPv4/IPv6 node.*
IPv6 In IPv4	*See IPv6 over IPv4 tunnels.*
IPv6 MTU	*The maximum IP packet size that can be sent over a link.*
IPv6 Node	*Node that implements IPv6; it can send and receive IPv6 packets. An IPv6 node can be an only IPv6 node or a dual IPv6/IPv4 node.*

This Appendix for Chapter 1 provides a basic glossary of IPv6 terms and concepts that are loosely based on references [IPV200501] and [MSD200401].

IPv6 Over IPv4 Tunnel	Sending IPv6 packets with an IPv4 header, so IPv6 traffic can be sent over an IPv4 infrastructure. In the IPv4 header, the protocol field value is 41.
IPv6 Prefixes	The prefix is the part of the address that indicates the bits that have fixed values or are the bits of the network identifier. Prefixes for IPv6 routes and subnet identifiers are expressed in the same way as the classless interdomain routing notation for IPv4. An IPv6 prefix is written in address/prefix-length notation. For example, 21DA:D3::/48 is a route prefix and 21DA:D3:0:2F3B::/64 is a subnet prefix. IPv4 implementations commonly use a dotted decimal representation of the network prefix known as the subnet mask. A subnet mask is not used in IPv6. Only prefix-length notation is used [MSD200401].
IPv6 Routing Table	Set of routes used to determine next node address and interface in IPv6 traffic sent by an equipment or redirected by a router.
IPv6/IPv4 Node	A node that has IPv4 and IPv6 implementations.
ISATAP	See Intra-Site Automatic Tunneling Addressing Protocol.
ISATAP Address	A |64-bit prefix|:0:5EFE:w.x.y.z address, where w.x.y.z is a public or private IPv4 address, that is allocated to an ISATAP device.
ISATAP Machine	A device to which an ISATAP address is assigned to.
ISATAP Name	The name solved by computers with Windows XP Service Pack 1 or Windows .NET Server 2003 operative systems to automatically discover the ISATAP router address. Windows XP equipments try to resolve the name "_ISATAP".
ISATAP Router	An IPv6/IPv4 router that answers to ISATAP equipments requests through tunnels and routes traffic between ISATAP equipments and nodes from another ISATAP network or subnetwork.
Jumbo Payload Option	An option in the hop-to-hop options header that shows the size of the jumbogram.
Jumbogram	An IPv6 packet that has a payload greater than 65,535 bytes. Jumbograms are identified with a 0 value in the payload length IPv6 header field, and including a Jumbo payload option in the hop-to-hop options header.
Lifetime In Preferred State	Time during a unicast address, obtained by means of stateless autoconfiguration mechanism, stays in the preferred state. This time is specified by the preferred lifetime field in routers advertisement messages prefix information option.
Link	A communication facility or medium over which nodes can communicate at the link layer, i.e., the layer immediately below IPv6. Examples are Ethernets (simple or bridged); PPP links; X.25, Frame Relay, or ATM networks; and internet (or higher) layer "tunnels", such as tunnels over IPv4 or IPv6 itself.
Link Maximum Transmission Unit (MTU)	The MTU is the number of bytes in the greatest IPv6 packet that can be sent through the link. Since the frame maximum size includes link layer headers, the link MTU does not equate with the link frame maximum size; rather, the link MTU matches the link layer technology payload maximum size.
Link State	Routing protocol technology that exchanges routes information, that consists of prefixes of networks connected to a router and its associated cost. Link state information is advertised in boot process, just as when changes are detected in the network topology.
Local Address	An IPv6 unicast address that is not reachable in IPv6 Internet. Local addresses include "link-local" and "site-local" addresses.
Local Area Network Segment	Link portion that consists of an only medium limited by bridges or layer 2 switches.
Local Interface	Internal interface that allows a node to send packets to itself.
Local Loop Address	IPv6::1 address, assigned to local interface.

This Appendix for Chapter 1 provides a basic glossary of IPv6 terms and concepts that are loosely based on references [IPV200501] and [MSD200401].

Local Site Address	*Local address identified by the 1111 1110 11 (FEC0::/10) prefix. The scope of these addresses is local sites (of an organization), without the necessity of a global prefix. Local site addresses (aka site-local addresses) are not accessible from other sites and routers should not direct local site traffic out of such site.*
MAC Address	*A link layer address of local network typical technologies such as Ethernet, Token Ring and FDDI. It is also known as physical address, hardware address, or network adapter address.*
Machine (Host)	*A node that cannot send datagrams not created by itself. A machine (device) is both the source and destination of IPv6 traffic, and will discard traffic that is not specifically addressed to it.*
Mapping IPv4 Address	*A 0:0:0:0:0:FFFF:w.x.y.z or ::FFFF:w.x.y.z address, where w.x.y.z is an IPv4 address. Mapped IPv4 addresses are used to represent an IPv4-only node in the presence of an IPv6 node.*
Maximum Transfer Unit (MTU)	*The longest Protocol Data Unit (PDU) that can be sent (unfragmented). Maximum transmission units are defined at the link layer (frame maximum size) and at the network or Internet layer (maximum IPv6 packet size).*
Maximum-Level Aggregation Identifier	*(aka Top-Level Aggregation Identifier -- TLA ID). 13 bits field inside the global unicast address reserved for large organizations or ISP by the IANA, hence it identifies the addresses range that they have delegated.*
Medium Access Control	*A sublayer of the link layer defined by the Institute of Electrical and Electronic Engineers. Its functionalities are the creation of frames and the management of the medium sharing (and access)*
MTU	*See Maximum Transmission Unit.*
Multicast Address	*An address that identifies several interfaces and is used to deliver data from one-to-several. By means of the multicast routing topology, packets to a multicast address will be delivered to all interfaces identified by it.*
Multicast Group	*Set of equipments listening to a specific multicast address.*
Multicast IPv4 Tunnel	*See 6over4.*
Name Resolution	*Procedure to obtain an address from a name. In IPv6, the resolution of names allows obtaining addresses from device names or domain names totally qualified (FQDN).*
ND	*See Neighbors Discovery.*
Neighbor	*Node connected to the same link.*
Neighbors Cache	*A cache supported by each IPv6 node that stores the IP address of its neighbors in the link, its corresponding link layer address, and an indication of its accessibility state. Neighbors cache is equivalent to ARP cache in IPv4.*
Neighbors Discovery (ND)	*A set of messages and ICMPv6 processes that fixes the relations between neighbors nodes. Neighbors discovery replaces ARP, ICMP routes discovery, and ICMP redirection messages used in IPv4. It also provides inaccessible neighbor detection.*
Neighbors Discovery Options	*Neighbors Discovery messages options that show link layer addresses, information about prefixes, MTU, routes, and configuration information for IPv6 mobility.*
Network Addresses Translator	*An IPv4 router that translate addresses and ports when sending packets between a network with private addresses and Internet.*
Network Prefix	*The fixed part of the address that is used to determine the subnetwork identifier, the route or the addresses range.*
Network Segment	*See Subnetwork.*
Next Hop Obtaining	*Process to obtain address- or next-hop-interface to facilitate sending a packet based on the routing table content.*

This Appendix for Chapter 1 provides a basic glossary of IPv6 terms and concepts that are loosely based on references [IPV200501] and [MSD200401].

Next-Level Aggregation Identifier (NLA ID)	24 bits field inside the global unicast aggregatable address that allows the creation of several hierarchical levels of addressing in its networks to organize addresses and routing to other ISPs, as well as to identify organization sites.
NLA ID	See Next-Level Aggregation Identifier.
No-Broadcast Multiple Access Link	A link layer technology that supports links with more than two nodes, but without allowing the sending of a packet to several destinations (broadcast). For example, X.25, Frame Relay and Asynchronous Transfer Mode (ATM).
Not Specified Address	0:0:0:0:0:0:0:0 (::) address is used to show the absence of any address, equivalent to IPv4 0.0.0.0 address.
Own Link	Home link. In mobile IP, the link in which the mobile node resides in its network. The mobile node uses the own link prefix to create its own address.
Packet	Protocol Data Unit (PDU) at Internet layer. In IPv6, a packet that consists of a header and an IPv6 payload.
Parameters Discovery	Neighbors Discovery process that allows equipments to know configuration parameters, including link MTU, and the default hops limit for outgoing packets.
Path Determination System	Procedure to select the route from the routing table the datagram will be forwarded through. That is, how the next router the datagram will be sent to is selected.
Path MTU	Maximum IPv6 packet size that can be sent without using fragmentation between a source and a destination over an IPv6 network route. The route MTU equates with the smallest link MTU for all links in such route.
Path Vector	A routing protocol's approach that involves the exchange of hops information sequences showing the path to follow in a route. For example, BGP-4 exchanges sequences of numbers of Autonomous Systems (ASs).
Path's MTU Discovery	Process relating to the use of Too Big message by means of ICMPv6 to discover the maximum IPv6 MTU value in all links between two devices.
PDU	See Protocol Data Unit.
Point-To-Point Protocol	Point-to-point network encapsulation method that provides frame delimiters, protocol identification, and integrity services at bit level.
Prefixes List	Link prefixes list supported by each host. Each entry defines the directly reachable IP addresses range, that is, neighbors.
Prefix-Length Notation	Notation used to represent network prefixes. It uses the address/prefix length form; this prefix length is the address' initial bits number that is employed to define the prefix.
Protocol Data Unit (PDU)	(aka datagram) The PDU is a fragment of a data sequence that is transmitted through the network. Data objects corresponding to a concrete layer in a network architecture consisting of layers. During transmission the data unit of the n layer turn into the payload of the n-1 layer (the lower layer).
Pseudo-Header	Provisional header that is built to calculate the needed checksum to associate the IPv6 header with the charge. IPv6 uses a new pseudo-header format to calculate UDP, TCP and ICMPv6 checksum.
Pseudo-Periodic	Event that is repeated at intervals of various lengths. For example, the routes advertisement sent by an IPv6 router is made at intervals that are calculated between a minimum and a maximum.
Reassembing	Procedure to rebuild the original charge of a datagram from several fragments.
Redirect	Procedure included in the neighbors discovery mechanisms to inform a host about the IPv6 address of another neighbor that is more appropriate as next hop to a destination.
Relay Router 6to4	An IPv6/IPv4 router that redirects traffic directed to 6to4 addresses between 6to4 routers in the Internet and IPv6 Internet devices.

This Appendix for Chapter 1 provides a basic glossary of IPv6 terms and concepts that are loosely based on references [IPV200501] and [MSD200401].

Router	*Node that can forward datagrams not specifically addressed to it. In an IPv6 network, a router is used to send advertisements related to its presence and its configuration information.*
Router Advertisement	*Neighbor discovery message sent by a router in a pseudo-periodic way or as a router solicitation message response. The advertisement includes, at least, information about a prefix that will be used later by the host to calculate its own unicast IPv6 address following the stateless mechanism.*
Router's Cache	*See Destination Cache.*
Routers Discovery	*Neighbors discovery process that allows for discovering of routers connected to a particular link.*
Routing Loop	*Undesirable situation in a network that causes the traffic was relayed over a closed loop, so it never reaches its destination.*
Scope	*For IPv6 addresses, the scope is the portion of the network to which the traffic will be propagated to.*
Scope ID	*The scope ID identifies a specific area within the reachability scope for nonglobal addresses. A node identifies each area of the same scope with a unique scope ID.*
Site Prefix	*A 48 bits prefix used to refer to all site addresses. Site prefixes are stored in a prefixes table that is used to confine traffic associated to these site prefixes.*
Site-Level Aggregation Identifier (SLA ID)	*16 bits field inside the global unicast address that uses an organization to identify subnetworks inside its own network.*
Solicited-Node Address	*Multicast address used by nodes during address resolution process. The solicited-node address facilitates efficient querying of network nodes during address resolution. IPv6 uses the Neighbor Solicitation message to perform address resolution. In IPv4, the ARP Request frame is sent to the MAC-level broadcast, disturbing all nodes on the network segment regardless of whether a node is running IPv4. For IPv6, instead of disturbing all IPv6 nodes on the local link by using the link-local scope all-nodes address, the solicited-node multicast address is used as the Neighbor Solicitation message destination. The solicited-node multicast address consists of the prefix FF02::1:FF00:0/104 and the last 24-bits of the IPv6 address that is being resolved.*
Static Routing	*Utilization of routes, introduced by hand, into routers routing tables.*
Subnet Anycast Router Address	*Anycast address (64 bits:: prefix) that is allocated to routers interfaces.*
Subnetwork	*One or more links that use the same 64 bits prefix in IPv6.*
Subnetwork Associated Path	*Path where the 64 bits prefix belongs to a concrete subnetwork.*
Suitable Path Selection	*The algorithm used by the routes selection procedure to choose the routes from the routing table that are nearer to the destination address the packet should be sent.*
TLA ID	*See Maximum-Level Aggregation Identifier.*
Top-Level Aggregation Identifier (TLA ID)	*See Maximum-Level Aggregation Identifier.*
Transition	*Conversion of IPv4-only nodes into dual-stack nodes or IPv6-only nodes.*
Tunnel	*An IPv6 over IPv4 tunnel, in which end points are specified by manual configuration.*

This Appendix for Chapter 1 provides a basic glossary of IPv6 terms and concepts that are loosely based on references [IPV200501] and [MSD200401].

Unicast Address

An address that identifies an only interface and allows network layer point-to-point communication. It identifies a single interface within the scope of the unicast address type. The following list shows the types of IPv6 addresses:

Aggregatable global unicast addresses

"Link-local" addresses

"Site-local" addresses

Special addresses, including unspecified and loopback addresses

Compatibility addresses, including 6to4 addresses

With the appropriate unicast routing topology, packets addressed to a unicast address are delivered to a single interface.

Appendix B: Basic Bibliography

IPv6 RFC Bibliography

4076   *Renumbering Requirements for Stateless Dynamic Host Configuration Protocol for IPv6 (DHCPv6), T. Chown, S. Venaas, A. Vijayabhaskar (May 2005)*

4074   *Common Misbehavior Against DNS Queries for IPv6 Addresses, Y. Morishita, T. Jinmei (May 2005)*

4057   *IPv6 Enterprise Network Scenarios, J. Bound, Ed. (June 2005)*

4038   *Application Aspects of IPv6 Transition, M-K. Shin, Ed., Y-G. Hong, J. Hagino, P. Savola, E. M. Castro (March 2005)*

4029   *Scenarios and Analysis for Introducing IPv6 into ISP Networks, M. Lind, V. Ksinant, S. Park, A. Baudot, P. Savola (March 2005)*

4007   *IPv6 Scoped Address Architecture, S. Deering, B. Haberman, T. Jinmei, E. Nordmark, B. Zill (March 2005)*

3956   *Embedding the Rendezvous Point (RP) Address in an IPv6 Multicast Address, P. Savola, B. Haberman (November 2004) (Updates RFC 3306)*

3919   *Remote Network Monitoring (RMON) Protocol Identifiers for IPv6 and Multiprotocol Label Switching (MPLS), E. Stephan, J. Palet (October 2004)*

3904   *Evaluation of IPv6 Transition Mechanisms for Unmanaged Networks, C. Huitema, R. Austein, S. Satapati, R. van der Pol (September 2004)*

3901   *DNS IPv6 Transport Operational Guidelines, A. Durand, J. Ihren (September 2004) (Also BCP91)*

3898   *Network Information Service (NIS) Configuration Options for Dynamic Host Configuration Protocol for IPv6 (DHCPv6), V. Kalusivalingam (October 2004)*

3849   *IPv6 Address Prefix Reserved for Documentation, G. Huston, A. Lord, P. Smith (July 2004)*

3831   *Transmission of IPv6 Packets over Fibre Channel C. DeSanti (July 2004)*

3810   *Multicast Listener Discovery Version 2 (MLDv2) for IPv6, R. Vida, Ed., L. Costa, Ed. (June 2004) (Updates RFC 2710)*

3776   *Using IPsec to Protect Mobile IPv6 Signaling Between Mobile Nodes and Home Agents, J. Arkko, V. Devarapalli, F. Dupont (June 2004)*

3775   *Mobility Support in IPv6, D. Johnson, C. Perkins, J. Arkko (June 2004)*

3769   *Requirements for IPv6 Prefix Delegation, S. Miyakawa, R. Droms (June 2004)*

3756   *IPv6 Neighbor Discovery (ND) Trust Models and Threats, P. Nikander, Ed., J. Kempf, E. Nordmark (May 2004)*

3750   *Unmanaged Networks IPv6 Transition Scenarios, C. Huitema, R. Austein, S. Satapati, R. van der Pol (April 2004)*

3736   *Stateless Dynamic Host Configuration Protocol (DHCP) Service for IPv6, R. Droms (April 2004)*

3701   *6bone (IPv6 Testing Address Allocation) Phaseout, R. Fink, R. Hinden (March 2004) (Obsoletes RFC 2471)*

3697   *IPv6 Flow Label Specification, J. Rajahalme, A. Conta, B. Carpenter, S. Deering (March 2004)*

3646   *DNS Configuration options for Dynamic Host Configuration Protocol for IPv6 (DHCPv6),R. Droms, Ed. (December 2003)*

3633   *IPv6 Prefix Options for Dynamic Host Configuration Protocol (DHCP) version 6, O. Troan, R. Droms (December 2003)*

3595   *Textual Conventions for IPv6 Flow Label, B. Wijnen (September 2003)*

3587   *IPv6 Global Unicast Address Format, R. Hinden, S. Deering, E. Nordmark (August 2003) (Obsoletes RFC 2374)*

3582   *Goals for IPv6 Site-Multihoming Architectures, J. Abley, B. Black, V. Gill (August 2003)*

41

IPv6 RFC Bibliography

3542	Advanced Sockets Application Program Interface (API) for IPv6, W. Stevens, M. Thomas, E. Nordmark, T. Jinmei (May 2003) Obsoletes RFC 2292)
3531	A Flexible Method for Managing the Assignment of Bits of an IPv6 Address, Block M. Blanchet (April 2003)
3513	Internet Protocol Version 6 (IPv6) Addressing Architecture, R. Hinden, S. Deering (April 2003) (Obsoletes RFC 2373)
3493	Basic Socket Interface Extensions for IPv6, R. Gilligan, S. Thomson, J. Bound, J. McCann, W. Stevens (February 2003) (Obsoletes RFC 2553)
3484	Default Address Selection for Internet Protocol version 6 (IPv6), R. Draves (February 2003)
3364	Tradeoffs in Domain Name System (DNS) Support for Internet Protocol version 6 (IPv6), R. Austein (August 2002) (Updates RFC 2673, RFC 2874)
3363	Representing Internet Protocol version 6 (IPv6) Addresses in the Domain Name System (DNS), R. Bush, A. Durand, B. Fink, O. Gudmundsson, T. Hain (August 2002) (Updates RFC 2673, RFC 2874)
3316	Internet Protocol Version 6 (IPv6) for Some Second and Third Generation Cellular Hosts, J. Arkko, G. Kuijpers, H. Soliman, J. Loughney, J. Wiljakka (April 2003)
3315	Dynamic Host Configuration Protocol for IPv6 (DHCPv6), R. Droms, Ed., J. Bound, B. Volz, T. Lemon, C. Perkins, M. Carney (July 2003)
3314	Recommendations for IPv6 in Third Generation Partnership Project (3GPP) Standards, M. Wasserman, Ed. (September 2002)
3307	Allocation Guidelines for IPv6 Multicast Addresses, B. Haberman (August 2002)
3306	Unicast-Prefix-based IPv6 Multicast Addresses, B. Haberman, D. Thaler (August 2002) (Updated by RFC 3956)
3266	Support for IPv6 in Session Description Protocol (SDP), S. Olson, G. Camarillo, A. B. Roach (June 2002) (Updates RFC 2327)
3226	DNSSEC and IPv6 A6 aware server/resolver message size requirements, O. Gudmundsson (December 2001) Updates RFC 2535, RFC 2874)(Updated by RFC 4033, RFC 4034, RFC 4035)
3178	IPv6 Multihoming Support at Site Exit Routers, J. Hagino, H. Snyder (October 2001)
3177	IAB/IESG Recommendations on IPv6 Address Allocations to Sites, IAB/IESG (September 2001)
3175	Aggregation of RSVP for IPv4 and IPv6 Reservations, F. Baker, C. Iturralde, F. Le Faucheur, B. Davie (September 2001)
3162	RADIUS and IPv6, B. Aboba, G. Zorn, D. Mitton (August 2001)
3146	Transmission of IPv6 Packets over IEEE 1394 Networks, K. Fujisawa, A. Onoe (October 2001)
3142	An IPv6-to-IPv4 Transport Relay Translator, J. Hagino, K. Yamamoto (June 2001)
3122	Extensions to IPv6 Neighbor Discovery for Inverse Discovery Specification, A. Conta (June 2001)
3111	Service Location Protocol Modifications for IPv6, E. Guttman (May 2001)
3089	A SOCKS-based IPv6/IPv4 Gateway Mechanism, H. Kitamura (April 2001)
3056	Connection of IPv6 Domains via IPv4, Clouds B. Carpenter, K. Moore (February 2001)
3053	IPv6 Tunnel Broker, A. Durand, P. Fasano, I. Guardini, D. Lento (January 2001)
3041	Privacy Extensions for Stateless Address Autoconfiguration in IPv6, T. Narten, R. Draves (January 2001)
2928	Initial IPv6 Sub-TLA ID Assignments, R. Hinden, S. Deering, R. Fink, T. Hain (September 2000)
2894	Router Renumbering for IPv6, M. Crawford (August 2000) (TXT = 69135 bytes)
2893	Transition Mechanisms for IPv6 Hosts and Routers, R. Gilligan, E. Nordmark (August 2000) (Obsoletes RFC 1933)
2874	DNS Extensions to Support IPv6 Address Aggregation and Renumbering, M. Crawford, C. Huitema (July 2000) (Updates RFC 1886) (Updated by RFC 3152, RFC 3226, RFC 3363, RFC 3364)
2740	OSPF for IPv6, R. Coltun, D. Ferguson, J. Moy (December 1999)

IPv6 RFC Bibliography

2732   *Format for Literal IPv6 Addresses in URL's, R. Hinden, B. Carpenter, L. Masinter (December 1999) (Obsoleted by RFC 3986) (Updates RFC 2396)*

2711   *IPv6 Router Alert Option, C. Partridge, A. Jackson (October 1999)*

2710   *Multicast Listener Discovery (MLD) for IPv6, S. Deering, W. Fenner, B. Haberman (October 1999) (Updated by RFC 3590, RFC 3810)*

2675   *IPv6 Jumbograms, D. Borman, S. Deering, R. Hinden (August 1999)(Obsoletes RFC 2147)*

2590   *Transmission of IPv6 Packets over Frame Relay Networks Specification, A. Conta, A. Malis, M. Mueller (May 1999)*

2553   *Basic Socket Interface Extensions for IPv6, R. Gilligan, S. Thomson, J. Bound, W. Stevens (March 1999) (Obsoletes RFC 2133) (Obsoleted by RFC 3493) (Updated by RFC 3152)*

2553   *Basic Socket Interface Extensions for IPv6, R. Gilligan, S. Thomson, J. Bound, W. Stevens (March 1999) (Obsoletes RFC 2133) (Obsoleted by RFC 3493) (Updated by RFC 3152)*

2529   *Transmission of IPv6 over IPv4 Domains without Explicit Tunnels, B. Carpenter, C. Jung (March 1999)*

2526   *Reserved IPv6 Subnet Anycast Addresses, D. Johnson, S. Deering (March 1999)*

2492   *IPv6 over ATM Networks, G. Armitage, P. Schulter, M. Jork (January 1999)*

2491   *IPv6 over Non-Broadcast Multiple Access (NBMA) networks, G. Armitage, P. Schulter, M. Jork, G. Harter (January 1999)*

2474   *Definition of the Differentiated Services Field (DS Field) in the IPv4 and IPv6 Headers, K. Nichols, S. Blake, F. Baker, D. Black (December 1998) (Obsoletes RFC 1455, RFC 1349)(Updated by RFC 3168, RFC 3260)*

2473   *Generic Packet Tunneling in IPv6 Specification, A. Conta, S. Deering (December 1998)*

2472   *IP Version 6 over PPP D. Haskin, E. Allen (December 1998) (Obsoletes RFC 2023)*

2471   *IPv6 Testing Address Allocation, R. Hinden, R. Fink, J. Postel (deceased) (December 1998) (Obsoletes RFC 1897) (Obsoleted by RFC 3701)*

2470   *Transmission of IPv6 Packets over Token Ring Networks, M. Crawford, T. Narten, S. Thomas (December 1998)*

2467   *Transmission of IPv6 Packets over FDDI Networks, M. Crawford (December 1998 (Obsoletes RFC 2019)*

2466   *Management Information Base for IP Version 6: ICMPv6 Group, D. Haskin, S. Onishi (December 1998)*

2465   *Management Information Base for IP Version 6: Textual Conventions and General Group, D. Haskin, S. Onishi (December 1998)*

2464   *Transmission of IPv6 Packets over Ethernet Networks, M. Crawford (December 1998 (Obsoletes RFC 1972)*

2463   *Internet Control Message Protocol (ICMPv6) for the Internet Protocol Version 6 (IPv6) Specification, A. Conta, S. Deering (December 1998) (Obsoletes RFC 1885)*

2462   *IPv6 Stateless Address Autoconfiguration, S. Thomson, T. Narten (December 1998) (Obsoletes RFC 1971)*

2461   *Neighbor Discovery for IP Version 6 (IPv6), T. Narten, E. Nordmark, W. Simpson (December 1998) (Obsoletes RFC 1970)*

2460   *Internet Protocol, Version 6 (IPv6) Specification, S. Deering, R. Hinden (December 1998)(Obsoletes RFC 1883)*

2428   *FTP Extensions for IPv6 and NATs,M. Allman, S. Ostermann, C. Metz (September 1998)*

2375   *IPv6 Multicast Address Assignments, R. Hinden, S. Deering (July 1998)*

2374   *An IPv6 Aggregatable Global Unicast Address Forma,t R. Hinden, M. O'Dell, S. Deering (July 1998) Obsoletes RFC 2073) (Obsoleted by RFC 3587)*

2373   *IP Version 6 Addressing Architecture, R. Hinden, S. Deering (July 1998) Obsoletes RFC 1884) (Obsoleted by RFC 3513)*

2292   *Advanced Sockets API for IPv6, W. Stevens, M. Thomas (February 1998)(Obsoleted by RFC 3542)*

2185   *Routing Aspects of IPv6 Transition, R. Callon, D. Haskin (September 1997)*

IPv6 RFC Bibliography

2147   TCP and UDP over IPv6 Jumbograms, D. Borman (May 1997)(Obsoleted by RFC 2675)

2133   Basic Socket Interface Extensions for IPv6, R. Gilligan, S. Thomson, J. Bound, W. Stevens (April 1997) Obsoleted by RFC 2553)

2080   RIPng for IPv6, G. Malkin, R. Minnear (January 1997)

2073   An IPv6 Provider-Based Unicast Address Format, Y. Rekhter, P. Lothberg, R. Hinden, S. Deering, J. Postel (January 1997) Obsoleted by RFC 2374)

2030   Simple Network Time Protocol (SNTP) Version 4 for IPv4, IPv6 and OSI, D. Mills (October 1996) (Obsoletes RFC 1769)

2019   Transmission of IPv6 Packets Over FDDI, M. Crawford (October 1996) (TXT = 12344 bytes)(Obsoleted by RFC 2467)

1972   A Method for the Transmission of IPv6 Packets over Ethernet Networks, M. Crawford (August 1996) (Obsoleted by RFC 2464)

1971   IPv6 Stateless Address Autoconfiguration, S. Thomson, T. Narten (August 1996) (Obsoleted by RFC 2462)

1970   Neighbor Discovery for IP Version 6 (IPv6), T. Narten, E. Nordmark, W. Simpson (August 1996) (Obsoleted by RFC 2461)

1933   Transition Mechanisms for IPv6 Hosts and Routers, R. Gilligan, E. Nordmark (April 1996) Obsoleted by RFC 2893)

1924   A Compact Representation of IPv6 Addresses, R. Elz (Apr-01-1996)

1897   IPv6 Testing Address Allocation, R. Hinden, J. Postel (January 1996) (Obsoleted by RFC 2471)

1888   OSI NSAPs and IPv6, J. Bound, B. Carpenter, D. Harrington, J. Houldsworth, A. Lloyd (August 1996)

1887   An Architecture for IPv6 Unicast Address Allocation, Y. Rekhter, T. Li, Eds. (December 1995)

1885   Internet Control Message Protocol (ICMPv6) for the Internet Protocol Version 6 (IPv6), A. Conta, S. Deering (December 1995)(Obsoleted by RFC 2463)

1884   IP Version 6 Addressing Architecture, R. Hinden, S. Deering, Eds. (December 1995) (Obsoleted by RFC 2373)

1883   Internet Protocol, Version 6 (IPv6) Specification, S. Deering, R. Hinden (December 1995) (Obsoleted by RFC 2460)

1881   IPv6 Address Allocation Management IAB/ESG (December 1995)

1809   Using the Flow Label Field in IPv6, C. Partridge (June 1995)

NAT RFC Bibliography

4008   Definitions of Managed Objects for Network Address Translators (NAT), R. Rohit, P. Srisuresh, R. Raghunarayan, N. Pai, C. Wang (March 2005)

3947   Negotiation of NAT-Traversal in the IKE, T. Kivinen, B. Swander, A. Huttunen, V. Volpe (January 2005)

3715   IPsec-Network Address Translation (NAT) Compatibility Requirements, B. Aboba, W. Dixon (March 2004)

3519   Mobile IP Traversal of Network Address Translation (NAT) Devices,H. Levkowetz, S. Vaarala (May 2003)

3235   Network Address Translator (NAT)-Friendly Application Design Guidelines, D. Senie (January 2002)

3022   Traditional IP Network Address Translator (Traditional NAT), P. Srisuresh, K. Egevang (January 2001) (Obsoletes RFC 1631)

2993   Architectural Implications of NAT, T. Hain (November 2000)

2766   Network Address Translation - Protocol Translation (NAT-PT), G. Tsirtsis, P. Srisuresh (February 2000) Updated by RFC 3152)

2709   Security Model with Tunnel-mode IPsec for NAT Domains, P. Srisuresh (October 1999)

2663   IP Network Address Translator (NAT) Terminology and Considerations, P. Srisuresh, M. Holdrege (August 1999)

1631   The IP Network Address Translator (NAT),K. Egevang, P. Francis (May 1994) (Obsoleted by RFC 3022)

SIP RFC Bibliography

4092   Usage of the Session Description Protocol (SDP) Alternative Network Address Types (ANAT) Semantics in the Session Initiation Protocol (SIP), G. Camarillo, J. Rosenberg (June 2005)

4083   Input 3rd-Generation Partnership Project (3GPP) Release 5 Requirements on the Session Initiation Protocol (SIP), M. Garcia-Martin (May 2005)

4032   Update to the Session Initiation Protocol (SIP) Preconditions Framework, G. Camarillo, P. Kyzivat (March 2005) (Updates RFC 3312)

4028   Session Timers in the Session Initiation Protocol (SIP), S. Donovan, J. Rosenberg (April 2005)

3976   Interworking SIP and Intelligent Network (IN) Applications, V. K. Gurbani, F. Haerens, V. Rastogi (January 2005)

3969   The Internet Assigned Number Authority (IANA) Uniform Resource Identifier (URI) Parameter Registry for the Session Initiation Protocol (SIP), G. Camarillo (December 2004) Updates RFC 3427) (Also BCP99)

3968   The Internet Assigned Number Authority (IANA) Header Field Parameter Registry for the Session Initiation Protocol (SIP), G. Camarillo (December 2004)(Updates RFC 3427) (Also BCP98)

3960   Early Media and Ringing Tone Generation in the Session Initiation Protocol (SIP), G. Camarillo, H. Schulzrinne (December 2004)

3959   The Early Session Disposition Type for the Session Initiation Protocol (SIP), G. Camarillo (December 2004)

3911   The Session Initiation Protocol (SIP) "Join" Header, R. Mahy, D. Petrie (October 2004)

3903   Session Initiation Protocol (SIP) Extension for Event State Publication, A. Niemi, Ed. (October 2004)

3893   Session Initiation Protocol (SIP) Authenticated Identity Body (AIB) Format, J. Peterson (September 2004)

3892   The Session Initiation Protocol (SIP) Referred-By Mechanism, R. Sparks (September 2004)

3891   The Session Initiation Protocol (SIP) "Replaces" Header,R. Mahy, B. Biggs, R. Dean (September 2004)

3857   A Watcher Information Event Template-Package for the Session Initiation Protocol (SIP), J. Rosenberg (August 2004)

3856   A Presence Event Package for the Session Initiation Protocol (SIP), J. Rosenberg (August 2004)

3853   S/MIME Advanced Encryption Standard (AES) Requirement for the Session Initiation Protocol (SIP), J. Peterson (July 2004) Updates RFC 3261)

3842   A Message Summary and Message Waiting Indication Event Package for the Session Initiation Protocol (SIP), R. Mahy (August 2004)

3841   Caller Preferences for the Session Initiation Protocol (SIP), J. Rosenberg, H. Schulzrinne, P. Kyzivat (August 2004)

3840   Indicating User Agent Capabilities in the Session Initiation Protocol (SIP), J. Rosenberg, H. Schulzrinne, P. Kyzivat (August 2004)

3824   Using E.164 numbers with the Session Initiation Protocol (SIP), J. Peterson, H. Liu, J. Yu, B. Campbell (June 2004)

3764   Enumservice registration for Session Initiation Protocol (SIP) Addresses-of-Record, J. Peterson (April 2004)

3725   Best Current Practices for Third Party Call Control (3pcc) in the Session Initiation Protocol (SIP), J. Rosenberg, J. Peterson, H. Schulzrinne, G. Camarillo (April 2004) (Also BCP85)

3702   Authentication, Authorization, and Accounting Requirements for the Session Initiation Protocol (SIP), J. Loughney, G. Camarillo (February 2004)

3680   A Session Initiation Protocol (SIP) Event Package for Registrations J. Rosenberg (March 2004)

3666   Session Initiation Protocol (SIP) Public Switched Telephone Network (PSTN) Call Flows, A. Johnston, S. Donovan, R. Sparks, C. Cunningham, K. Summers (December 2003) (Also BCP76)

3665   Session Initiation Protocol (SIP) Basic Call Flow Examples, A. Johnston, S. Donovan, R. Sparks, C. Cunningham, K. Summers (December 2003) (Also BCP75)

SIP RFC Bibliography

3608   Session Initiation Protocol (SIP) Extension Header Field for Service Route Discovery During Registration, D. Willis, B. Hoeneisen (October 2003)

3603   Private Session Initiation Protocol (SIP) Proxy-to-Proxy Extensions for Supporting the PacketCable Distributed Call Signaling Architecture, W. Marshall, Ed., F. Andreasen, Ed. (October 2003)

3581   An Extension to the Session Initiation Protocol (SIP) for Symmetric Response Routing, J. Rosenberg, H. Schulzrinne (August 2003)

3578   Mapping of Integrated Services Digital Network (ISDN) User Part (ISUP) Overlap Signalling to the Session Initiation Protocol (SIP), G. Camarillo, A. B. Roach, J. Peterson, L. Ong (August 2003)

3515   The Session Initiation Protocol (SIP) Refer Method, R. Sparks (April 2003)

3487   Requirements for Resource Priority Mechanisms for the Session Initiation Protocol (SIP) H. Schulzrinne (February 2003)

3486   Compressing the Session Initiation Protocol (SIP), G. Camarillo (February 2003)

3485   The Session Initiation Protocol (SIP) and Session Description Protocol (SDP) Static Dictionary for Signaling Compression (SigComp), M. Garcia-Martin, C. Bormann, J. Ott, R. Price, A. B. Roach (February 2003)

3455   Private Header (P-Header) Extensions to the Session Initiation Protocol (SIP) for the 3rd-Generation Partnership Project (3GPP), M. Garcia-Martin, E. Henrikson, D. Mills (January 2003)

3428   Session Initiation Protocol (SIP) Extension for Instant Messaging, B. Campbell, Ed., J. Rosenberg, H. Schulzrinne, C. Huitema, D. Gurle (December 2002)

3427   Change Process for the Session Initiation Protocol (SIP), A. Mankin, S. Bradner, R. Mahy, D. Willis, J. Ott, B. Rosen (December 2002) (Updated by RFC 3968, RFC 3969)(Also BCP67)

3398   Integrated Services Digital Network (ISDN) User Part (ISUP) to Session Initiation Protocol (SIP) Mapping, G. Camarillo, A. B. Roach, J. Peterson, L. Ong (December 2002)

3372   Session Initiation Protocol for Telephones (SIP-T): Context and Architectures, A. Vemuri, J. Peterson (September 2002) (Also BCP63)

3361   Dynamic Host Configuration Protocol (DHCP-for-IPv4) Option for Session Initiation Protocol (SIP) Servers, H. Schulzrinne (August 2002)

3351   User Requirements for the Session Initiation Protocol (SIP) in Support of Deaf, Hard of Hearing and Speech-impaired Individuals, N. Charlton, M. Gasson, G. Gybels, M. Spanner, A. van Wijk (August 2002)

3329   Security Mechanism Agreement for the Session Initiation Protocol (SIP), J. Arkko, V. Torvinen, G. Camarillo, A. Niemi, T. Haukka (January 2003)

3327   Session Initiation Protocol (SIP) Extension Header Field for Registering Non-Adjacent Contacts, D. Willis, B. Hoeneisen (December 2002)

3326   The Reason Header Field for the Session Initiation Protocol (SIP), H. Schulzrinne, D. Oran, G. Camarillo (December 2002)

3325   Private Extensions to the Session Initiation Protocol (SIP) for Asserted Identity within Trusted Networks, C. Jennings, J. Peterson, M. Watson (November 2002)

3324   Short Term Requirements for Network Asserted Identity M. Watson (November 2002)

3323   A Privacy Mechanism for the Session Initiation Protocol (SIP) J. Peterson (November 2002)

3322   Signaling Compression (SigComp) Requirements and Assumptions, H. Hannu (January 2003)

3321   Signaling Compression (SigComp) - Extended Operations, H. Hannu, J. Christoffersson, S. Forsgren, K.-C. Leung, Z. Liu, R. Price (January 2003)

3320   Signaling Compression (SigComp), R. Price, C. Bormann, J. Christoffersson, H. Hannu, Z. Liu, J. Rosenberg (January 2003)

3319   Dynamic Host Configuration Protocol (DHCPv6) Options for Session Initiation Protocol (SIP) Servers, H. Schulzrinne, B. Volz (July 2003)

3313   Private Session Initiation Protocol (SIP) Extensions for Media Authorization, W. Marshall, Ed. (January 2003)

SIP RFC Bibliography

3312   *Integration of Resource Management and Session Initiation Protocol (SIP), G. Camarillo, Ed., W. Marshall, Ed., J. Rosenberg (October 2002) (Updated by RFC 4032)*

3311   *The Session Initiation Protocol (SIP) UPDATE Method, J. Rosenberg (October 2002)*

3265   *Session Initiation Protocol (SIP)-Specific Event Notification, A. B. Roach (June 2002) (Obsoletes RFC 2543) (Updates RFC 3261)*

3263   *Session Initiation Protocol (SIP): Locating SIP Servers, J. Rosenberg, H. Schulzrinne (June 2002) (Obsoletes RFC 2543)*

3262   *Reliability of Provisional Responses in Session Initiation Protocol (SIP), J. Rosenberg, H. Schulzrinne (June 2002) (Obsoletes RFC 2543)*

3261   *SIP: Session Initiation Protocol, J. Rosenberg, H. Schulzrinne, G. Camarillo, A. Johnston, J. Peterson, R. Sparks, M. Handley, F. Schooler (June 2002) (Obsoletes RFC 2543) (Updated by RFC 3265, RFC 3853)*

3087   *Control of Service Context using SIP Request-URI, B. Campbell, R. Sparks (April 2001)*

3050   *Common Gateway Interface for SIP, J. Lennox, H. Schulzrinne, J. Rosenberg (January 2001)*

2976   *The SIP INFO Method, S. Donovan (October 2000)*

2848   *The PINT Service Protocol: Extensions to SIP and SDP for IP Access to Telephone Call Services, S. Petrack, L. Conroy (June 2000)*

2543   *SIP: Session Initiation Protocol, M. Handley, H. Schulzrinne, E. Schooler, J. Rosenberg (March 1999) (Obsoleted by RFC 3261, RFC 3262, RFC 3263, RFC 3264, RFC 3265)*

Basic VoP/VoIP Concepts

Considerable interest has been shown during the past quarter century in supporting Voice over Packet (VoP)-based networks. Several packet technologies have been proposed, tested and deployed over the years, including Voice over X.25 (VoX25) networks, Voice over Frame Relay (VoFR) networks, Voice over Asynchronous Transfer Mode (VoATM) networks, Voice over IP (VoIP) networks, Voice over Wi-Fi/Hotspot networks, and Voice over Multiprotocol Label Switching (VoMPLS) networks (e.g., see [MIN199801], [MIN199802], [MIN200102], [MIN200203], among other references.) Of these, VoIP has seen the largest market penetration in recent years, and it has stirred the most interest as a technology to perhaps replace the traditional circuit-switched mechanisms of the public-switched telephone network. Hence, there is keen research and commercial interest in VoIP at this time. Issues surrounding VoP/VoIP include the following six major areas:

1. Voice digitization/compression
2. Standards in the user plane (information flow) and in the control plane (call/session signaling/management)
3. Signaling
4. Numbering
5. Applications
6. Wireless deployment

This chapter provides an overview of these topics. We are not deliberately focusing on IPv6-related issues in this chapter; rather, we simply aim at establishing a baseline of the VoP space and supportive technologies before proceeding with the major topic at hand. Also, note that in Chapter 1 we have positioned the discussion (motivation) of IPv6 in part as a solution to scalability limitations of IPv4-based VoIP; hence, the discussion here and in the chapters that follow partially takes on the perspective of organizations that may have to deal with such scalability considerations, namely, service providers and telecommunication carriers (and, also equipment providers to these industries); this approach also applies to large firms with 10,000–50,000 users or more.

2.1 Introduction and Background

There are conceivably several possible motivations for considering a packetized approach to voice, as advocated by various industry constituencies. Ranking these publicly-stated industry motivations by the order that would make the most sense, one has:

- New applications become possible with VoP, thereby generating opportunities for new services and new revenues. New applications could include, for example, the following, among others: mobility, unified messaging, converged-delivery of any kind of information service, Computer Telephony Integration (CTI) applications for Contact Center, including virtual contact centers and, hosted "IP PBX/Centex" services without traditional distance limitations.

- Cost savings can be achieved by the carrier by using a packetized technology for convergence in some or all of the following: the operations budget, the equipment budget, or the transmission budget. These savings can then be passed on to the users.
- Voice still brings in about 80% of the revenues for carriers; the U.S. voice revenues were around $200B per year at press time, and the worldwide revenues were around $800B, including mobile services[1,2]. Therefore, voice is a desirable market to optimize (with new technologies) and/or penetrate (the data revenue figure being partitioned as 13% corporate and 7% Internet[3].)
- An elegant new integrated (converged—"triple play") architecture becomes possible with a connectionless packet: a new does-it-all network that is a based on a single approach supporting a gamut of services.
- IP has become ubiquitous in the data arena. It is desirable, therefore (according to proponents), to use IP for everything, and IP can be made good enough to support anything and everything.

It appears at face value that new applications have taken somewhat of a back seat during the past few years of advocacy, while transfer of market share (from traditional equipment providers to newer equipment providers and from traditional carriers to new carriers; for example, Multiple System Operators [MSOs]) appears to have been a key driver for proponents. VoP conserves bandwidth, and this has been the secondary key focus of the proponents. However, backbone bandwidth has become a near-commodity of late and bandwidth conservation is of limited interest in terms of the overall benefits that VoP can afford—optical transmission links now can carry in the range of Tbps. Time Division Multiplexing (TDM) trunk replacement with Statistical TDM (STDM) trunks does little to change this revenue picture of the service providers. Also, transmission costs tend to represent only 10–15% of the total budget of carriers. Hence, decreasing this cost by even 95% has limited impact to the bottom line. After all, we have had more than 125 years to optimize voice transport. If VoP only does something to transport, then it will have rather limited success in the general market. And the "packet" portion of VoP has very little to do with the bandwidth savings itself: that credit must go to the vocoding technology. This compression could well be accomplished without any packetization of any kind (IP, ATM, frame, or MPLS.)

Traditionally, there are two business approaches to carrier operations: (1) introduce new technology to reduce costs—this is most effective when the technology aims at reducing people's costs (for example, automation) and in improving Operations, Administration, Maintenance, and Provisioning (OAM&P); and, (2) introduce new technology to bring about new services and new applications. Although a hybrid strategy that pursues both approaches is ideal, if one had to pick only one, then from a macro-economics point-of-view, the latter strategy would appear to be better.

In a greenfield environment, the planner in a carrier might look at deploying a VoP architecture rather than a traditional Class 5 TDM switch for transmission savings and switch cost reduction, when small-to-medium switches are needed[4]. But, in existing environments, the advantages to VoP have to be secured through new applications; obviously, at some late stage such deployment may be based on *technology pull*; however, that

---

[1] Internet, and private lines represented a $100B international market in 2002. The majority of this figure, however, is from private line services.

[2] The 80% figure is an often-quoted number. However, actual numbers showed a $900B market in 2002 with about $50B in data and Internet (the remainder of the $120B "other" bucket is for private line services). This actually makes the data and Internet figure about 5.5% (= 50/900).

[3] Source: Nortel Networks, Lehman Brothers, Merrill Lynch, Nortel.

[4] The cost per line of a traditional switch is in the range of $300–350.

is not yet the case today. In recent years, the competitive pendulum has shifted against the formation of new (greenfield) carriers such as Competitive Local Exchange Carriers (CLECs), Data Local Exchange Carriers (DLECs), Rural Local Exchange Carriers (RLECs), Ethernet Local Exchange Carriers (ELECs), and Building Local Exchange Carriers (BLECs). And, while some pure-play carriers in Asia and elsewhere may in fact utilize VoP/softswitch technology, the case is not yet so to any significant extent in North America, with only a handful of such pure-play carriers achieving brand name recognition.

Several technical factors that have held back the deployment of VoP/VoIP on a broad scale are (1) Quality of Service (QoS) considerations for packet networks, especially for network-to-network (carrier-to-carrier) environments; (2) robust signaling in support of interworking with the embedded PSTN, which is not going away any time in the foreseeable decade or two; and, (3) security considerations including coexistence with typical firewall environments (for example, questions may arise if to deploy various network elements supporting VoIP in the uncontrolled domain, the controlled domain, the restricted domain, or the secure domain—such decision clearly impacts firewall/DMZ architecture considerations—this topic is revisited in Chapter 5.)

MPLS is a relatively new technology that is expected to be utilized in the immediate future for core networks, including converged data and voice networks, possibly prior to IPv6, but that remains to be seen in the final analysis. The promise of MPLS is to: (1) provide a connection-oriented protocol for Layer 3 IP; (2) support the ability to traffic engineer the network; and, (3) support wire-speed forwarding of protocol data units. MPLS enhances the services that can be provided by IP networks, offering Traffic Engineering (TE, that is, specified routes through the network), "guaranteed" QoS, and Virtual Private Networks (VPNs). MPLS enjoys certain attributes that, prima facie, make it a better technology than pure IP to support packetized voice applications. Proponents see MPLS as a key development in IP/Internet technologies that will assist in adding a number of essential capabilities to today's best effort packet networks, including traffic engineering capabilities, providing traffic with different qualitative Classes of Service (CoS), providing traffic with different quantitative QoS and, providing IP-based VPNs. The improved traffic management, the QoS capabilities, and the expedited packet-forwarding via the label mechanism could offer some technical advantage to voice. Other "purists" prefer a basic IP approach due to its perceived simplicity and broad applicability. However, there are fundamental scalability and end-to-end protocol issues with pre-IPv6 technologies, as we discussed in Chapter 1. Hence, the longer term solution (2–5 years out) is to plan the use of IPv6 architectures.

This chapter covers both VoIP and VoMPLS (we do not cover VoATM or VoFR[6]). Table 2.1 provides a small (far from exhaustive) set of VoIP-related terms [VOV200501].

Table 2.1: Small set of of VoIP-related terms.

Term	Definition
Common Open Policy Service (COPS)	*A query and response protocol that can be used to exchange policy information between a policy server and its clients. Defined in IETF RFC 2748.*
COPS	*See Common Open Policy Service.*
GDOI	*See Group Domain of Interpretation Group Keying.*

---

[6]   The reader may refer to reference [MIN199801] for a treatment of that topic.

Term	Definition
Group Domain of Interpretation (GDOI) Group Keying	Process that provides a means for a group of users or devices to share cryptographic keys, get efficient key updates, and efficient remove group members. Defined in IETF RFC 3547 and tracked by the IETF MSEC Working Group.
Media Gateway Control Protocol (MGCP)	A protocol used for controlling VoIP Gateways from external call control elements. Described in IETF RFC 2705 and tracked by the International Softswitch Consortium (ISC) Device Control Working Group.
MGCP	See Media Gateway Control Protocol.
RADIUS	See Remote Authentication Dial-In User Service.
Real-Time Streaming Protocol (RTSP)	An application-level protocol for control over the delivery of data with real-time properties. Described in IETF RFC 2326.
Real-Time Transport Protocol (RTP)	A protocol used to carry streaming real-time multimedia data over IP Networks. Defined in IETF RFC 1889.
Remote Authentication Dial-In User Service (RADIUS)	Protocol used for authentication and authorization, as well as billing information that details the service or services being delivered to the end user. Described in IETF RFC 2138 and RFC 2139.
RTP	See Real-Time Transport Protocol.
RTSP	See Real-Time Streaming Protocol.
Secure Real-Time Transport Protocol (SRTP)	Secure real-time transport protocol that provides confidentiality, message authentication, and replay protection for Real-time Transport Protocol (RTP).
Session Initiation Protocol (SIP)	An application-layer control protocol defined by Internet Engineering Task Force (IETF) that can establish, modify and terminate multimedia sessions or calls. These multimedia sessions include invitations to both unicast and multicast conferences and Internet telephony applications. SIP can be used in conjunction with other call setup and signaling protocols, and is originally described in IETF RFC 2543 (March 1999), now obsoleted by RFC 3261, RFC 3262, RFC 3263, RFC 3264, and RFC 3265 – June 2002). IETF SIP Working Group tracks issue.
SIP	See Session Initiation Protocol.
SRTP	See Secure Real-time Transport Protocol.
Telephony Routing over IP (TRIP)	A policy-driven, dynamic routing protocol used for advertising a range of possible telephony destinations and their routing attributes. TRIP can serve as the telephony routing protocol for any signaling protocol. Described in IETF RFC 2871 and tracked by the IETF IPTEL Working Group.
TRIP	See Telephony Routing over IP.

2.1.1 Carriers' Voice Networks

Figure 2.1 depicts a boiled down network for a traditional Interexchange Carrier (IXC) where VoP would need to coexist and support connectivity and information flows (note that with the recent mergers and acquisitions in the telecom industry, the lines of demarcation between an interexchange carrier and a local exchange carrier are blurring considerably; however, nearly all of the equipment in place remains unaffected). While some believe that one should use IP (or, more specifically, packet) for everything, it has always been known that multiplexing in general, and statistical multiplexing in particular, only pay for themselves at the transmission level when the cost of the link to be shared is high (e.g., national or international applications). Therefore, if the main tenet of the VoP industry is that VoP is bandwidth-efficient, then it will simply have an extremely limited ILEC/local service market; the converse would be true if the focus of VoIP were in new services, which also addresses the converged environment rubric. As a further challenge to the VoP industry, the cost of long-haul bandwidth is at a historical low. The fact is that engineers designing new technology must move beyond the dry clinical abstraction offered by packet disciplines, queue management, and protocol state machine, and offer the buyers of their gear (the carriers) and the clients of the buyers of their gear (the end users) new solutions, not just functionally-equivalent new technology.

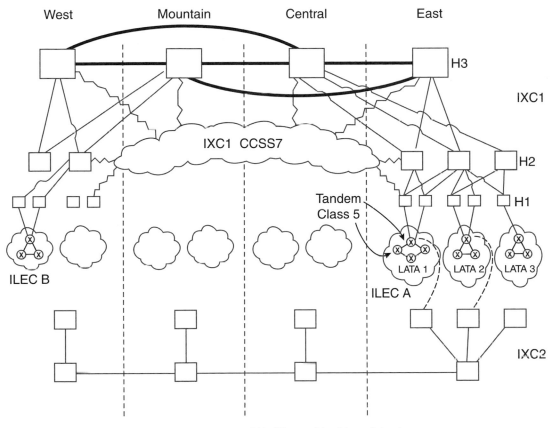

H1: Hierarchical Level 1, etc.

Note: Cable companies, CLECs, cellular providers,
and hotspot service providers are not shown for simplicity.

Figure 2.1: Typical traditional IXC network.

Indications are that carriers are expected to deploy VoP when:

- it provides major new revenue opportunities over and above the revenue stream the carriers currently have;
- or, it provides major savings in the OAM&P side of the house.

A major breakthrough would be the introduction of an entire set of new data-enriched voice applications. For example, follow-me/find-me, distributed call centers, dynamic Interactive Voice Response (IVR), personal voice mail, unified messaging, advanced presence services, and others.

2.1.2 VoIP in Cable TV Environments

The cable industry is now expanding its competitive offerings to include business and residential telephone services delivered over its fiber optic infrastructure. Cable-delivered telephone service is a natural extension of a network already capable of delivering services and products not possible just a few short years ago. Once upgraded to two-way fiber optics, a cable system can offer telephone service over the same cable line that already carries digital video, high-speed Internet, and other advanced services to consumers. Many of the nation's largest Multiple System Operators (MSOs) (also known as *cable TV providers*, or simply, *cable companies*) now offer residential and/or commercial phone service based on VoIP.

According to data from the National Cable & Telecommunications Association (NCTA), about 73 million households had cable service in the U.S. at the beginning of 2002, which represents about 70% of TV households; these numbers have continued to increase since then—the industry has enjoyed a good growth rate during the past decade or more. NCTA notes that though still a new business, cable telephony is a key component of the cable industry's business strategy in the coming years: with the continued improvements in IP telephony, cable-delivered telephone service could evolve into a simple telecommunications "after thought" of consumers, rather than a separate, independent service. Number portability also helps this transition. Delivering telephone service over traditional circuit-switched facilities can be relatively expensive, can be limiting in terms of future follow-on services and, can impose certain complexities in the interworking of the cable and copper plant. There is a general consensus that VoIP/VoMPLS may be a better approach for the cable TV space, particularly for operators just getting into the space. Recent statistics show that a typical U.S. household spends about $1,000 a year in telephone service. This opens up the opportunity of another $100–150B or so in revenues for the MSOs just in the U.S. A study by RHK Inc. indicated that voice-over-broadband (cable) was expected to reach about $1B by 2005, a 60% compound annual growth rate. RHK foresees incumbent Local Exchange Carriers (ILECs), which have recently all but effectively morphed into carriers providing end-to-end services, and MSOs starting to offer bundled services to consumers and small- and medium-sized businesses. Anecdotal information shows that around 15% of households in the U.S. have replaced a traditional telephony line with another form of access, usually cable TV. Major work in this arena is undertaken by Cable Television Laboratories (Cable Labs) and PacketCable. CableLabs, the industry's standards advocacy group, published the spec Data-Over-Cable Service Interface Specification (DOCSIS) for the purpose of supporting VoIP, multimedia services, and advanced security. PacketCable is a collaboration initiative between MSOs and VoIP vendors to develop system specs for VoIP-over-cable.

2.2 Voice Digitization and Encoding

This section discusses voice digitization and compression methodologies that are relevant to VoP applications. Low-bit-rate voice methods, namely, methods that provide voice compression at 8 kbps or less, are of specific interest to packet network implementations. At a macro level, one needs to collect speech at the source, digitize it in a way that is perceptually acceptable (as measured by a technique known as Mean Opinion Score (MOS) ([MOO199701], [KAR198501], [SPO199701], [ITU199201], [STO199301], [ZWI199101])), or equivalent, and deliver the speech samples at the remote end with low delay, low jitter, low packet loss, and low mis-sequencing, while avoiding the creation of echo or other impairments for the speaker. Figure 2.2 depicts this concept pictorially. One also needs to be able to support a full-feature call model that entails user-to-network and network-node signaling, so that the user can request specific services from the network. Also, supplementary services, including but not limited to: three-way calling, conferencing, transferring, unified messaging, presence indication, instant messaging, centrex/virtual switch, and so forth, need to be supported. Finally, interworking with the hundreds of millions of other existing telephones in the world must be supported by any VoP system.

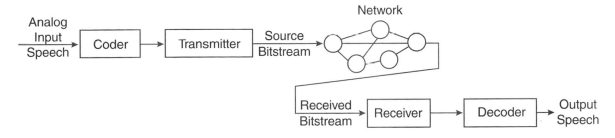

Figure 2.2: Use of coders in a VoP application.

2.2.1 Overview of Speech Encoding Methods

It was the speech encoding advancements of the early 1990s that gave VoP/VoIP a key (but not the only) technical "push." There are two families of techniques used in speech analysis: waveform coding (e.g., see [JAY198401]) and vocoding (also called *coding*) (e.g., see [BEL200001]). Table 2.2 depicts these two techniques and some examples of specific algorithms. Waveform coding is applicable to traditional voice networks, VoATM, and some implementations of VoMPLS; vocoding is applicable to VoIP/cable/cellular applications. Waveform coding was developed (at the practical level) in the early 1960s; vocoding was developed (at the practical level) in the early 1990s (starting with GSM cellular services), although the research goes back more than half a century.

A lossless[7] (also known as *noiseless*) coding system is able to reconstruct perfectly the samples of the original signal from the coded (compressed) representation; waveform coding schemes are nearly-lossless. On the other hand, a coding scheme incapable of perfect reconstruction from the coded representation is called *lossy*. Vocoding schemes are lossy. Lossy schemes offer the advantage of lower bit rates (e.g., less than 1 bit per traditional sample) relative to lossless schemes (e.g., eight or more bits per traditional sample). So far, however, the PSTN has used lossless methods.

Table 2.2: Speech digitization methods and some illustrative examples.

Method	Aspect
Waveform coders: Utilize algorithms to produce an output that approximates the input waveform.	*PCM (Pulse Code Modulation): Standard telephony method for "toll" quality voice. Typically used at 64 kbps.*
	ADPCM (Adaptive Differential PCM): Adaptive coding for rates of 40-, 32-, 24-, and 16 kbps. Uses a combination of adaptive quantization and adaptive prediction.
Vocoding: Digitize a compact description of the voice spectrum in several frequency bands, including extraction of the pitch component of the signal.	*Adaptive subband coding (e.g., see [AKA199601], [AKA199201]): Supports rates of 16- and 8 kbps. Speech is separated into frequency bands and each is coded using different strategies. The strategies are selected to suit properties of hearing and some predictive measure of the input spectrum.*
	(hybrid) Multipulse linear predictive coding: Support rates of 8- and 4 kbps. A suitable number of pulses are utilized to optimize the excitation information for a speech segment and to supplement linear prediction of the segments.
	Stochatically-excited Linear Predictive Coding (LPC): Supports rates of 8- to 2 kbps. The coder stores a repository of candidate excitations, each a stochastic sequence of pulses, and the best is matched.

---

[7] The loss is determined by the quantization process.

In waveform coding, one attempts to code, transmit, and then reproduce the analog voice time-amplitude curve by modeling its physical shape in the amplitude-time domain. The number of bits per second to represent the voice with this method is "high": 64-, 32-, or 16-kbps, depending on the technology. Vocoding attempts to reproduce the analog voice curve by performing a mathematical analysis (spectral transformation), which "identifies" abstractly the type of curve at hand; what is transmitted is a small set of parameters describing the nature of the curve. The number of bits per second to represent the voice with this method is low: 9.6-, 6.3-, 5.3-kbps and even lower, depending on the technology. However, voice quality becomes degraded as the digitization rate becomes small(er) than 4.8 kbps. Low bit rates are realized by methods that reduce the redundancies in speech, provide adaptive quantization and pitch extraction, and then code the processed signal in a perceptually-optimized manner. Quality, bit rate, complexity, and delay are all impacted by the processing and coding. Quality is negatively impacted as the bit rate goes down, but this effect can be decreased to some extent by adding complexity of processing (at an increased cost, however); in turn, processing increases the delay. In general, except for Linear Predictive Coding (LPC), as the bit rate decreases by a binary order of magnitude, the complexity of the vocoder (also known as *coders*) increases by approximately a decimal order of magnitude. At the same time the delay increases and the quality deteriorates. Most VoIP applications to date have operated in the 8 kbps range, although higher rates are not precluded, particularly for intraenterprise applications.

In spite of the fact that an extensive body of research on vocoding methods has evolved in the past 25 years, historically, the technology as noted has not experienced major deployment in the PSTN. However, with use of these techniques in wireless networks and in VoP networks, increased penetration even in the PSTN itself is expected in the next few years. If the network planner is going to introduce major new digitization and coding systems in an existing network, like the PSTN, they may as well go "all the way" and deploy the 10-fold compression methods embodied in modern vocoders rather than the 2-to-1 compression of newer waveform coding schemes.

Nyquist theory specifies that to properly waveform-code an analog signal of bandwidth W with basic Pulse Code Modulation (PCM) techniques, one needs 2W samples per second. For voice, when band-limited to a nominal 4,000 Hz bandwidth, one needs 8,000 samples per second. The dynamic range of the signal (and, ultimately, the signal-to-noise ratio) dictates the number of quantizing levels required. For telephonic voice, 256 levels suffice based on psychoacoustic studies conducted in the 1950s and early 1960s; it follows that 8 bits are needed to uniquely represent these many levels. In turn, this implies than one needs 64,000 bps to encode telephonic human speech in digital form (for the 4,000 Hz spectrum). PCM does not require sophisticated signal processing techniques and related circuitry; hence, it was the first method to be employed, and is the prevalent method used today in the telephone plant (PCM was first deployed in the early 1960s). PCM provides excellent quality. This is the method used in modern Compact Disc (CD) music recording technology, although the sampling rate is higher and the coding words are longer, to guarantee a frequency response to 22 kHz. The problem with PCM is that it requires a fairly high bandwidth to represent a voice signal.

One way to reduce the bit rate in a waveform coding environment is to use differential encoding methods. The problem with these voice coding methods, however, is that if the input analog signal varies rapidly between samples, the differential technique is not able to represent with sufficient accuracy the incoming signal. Just as in the PCM technique, clipping can occur when the input to the quantizer is too large; in this case, the input signal is the change in signal from the previous sample. The resulting distortion is known as slope-overload distortion. This issue is addressed by the Adaptive Differential Pulse Code Modulation (ADPCM) scheme. ADPCM provides "toll quality voice with minimal (voice) degradation" at 32 kbps. In ADPCM, the coder can be made to adapt to slope overload by increasing the range represented by the 4 bits used per sample. In principle, the range implicit in the 4 bits can be increased or decreased to match different situations; this will reduce the quantizing noise for large signals, but will increase noise for normal signals. In practice, the ADPCM coding device accepts the PCM-coded signal and then applies a special algorithm to reduce the 8-bit samples to 4-bit words using only 15 quantizing levels. These 4-bit words no longer represent sample amplitudes; instead, they contain only enough information to reconstruct the amplitude at the distant end. The adaptive predictor predicts the value of the next signal based on the level of the previously sampled signal. A feedback loop ensures that voice variations are followed with minimal deviation. The deviation of the predicted value measured against the actual signal tends to be small and can be encoded with 4 bits. In the event that successive samples vary widely, the algorithm adapts by increasing the range represented by the 4 bits through a slight increase in the noise level over normal signals [BEL200001]. Low bit rate voice methods such as ADPCM reduce not only the capacity needed to transmit digital voice but also for voiceband data (e.g., fax and dial-up Internet access). ADPCM encoding methods can and have been utilized in ATM and frame relay environments; vocoding is now used more prevalently in IP-based voice and in cellular networks.

Sophisticated voice coding methods have become available in the past decade due to the evolution of VLSI/DSP technology. As noted earlier, coding rates of 32 kbps, 16 kbps, and even "vocoder" methods requiring 6,300 bps, 5,300 bps, 4,800 bps, 2,400 bps, and even less, have evolved in recent years, while the quality of the synthetized voice has increased considerably. There is interest for pursuing these new coding schemes, particularly, in VoP environments, since the implication is that one can increase the voice carrying capacity of the network in place, up to ten times without the introduction of new transmission equipment. Unfortunately, current switching technology is based on DS0 (64 kbps) channels. And, as a rough figure, we estimate that there is an embedded base of about $50B of traditional Class 5 switches in North America. Given this predicament, a carrier can either: (a) ignore these new coding methods; or, (b) use trunk gateway hardware outside the switch to achieve the trunk-level voice compression; or, (c) introduce new switching technology that uses these schemes directly.

Standards for voice digitization/encoding are critical, if one is to be able to interconnect and interwork any two telephones in the world. In the International Telecommunications Union – Telecommunications (ITU-T) (the sector dedicated to telecommunications), Study Group 15 (SG15) is charged with making recommendations related to speech and video processing. Table 2.3 identifies the collection of ITU-T standards and specifications.

Table 2.3: ITU-T standards related to voice coding.

Coder	Description
G.711	Pulse Code Modulation (PCM) of voice frequencies.
G.712	Transmission performance characteristics of pulse code modulation channels. Replaces G.712, G.713, G.714, and G.715.
G.713	[Withdrawn] Performance characteristics of PCM channels between two-wire interfaces at voice frequencies. The content of this Recommendation is now covered by ITU-T G.712.
G.714	[Withdrawn] Separate performance characteristics for the encoding and decoding sides of PCM channels applicable to four-wire voice-frequency interfaces. The content of this Recommendation is now covered by ITU-T G.712.
G.715	[Withdrawn] Separate performance characteristics for the encoding and decoding side of PCM channels applicable to two-wire interfaces. The content of this Recommendation is now covered by ITU-T G.712.

G.720	Characterization of low-rate digital voice coder performance with nonvoice signals.
G.721	*[Withdrawn]* 32 kbps Adaptive Differential Pulse Code Modulation (ADPCM). The content of this Recommendation is now covered by ITU-T G.726
G.722	7 kHz audio-coding within 64 kbps.
G.722.1	Coding at 24 and 32 kbps for hands-free operation in systems with low frame loss.
G.722.2	Wideband coding of speech at around 16 kbps using Adaptive Multirate Wideband (AMR-WB).
G.723	*[Withdrawn]* Extensions of Recommendation G.721 adaptive differential pulse code modulation to 24 and 40 kbps for digital circuit multiplication equipment application. The content of 1988 edition of ITU-T G.723 is now covered by ITU-T G.726
G.723.1	Dual-rate speech coder for multimedia communications transmitting at 5.3 and 6.3 kbps.
G.724	Characteristics of a 48-channel low bit rate encoding primary multiplex operating at 1544 kbps.
G.725	System aspects for the use of the 7 kHz audio codec within 64 kbps.
G.726	40, 32, 24, 16 kbps Adaptive Differential Pulse Code Modulation (ADPCM).
G.727	5-, 4-, 3- and 2-bit/sample embedded Adaptive Differential Pulse Code Modulation (ADPCM).
G.728	Coding of speech at 16 kbps using low-delay code excited linear prediction.
G.Imp728	Implementors' Guide for ITU-T Recommendation G.728 ("Coding of speech at 16 kbps using low-delay code excited linear prediction").
G.729	Coding of speech at 8 kbps using Conjugate-Structure Algebraic-Code-Excited Linear-Prediction (CS-ACELP).

Table 2.4 depicts some of the key technical facets of the relevant to vocoders that are applicable to VoP networks that have emerged in the past decade [RAD200401]. One of the key questions regarding vocoding relates to speech quality. As alluded to earlier, MOS is a popular method to assess subjective quality measures (other methods also exist). Figure 2.3 depicts the MOS of a number of popular vocoder technologies.

Table 2.4: Coder/vocoder technology at a snapshot.

Coder	Description
G.711 Speech Coder	A-Law/μ-Law Pulse Code Modulation (PCM) coding of speech at 64 kbps. The speech coder is implemented as an encoder and decoder with an option to select A-Law/μ-Law and multiple frame sizes at compilation/run time.
G.722 Speech Coder	Variable-rate wide-band audio coder. The encoder compresses 16 kHz linear-PCM input data to 48/56/64kbps and decodes into one of three bit rates. G.722 is a mandatory coding scheme for wideband audio for videoconferencing.
G.723.1 Dual Rate Speech Coder with Annex A	Encoding 8 kHz sampled speech signals for transmission at a rate of either 6.3 kbps or 5.3 kbps. G.723.1 provides near toll quality performance under clean channel conditions. Coder operates on 30 ms frames with 7.5 ms of look-ahead. The coder offers good speech quality in network impairments such as frame loss and bit errors, and is suitable for applications such as voice over frame relay, teleconferencing or visual telephony, wireless telephony, and voice logging. Additional bandwidth savings are possible via voice activity detection and comfort noise generation.

G.726 ADPCM Waveform Coder	G.726 Adaptive Differential PCM (ADPCM) compresses speech and other audio signal components of multimedia. This coder accepts A-law or μ-law PCM speech samples and compresses it at rates of 40, 32-, 24- or 16- kbps. The G.726 algorithm has been optimized to compress speech to the highest quality. The coder is based on adaptive differential waveform quantization and passes fax, Dual-Tone MultiFrequency (DTMF) and other telephony tones. The primary applications of this coder are in Digital Circuit Multiplex Equipment (DCME), satellite telephony and wireless standards such as PACS, DECT and PHP (Japan).
G.729 with Annex-B CS-ACELP Voice Coder	Conjugate-Structure Algebraic Code Excited Linear Prediction, encoding 8 kHz sampled speech signals for transmission over 8kbps channels. Also includes G.729B implementation for fixed rate speech coders. G.729 encodes 80 sample frames (10 ms) of 16-bit linear PCM data into ten 8-bit code words, and provides near toll quality performance under clean channel conditions. Codec operates on 10 ms frames with 5 ms of look-ahead, allowing low transmission delays. The coder offers good speech quality in network impairments such as frame loss and bit errors, and is suitable for applications such as voice over frame relay, teleconferencing or visual telephony, wireless telephony and voice logging. Additional bandwidth savings are possible via voice activity detection (silence suppression) (G.729 Annex B).
GSM-FR Speech Coder	The ITU-T RPE/LTP supports encoding 8 kHz sampled speech signals for transmission at a rate of 13 kbps. The encoder compresses linear-PCM narrow band speech input data, and uses Regular Pulse Excitation with Long-Term Prediction (RPE-LTP) algorithm. The GSM-FR coder has been optimized to compress speech to the highest quality. The primary applications of this coder are in Digital Circuit Multiplex Equipment (DCME), satellite telephony and wireless standards such as PACS, DECT and PHP (Japan).

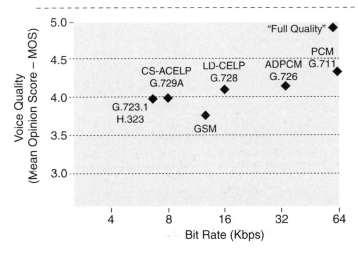

Speech Coder	MOS
Original Speech (No Coding)	5.0
ITU Rec. G.711, 64 Kbps μ-law PCM	4.1
ITU Rec. G.726, 32 Kbps μ-law ADPCM	3.8
ITU Rec. G.729, 8 Kbps CS-CELP	3.9
ITU Rec. G.728, 16 Kbps LD-CELP	4.0
ITU Rec. G.723.1, 6.4 Kbps MP-MPQ	3.9
GSM, 13 Kbps RPE-LTP	~3.7

Figure 2.3: MOS of various vocoding schemes.
(Courtesy of Radisys Corporation)

Having examined the issues and opportunities afforded by speech encoding methods in preliminary form, we now look at various technologies in more detail.

2.2.1.1 Waveform Coding

Speech coding or speech compression algorithms are used to obtain compact digital representations of (wideband) speech (audio) signals for the purpose of efficient transmission or storage (say, storage in a voice mail system). The key objective in coding is to represent the signal with a minimum number of bits while achieving transparent signal reproduction, that is, generating output audio that cannot be distinguished from the original input, even by a sensitive listener [SPA200001]. Two processes are required to digitize an analog signal, as follows:

1.   Sampling: this discretizes the signal in time, and,
2.   Quantizing: this discretizes the signal in amplitude.

The devices that accomplish speech analysis (digitization) are called *codecs* (for coder/decoder). Coders include Analog-to-Digital (A/D) converters that typically perform a digitization function, and "analysis modules" that further process the speech to reduce its data rate and prepare it for transmission. The reverse process uses synthesis modules to decode the signal and Digital-to-Analog (D/A) converters that reconvert the signal back to analog format.

Naturally, the goal of the entire digitizing process is to derive from an analog waveform a digital waveform that is a faithful facsimile (at the acoustical perception level) of the original speech. The sampling theorem, indicates that if the digital waveform is to represent the analog waveform in useful form, then the sampling rate must be at least twice the highest frequency present in the analog signal. Waveform coding methods are driven by this theorem. Analog telephonic speech is filtered before digitization to remove higher frequencies. The human speech spectrum contains frequencies beyond 12,000 Hz, but for telephony applications, higher frequencies can be safely filtered out. Specifically, in traditional telephone networks, the channel bank and digital loop carrier equipment in telephone networks is designed to eliminate frequencies above 3.3 kHz[8]. Consequently, analog speech signals are sampled at 8,000 Hz for PCM applications. PCM, as specified in the ITU G.711 recommendation, is currently the most often used digitization in telephony. Today, nearly every wireline telephone call in the U.S. is digitized at some point along the way using PCM.

As noted, sampling used in waveform coding of voice makes an analog waveform discrete in time; quantizing makes the signal discrete in amplitude. This discreteness is a direct consequence of the fact that computers are digital devices, where the values that are allowed for variables are discrete. The digitization process measures the analog signal at each sample time and produces a digital binary code value representing the instantaneous amplitude.

Optimizing speech quality means production of a digital waveform that can be reconverted to analog with as small an error as possible. Quantization is the process that maps a continuum of amplitudes into a finite number of discrete values. This results in a (small) loss of information and the ensuing introduction of noise, called *quantization noise* or *quantization error*. In waveform coding, this loss of information is "small" and the results in called *(nearly) lossless*; vocoding methods discard much more information and are therefore called *lossy*. Signal-to-Noise Ratio (SNR) expressed in decibels (dB) is a measure used to describe voice quality. For telephony application, speech coders are designed to have a signal-to-noise ratio above 30 dB over most of its range.

PCM can reproduce any signal to any desired level of quality, and has applications beyond telephony. For example, the introduction of the CD in the early 1980s brought to the forefront all of the advantages of

---

[8]   Nominally, the voice band is actually 4 kHz.

digital audio representation, including high fidelity, dynamic range, and robustness. These advantages, however, came at the expense of high data rates. Conventional CD and Digital Audio Tape (DAT) systems are typically sampled at either 44.1 or 48 kHz using PCM with a 16-bit sample resolution. This results in uncompressed data rates of 705.6/768 kbps for a monaural channel, or 1.41/1.54 Mbps for a stereo pair at 44.1/48 kHz, respectively, [SPA200001]. Compression techniques other than PCM are now being sought for high-fidelity music and have already been deployed in iPods and the like, under MPEG-1, Layer 3 (MP3) encoding; MP3 is the audio codec specified in Moving Pictures Expert Group 1 (MPEG-1) that, for a number of reasons, became the most popular format for digital music as its popularity hit critical mass (at this juncture "MP3" is commonly used as shorthand for any type of digital music file.)

Uniform Quantization

In a basic PCM system, input to the quantizer hardware comes in the form of an analog voltage provided by the sampler circuit. The simplest approach would be to use a uniform quantization method. Here, the range of input voltages is divided into 2^n segments, and a unique code word of n bits is associated with each segment. The width of each segment is known as the *step size*. The range, R, of an n-bit quantizer with step size s is:

$$R = (s)(2^n).$$

This implies that if the input voltage were to exceed R, clipping would result. To address this issue, logarithmic quantization is used.

Logarithmic Quantization

The goal of logarithmic quantization is to maintain a reasonably constant signal-to-noise ratio over a range of analog amplitudes; using this technique the signal-to-noise ratio will not vary with incoming signal amplitude. To accomplish this, one quantizes (not the incoming signal), but the log value of the signal, for example, for analog values, w, the equation $y = h + k \log(w)$ with h and k constants provides such a logarithmic function[9]. Logarithmic quantization is a compression process: it reduces the dynamic range of a signal according to a logarithmic function. After compression, a reverse process, exponentiation, is required to recover a facsimile of the original; the entire cycle is often referred to as companding (for compressing/expanding) [PEL199301]. In North America, a specific logarithmic scheme called μ-Law is used; in Europe a similar but not identical approach called *A-Law*; both methods employ 8-bit logarithmic quantization with sixteen regions and sixteen steps per region.

Adaptive Quantization

Speech signals contains a significant amount of redundant information. By making use of this fact and by removing some of these redundancies through processing, one is able to produce data parameters describing the waveform with a lower data rate than otherwise possible, and still be able to make a reasonably faithful reconstruction of the original. Speech samples generated at the Nyquist rate are correlated from sample to sample (actually they remain moderately correlated over a number of consecutive samples). This implies that values of adjacent samples do not differ significantly. Consequently, given some number of past samples, it is possible to predict with a degree of accuracy the value of the next sample.

One can achieve further reductions in voice bit rate in a waveform coding environment by employing analysis algorithms that make use of the technique of dynamically adapting the quantizer step size in response to variations in input signal amplitude. The goal is to maintain a quantizer range that is matched to the input

---

[9] This function is applicable when w > 0. A piecewise-linear approximation to the function can be utilized that is valid both for the value zero and for negative values.

signal's dynamic range. This discussion has mainly historical value since VoP systems have generally not utilized encoding other than PCM (e.g., VoATM) or vocoding (discussed in the next section)[10].

PCM techniques that adapt step size are referred to as Adaptive PCM (APCM). The technique can be applied to both uniform and logarithmic (nonuniform) quantizers. There are several adaptation algorithms, but all aim at estimating the slowly-varying amplitude of the input signal, while at the same time balancing the need to increase step size to attain appropriate range against the worsening signal-to-noise ratio that results from larger step sizes. For syllabic companding techniques, the quantization characteristics change at about the same rate as syllables occur in speech. Other methods use instantaneous companding. Yet other methods calculate signal amplitude statistics over a relatively short group of samples and adjust the step size accordingly (for example, *feed forward adaptive PCM* and *feedback adaptive PCM*). Some of these adaptive techniques are discussed next.

In the differential coding technique (also called *linear prediction*[11]), for example ADPCM, rather than coding the input waveform directly, one codes the difference between that waveform and one generated from linear predictions of past quantized samples. At sample time j, this encoder codes $e(j)$, the prediction errors at time j, where,

$$e(j) = y(j) - [a_1 y(n-1) + a_2 y(n-2) + ... + a_p y(j-p)]$$

and where $y(j)$ is the input sample and the term in square brackets is a predicted value of the input, based on previous values. The terms a_i are known as *prediction coefficients*. The output values $e(j)$ have a smaller dynamic range than the original signal, hence, they can be coded with fewer bits[12].

This method entails linear predictions because, as the preceding equation shows, the error predictions involve only first-order (linear) functions of past samples. The prediction coefficients a_i are selected so as to minimize the total squared prediction error, E where,

$$E = e^2(0) + e^2(1) + ... + e^2(n)$$

and n is the number of samples. Once computed, the coefficients are used with all samples until they are recalculated. In differential coding, a trade-off can be made by adapting the coefficients less frequently in response to a slowly changing speech signal. In general, predictor coefficients are adapted every 10 to 25 milliseconds.

As is the case with adaptive quantization, adaptive prediction is performed with either a feedback or feedforward approach. In the case of feedback predictive adaptation, the adaptation is based on calculations involving the previous set of n samples; with feed-forward techniques, a buffer is needed to accumulate n samples before the coefficients can be computed (this, however, introduces a delay, because the sample values have to be accumulated) [PEL199301]. Values of $n = 4$ to $n = 10$ are used. For $n \geq 4$, adaptive predicators achieve signal-to-noise ratios of 3 or 4 dB better than the nonadaptive counterparts, and more than 13 dB over PCM.

A basic realization of linear prediction can be found in DPCM coding, where, rather than quantizing samples directly, the difference between adjacent samples is quantized. This results in one less bit being needed per sample compared to PCM, while maintaining the signal-to-noise ratio. Here, if $y(j)$ is the value of a sample at a time j for a PCM waveform, then the DPCM sample at time j is given by $e(j)$, where,

$$e(j) = y(j) - [a_1 y(n-1) + a_2 y(n-2) + ... + a_p y(j-p)]$$

---

[10] Some frame relay implementations of the early 1990s did use ADPCM.

[11] Devices that use this technique are referred to as adaptive predictive coders (APC).

[12] Alternatively, one can achieve a higher signal-to-noise ratio with the same number of bits.

and where a_1 is a scaling factor, while $a_2 = a_3 = ... = a_p = 0$; namely,

$$e(j) = y(j) - a_1y(n-1)$$

Further gains over PCM and DPCM are obtained by including adaptation (as used in ADPCM). This is done either by incorporating adaptive quantization or by adjusting the scale factor (at syllabic rate), or both.

2.2.1.2 Vocoding (Analysis/Synthesis) in the Frequency Domain

We now shift our attention to vocoding. Advances in VLSI/DSP technology that took place in the late 1980s and early 1990s have permitted a wide variety of applications for speech coding, including digital voice transmissions over telephone channels. The processing can be done by a digital signal processor or by a general-purpose microprocessor, although the former is preferred. Transmission can either be online (real time) as in normal telephone conversations, or offline, as in storing speech for electronic mail of voice messages or automatic announcement devices. Many of the low bit rate voice methods make use of the features of human speech, in terms of the properties than can be derived from the vocal track apparatus. The coders typically segment input signals into quasistationary frames ranging from 2 to 50 ms in duration. Then, a time-frequency analysis section estimates the temporal and spectral components on each frame. Often, the time-frequency mapping is matched to the analysis properties of the human auditory system, although this is not always the case. Either way, the ultimate objective is to extract from the input audio a set of time-frequency parameters that is amenable to quantization and encoding in accordance with a perceptual distortion metric [SPA200001].

While, as noted, nearly all the traditional PSTN voice is still carried via waveform methods, specifically PCM, almost invariably all new applications such as cellular, voice over cable, VoWi-Fi, enterprise VoIP, and end-to-end VoIP (enterprise + pure-play-carrier + enterprise) applications are now vocoder-based.

The human vocal tract is excited by air from the lungs (see Figure 2.4). The excitation source is either voiced or unvoiced. In voiced speech, the vocal cords vibrate at a rate called the *fundamental frequency*; this frequency is what we experience as the pitch of a voice. Unvoiced speech is created when the vocal chords are held firm without vibrations and either the air is aspirated through the vocal tract or is expelled with turbulence through a constriction at the glottis, tongue, teeth or lips.

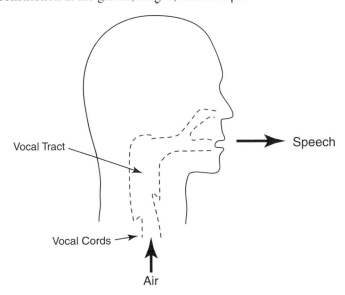

Figure 2.4: Model of human speech apparatus.

Two techniques play a role in speech processing:

1. Speech analysis is that portion of voice processing that converts speech to digital forms suitable for storage on computer systems and transmission on digital (data or telecommunications) networks;
2. Speech synthesis is that portion of voice processing that reconverts speech data from a digital form to a form suitable for human usage. These functions are essentially the inverse of speech analysis.

Speech analysis processes are also called *digital speech encoding* (or *coding*), and speech synthesis is also called *speech decoding*. The objective of any speech-coding scheme is to produce a string of voice codes of minimum data rate, so that a synthesizer can reconstruct an accurate facsimile of the original speech in an effective manner, while optimizing the transmission (or storage) medium. Figure 2.5 depicts a generic audio encoder [SPA200001].

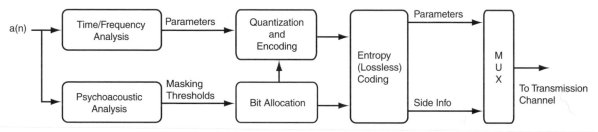

Figure 2.5: Generic audio encoder.

The waveform methods discussed earlier relate to time-domain (signal amplitude versus time) representation of the speech signal. Vocoding looks at the signal in the frequency domain; the spectrum represents the frequency distribution of energy present in speech over a period of time. Frequency domain coders attempt to produce code of minimum data rate by exploiting the resonant characteristics of the vocal tract. There is a lot of information that can be extracted and exploited in the speech spectrum. Different vocoder technologies have different designs, as identified in Table 2.5 (just a small set of vocoder technologies are shown in the table).

Table 2.5: Vocal track mechanism for various vocoders.

Vocoder	**Vocal Track Mechanism**
Formant vocoder	*Reproduces the formants; a filter for each of the first few formants is include, then all higher formants are lumped into one final filter.*
Channel vocoder	*Filters divide the spectrum into a number of bands.*
LPC vocoders (*)	*Models track based on concatenated acoustic tubes.*

(*) Popular vocoders now commercially deployed are derivatives of LPC.

Parametric Vocoders

Parametric vocoders model speech production mechanisms. They do so by taking advantage of the slow rate of change of the signals originating in the vocal tract, allowing one set of parameters to approximate the state over a period up to about 25 ms. Most vocoders aim at characterizing the frequency spectrum and the vocal tract excitation source (lungs and vocal chords) with only a small set of parameters. These parameters (called a *data frame*) include:

- about a dozen coefficients that define vocal tract resonance characteristics,
- a binary parameter specifying whether the excitation source is voiced or unvoiced,
- a value for the excitation energy and,
- a value for pitch (during voicing only).

The vocal tract state is approximated by analyzing the speech waveform every 10–25 milliseconds and calculating a new set of parameters at the end of the period. A sequence of data frames is used remotely (or on playback from storage) to control synthesis of a mirror waveform. Because only a handful of parameters are transmitted, the voice data rate is low. One of the advantages of vocoders is that they often separate excitation parameters: pitch, gain, and voiced/unvoiced indications are carried individually in the data frame, so each of these variables can be modified separately before or during synthesis. Commercial-level vocoder data rates range from about 13,000 bps to 1,200 bps, with 8,000 bps being a common rate. The rate is dependent upon the frame rate, the number of parameters in the frame and, upon the accuracy with which each parameter is coded [BEL200001].

In a typical coder there are excitation sources (voice/unvoiced), loudness controls, and a vocal track filter. The excitation source for voiced speech consists of a periodic impulse generator and a pulse-shaping circuit. The impulse period adjusts to follow the original pitch according to the pitch frequency parameter being fed to it from the data frame. The vocal tract filter network emulates resonance characteristics of the original vocal tract. The synthetic glottal waveform entering this section of the synthesizer is transformed to a speech waveform approximating the original [PEL199301].

Linear Predictive Coding

Linear Predictive Coding (LPC) is a parametric vocoding technique that utilizes linear prediction methods. LPC is one of the most powerful speech analysis techniques, and one of the most useful methods for encoding reasonable quality speech at a low bit rate. It provides accurate estimates of speech parameters, and is relatively efficient for computation. The term is applicable to those vocoding schemes that represent the excitation source parametrically (as just discussed) and that use a higher-order linear predictor ($n > 1$). LPC analysis enjoys a number of desirable features in the estimation of speech parameters such as spectrum, formant frequencies, pitch, and other vocal-tract measures. LPC analysis is conducted as a time-domain process.

LPC coding produces a data frame at a rate of about 40–100 frames per second (lower frame rates produce lower-quality speech). As should be clear, the data rate originated by a frame depends on the number of coefficients (e.g., the order of the predictor) and on the accuracy to which each of the parameters is quantized. It should be noted that speech synthesized from LPC coders is most sensitive to the first few coefficients; this, in turn, implies that the coefficients need not necessarily all be quantized with the same accuracy.

The analog model that is solved by LPC is an approximation of the vocal tract (glottis and lips, but no nasal cavities) using concatenated acoustic tubes. If the number of cylinders is appropriately selected in the model, the frequency domain mathematics of the concatenated tubes-problem solves approximately the vocal tract problem. LPC allows one to estimate frequency-domain acoustic tube parameters from the speech waveform, as described next.

The LPC prediction coefficients obtained from the time-domain signal can be converted to reflection coefficients representing the set of concatenated tubes. This implies that frequency-domain estimations that approximately describe the vocal tract can be obtained (with this methodology) from time-domain data using linear algebra. Specifically, the n prediction coefficients of an nth order predictor can be calculated by solving a system of n linear equations in n unknowns; the n reflection coefficients that are present in equations describing resonances in a concatenated acoustic tube on $0.5 * (n - 1)$ sections, can be calculated from the *n* prediction coefficients. Hence, LPC analysis generates a set of reflection coefficients, excitation energy, voice/unvoiced indication bit, and fundamental frequency (if signal is voiced).

LPC starts with the assumption that the speech signal is produced by a buzzer at the end of a tube. The glottis (the space between the vocal cords) produces the buzz, which is characterized by its intensity (loudness) and frequency (pitch). The vocal tract (the throat and mouth) forms the tube, which is characterized by its resonances (the *formants*). LPC analyzes the speech signal by estimating the formants, removing their effects from the speech signal, and estimating the intensity and frequency of the remaining buzz. The process of removing the formants is called *inverse filtering*, and the remaining signal is called the *residue*. The numbers that describe the formants and the residue can be stored or can be transmitted. LPC synthesizes the speech signal by reversing the process—use the residue to create a source signal, use the formants to create a filter (which represents the tube), and run the source through the filter, resulting in speech. Because speech signals vary with time, this process is done on short sections of the speech signal, which are called *frames* [HOW200401].

The basic problem of the LPC system is to determine the formants from the speech signal. The basic solution is a difference equation (called a *linear predictor*), which expresses each sample of the signal as a linear combination of previous samples. The coefficients of the difference equation (the prediction coefficients) characterize the formants; hence, the LPC system needs to estimate these coefficients. The estimate is done by minimizing the mean-square error between the predicted signal and the actual signal. This is a straightforward problem, in principle. In practice, it involves: (1) the computation of a matrix of coefficient values, and, (2) the solution of a set of linear equations. Several methods (for example, autocorrelation, covariance, recursive lattice formulation) can be utilized to assure convergence to a unique solution with efficient computation.

It may seem surprising that the signal can be characterized by a simple linear predictor. It turns out that, in order for this to work, the tube must not have side branches. For ordinary vowels, the vocal tract is well represented by a single tube. However, for nasal sounds, the nose cavity forms a side branch. Theoretically, therefore, nasal sounds require a different and more complex algorithm. In practice, this difference is partly ignored and partly dealt with during the encoding of the residue [HOW200401].

If the predictor coefficients are accurate, and everything else works correctly, the speech signal can be inverse-filtered by the predictor, and the result will be the pure source (buzz). For such a signal, it is fairly easy to extract the frequency and amplitude and to encode them. However, some consonants are produced with turbulent airflow, resulting in a hissy sound (fricatives and stop consonants). Fortunately, the predictor equation does not care if the sound source is periodic (buzz) or chaotic (hiss). This means that for each frame, the LPC encoder must decide if the sound source is buzz or hiss—if buzz, estimate the frequency. In either case, estimate the intensity, and encode the information so that the decoder can undo all these steps.

This methodology is how LPC-10e (LPC-10 enhanced), the algorithm described over a decade ago in Federal Standard 1015 works. LPC-10e uses one number to represent the frequency of the buzz, and the number zero is understood to represent hiss. LPC-10e provides intelligible speech transmission at 2,400 bps[13]. 2,400 is not a toll-quality technology, but it does establish a lower target for reasonably-intelligible speech over bandwidth-constrained networks (e.g., in some military applications.)

Some enhancements are needed to improve quality. One reason is that there are speech sounds that are made with a combination of buzz and hiss sources (for example, the middle consonant in "azure"). Speech sounds like this will not be reproduced accurately by a simple LPC encoder. Another problem is that, inevitably, any inaccuracy in the estimation of the formants means that more speech information gets left in the residue. The aspects of nasal sounds that do not match the LPC model (as discussed previously, for example), will end up in the residue. There are other aspects of the speech sound that does not match the LPC model; side branches introduced by the tongue positions of some consonants, and tracheal (lung) resonances are some examples [HOW200401].

Because of the issues discussed in the previous paragraph, the residue contains important information about how the speech should sound, and LPC synthesis without this information will result in poor quality speech. For the best quality results, one could just send the residue signal, and the LPC synthesis would sound fine. Unfortunately, the motivation for using this technique is to compress the speech signal, and the residue signal requires just as many bits as the original speech signal, so this would not provide any compression.

Various attempts have been made to encode the residue signal in an efficient way, providing better quality speech than LPC-10e, mentioned above without excessively increasing the bit rate. The most successful methods use a "codebook." The codebook is a table of typical residue signals, which is setup by the system designers. In operation, the analyzer compares the residue to all the entries in the codebook, chooses the entry that is the closest match, and just sends the code for that entry. The synthesizer receives this code, retrieves the corresponding residue from the codebook, and uses that to excite the formant filter. Schemes of this kind are called *Code Excited Linear Prediction (CELP)*, which we cover in more detail below. For CELP to work well, the codebook must be big enough to include all the various kinds of residues. But if the codebook is too big, it will be time consuming to search through, and will require large codes to specify the desired residue. The biggest problem is that such a system would require a different code for every frequency of the source (pitch of the voice), which would make the codebook extremely large [RAB197801], [CAM199001], [CAM199002], [HOW200401].

The problem just identified can be solved by using two small codebooks instead of a very large one. One codebook is fixed by the designers, and contains just enough codes to represent one pitch period of residue. The other codebook is adaptive; it starts out empty, and is filled in during operation, with copies of the previous residue delayed by various amounts. Therefore, the adaptive codebook acts like a variable shift register, and the amount of delay provides the pitch. This is the CELP algorithm described in *Federal Standard 1016*. It provides reasonable quality, natural sounding speech at 4,800 bps.

Code Excited Linear Prediction

This section expands on the description of code excited linear predictive coding as described above. CELP synthesizes speech using encoded excitation information to excite an LPC filter. This excitation information is found by searching though a table of candidate excitation vectors on a frame by frame basis. LPC analysis is performed on input speech to determine the LPC filter parameters. The analysis includes comparing the outputs of the LPC filter when it is excited by the various candidate vectors from the table or codebook. The

---

[13] Federal Standard 1016, Telecommunications: Analog to Digital Conversion of Radio Voice by 4,800 bit/second Code Excited Linear Prediction (CELP), FS 1016 (1991)" is a 4,800 bps code excited linear prediction voice coder. LPC10 compression uses Federal Standard 1015, it provides intelligible speech transmission at only 2400 bit per second, with a compression ratio more than 26. LPC10 compression is sensitive to noise. To get better result, one needs to adjust the microphone input level to avoid overly-loud signals, and eliminate background noise that can interfere with the compression process. The U.S. DoD's Federal-Standard-1015/NATO-STANAG-4198 based 2,400 bps linear prediction coder (LPC-10) was republished as a Federal Information Processing Standards Publication 137 ("Analog to Digital Conversion of Voice by 2400 bit/second Linear Predictive Coding," FIPS Pub 137 (1984)). The U.S. Federal Standard 1015 (NATO STANAG 4198) is described in: Thomas E. Tremain, "The Government Standard Linear Predictive Coding Algorithm: LPC-10," Speech Technology Magazine, April 1982, p. 40–49. The voicing classifier used in the enhanced LPC-10 (LPC-10e) is described in: Campbell, Joseph P., Jr. and T. E. Tremain, "Voiced/Unvoiced Classification of Speech with Applications to the U.S. Government LPC-10E Algorithm," Proceedings of the IEEE International Conference on Acoustics, Speech, and Signal Processing, 1986, p. 473–6. The following article describes the FS 1016 4.8-kbps CELP coder: Campbell, Joseph P. Jr., Thomas E. Tremain and Vanoy C. Welch, "The Proposed Federal Standard 1016 4800 bps Voice Coder: CELP," Speech Technology Magazine, April/May 1990, p. 58–64.

best candidate is chosen based on how well its corresponding synthesized output matches the input speech frame. After the best match has been found, information specifying the best codebook entry and the filter are transmitted to the speech synthesizer. The speech synthesizer has the same codebook and accesses the appropriate entry in that codebook, using it to excite the same LPC filter to reproduce the original input speech frame [USP199401].

The codebook is made up of vectors whose components are consecutive excitation samples. Each vector contains the same number of excitation samples as there are speech samples in a frame. In typical CELP coding techniques (see Figure 2.6 [USP199401]), each set of excitation samples in the codebook must be used to excite the LPC filter and the excitation results must be compared using an error criterion. Normally, the error criterion used determines the sum of the squared differences between the original and the synthesized speech samples resulting from the excitation information for each speech frame. These calculations involve the convolution of each excitation frame stored in the codebook with the perceptual weighting impulse response. Calculations are performed by using vector and matrix operations of the excitation frame and the perceptual weighting impulse response [USP199401].

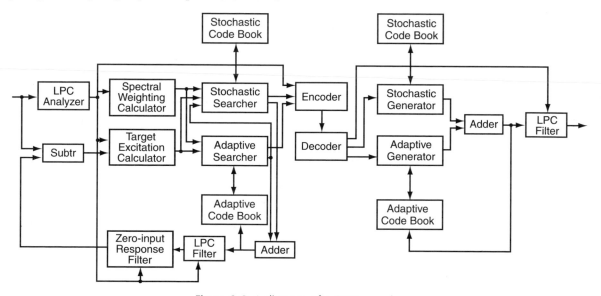

Figure 2.6: A diagram of a CELP vocoder.

A large number of computations must be performed as the initial versions of CELP required approximately 500 million multiply-add operations per second for a 4.8 kbps voice encoder. In addition, the search of the stochastic codebook for the best entry is computationally complex. The search process is the main source of the high computational complexity. Since the original appearance of CELP coders, the goal has been to reduce the computational complexity of the codebook search so that the number of instructions to be processed can be handled by inexpensive digital signal processing chips. Newer, low-complexity CELP speech coders:

- accurately and efficiently digitally code human speech using a CELP speech processor;
- optimize processing of a speech residual in the CELP speech processor using an algebraic, deterministic codebook;

- substantially reduce the computational complexity of processing the speech residual in the CELP speech processor through use of the codebook;
- construct the codebook by uniformly distributing a number of vectors over a multidimensional sphere.

Low-complexity CELP speech processors receive a digital speech input representative of human speech (Figure 2.7 [USP199401]) and performs linear predictive code analysis and perceptual weighting filtering to produce short- and long-term speech information. It uses an organized, nonoverlapping, deterministic algebraic codebook containing a predetermined number of vectors, uniformly distributed over a multidimensional sphere to generate a remaining speech residual. The short- and long-term speech information and remaining speech residual are combinable to form a quality reproduction of the digital speech input [USP199401].

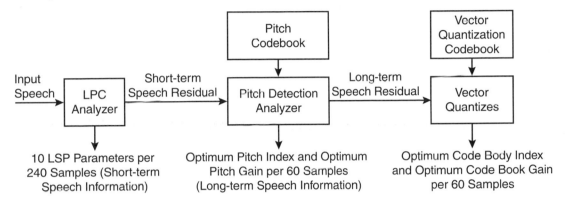

Figure 2.7: A diagram of the multistage extraction of information from the input speech frame signal.

The codebook is constructed by uniformly distributing a number of vectors over a multidimensional sphere. This is accomplished by constructing ternary valued vectors (where each component has the value -1, 0, or $+1$), having 80% of their components with value zero, and fixed nonzero positions. The fixed position of the nonzero elements is uniquely identifiable with this coder in comparison with other schemes [USP199401].

2.2.2 Technology and Standards for Low Bit Rate Vocoding Methods

As noted in the previous sections, during the past quarter-century, there has been a significant level of research and development in the area of vocoder technology and compressed speech. During the early- to-mid-1990s, the ITU-T (specifically, SG14 and SG15) standardized several vocoders that are applicable to low bit rate multimedia communications in general, and to VoP in intranets, Internet, and private-label IP networks in particular. Standardization is critical for interoperability and assurance of ubiquitous end-to-end connectivity. This standardization provides end-to-end compatibility at the physical level; the theme of this text, however, is total end-to-end compatibility—this requires compatibility at any number of layers including the network layer. Otherwise if the protocols are not directly compatible, an internetworking function is required and unfortunately this adds cost and complexity.

The standards developed in the early- to mid-1990s are ITU-T G.728, G.729, G.729A, and G.723.1. For some applications, the dominant factor is cost; for other applications, quality is critical. This is part of the reason why several standards have evolved in the recent past. However, to be ultimately successful, VoP will have to narrow down to one choice (or a small set of choices) so that anyone can call anyone else (as we do today with modems; or, with another example, with traditional TDM telephone instruments[14]), without worrying what technology the destination party may be using.

Corporate enterprise networks and intranets are often congested. Hence, for voice over IP take off, it is necessary to trade-off high computational power required for compressing speech down to the lowest possible rates, in order to to keep congestion low; however, one needs to do this without compromising the delay budget, especially for softphones that make use of desktop/PC's own computation power. Most planners have come to realize that a nontrivial amount of overhaul is needed for the intranets (and also legacy carriers' networks) to actually support VoIP (for example, QoS, power over Ethernet, architecture/number of hops, architecture/capacity, and so on.)

The vocoders discussed in the rest of this chapter require between 10 and 20 Millions of Instructions Per Second (MIPS). See Table 2.6 for a sample of processors' computing power (vocoders are typically implemented in digital signal processing chips, but the table and figure provide an intuitive sense of the required computing power). Note that a gateway that transcodes multiple speech systems would require the sum total of the MIPS (e.g., a gateway supporting 25 simultaneous users would need about 500 MIPS).

Table 2.6: MIPS of various systems example.

MC 68000 (8 MHz, 68,000 transistors)	1 MIPS
StrongARM (Newton MessagePad 2100) (2.5 million transistors)	185 MIPS
SGI Indy-R4400 (2.3 million transistors)	250 MIPS
PowerPC 604e (300 MHz) (5 million transistors)	500 MIPS
PowerPC G3 (750/300 MHz) (6.4 million transistors)	750 MIPS
Pentium II (7.5 million transistors)	500 MIPS
SGI Octane R10000 (6.8 million transistors)	800 MIPS

This discussion focuses principally on G.729, G.729A, and G.723.1; G.728's data rate (16 kbps) may be too high for (enterprise) VoP applications (although it would not necessarily be so for carrier applications). ITU-T G.729 is an 8 kbps Conjugate-Structure Algebraic Code Excited Linear Prediction (CS-ACELP) speech algorithm providing "good" speech quality. G.729 was originally designed for wireless environments, but it is applicable to IP/multimedia communications as well. Annex A of Rec. G.729 (also called *G.729A*) describes a reduced-complexity version of the algorithm that has been designed explicitly for integrated voice and data applications that are prevalent in Small Office/Home Office (SoHo) low bit rate multimedia communications. These vocoders use the same bitstream format and can interoperate with one another[15]. The basic concept of CELP was discussed in generality earlier in the chapter. Additional details are provided herewith. Figure 2.8 shows at a very high level the operation of a (G.729) CELP coder.

---

[14] Unfortunately (by design), the phone behind the PBX of vendor X usually does not work with the PBX of vendor Y—hopefully VoIP in general, and SIP in particular, will alleviate this issue.

[15] A signal analyzed with the G.729A coder can be reconstructed with the G.729 decoder, and vice versa. The major complexity reduction in G.729A is obtained by simplifying the codebook search for both the fixed and adaptive codebooks. By doing this, the complexity is reduced by nearly 50%, at the expense of a small degradation in performance.

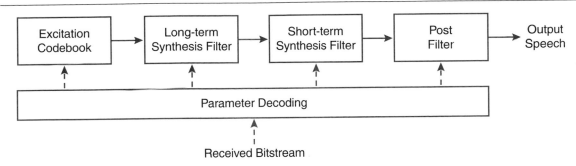

Figure 2.8: Block diagram of conceptual CELP synthesis model.

2.2.2.1 Overview

As noted earlier, the design goal of vocoders is to reduce the bit rate of speech for transmission or storage, while maintaining a quality level acceptable for the application at hand. On intranets and the Internet, voice applications may be standalone or multimedia-based. Since multimedia implies the presence of a number of media, speech coding for multimedia applications implies that the speech bitstream shares the communication link with other signals.

In principle, the use of a uniquely-specified vocoder might be desirable. Unfortunately, short-term local optimization considerations have lead developers to the conclusion that it is more economical to tailor the vocoder to each application. Consequently, a number of vocoders were standardized during the mid 1990s. Specifically, three "international" standards (ITU-G.729, G.729A, and G.723.1), and three regional standards (enhanced full-rate vocoders for North American (IS-54 operating at 8 kbps) and European (GSM RPE-LTP operating at 13 kbps) mobile systems have emerged. As a consequence of this over-abundance of standards, making an appropriate choice can be challenging. Vocoder attributes can be used to make trade-off analysis during the vocoder selection process that the developer of carrier, intranet or Internet multimedia or telephony application needs to undertake.

Vocoder Attributes

Vocoder speech quality is a function of bit rate, complexity, and processing delay. Developers of carrier, intranet, or Internet telephony products must review all these attributes. There usually is an interdependence between all these attributes and that they may have to be traded off against each other. For example, low bit-rate vocoders tend to have more delay than higher bit-rate vocoders. Low bit rate vocoders also require higher VLSI complexity to implement. As might be expected, often low bit-rate vocoders have lower speech quality than the higher bit-rate vocoders[16].

Bit Rate

Bandwidth efficiency is always at the top of the list for design engineers. Their thinking is that since the vocoder is sharing the access communications channel or the likely-overloaded enterprise network/Internet with other information streams, the peak bit rate should be as low as possible. Bandwidth limitations may not be an issue for carrier networks that are designed from the bottom up to support VoP, VoMPLS, or VoIPv6. Today, most vocoders operate at a fixed-bit rate regardless of the input signal characteristics;

---

[16] Additional factors that influence the selection of a speech vocoder are availability, licensing conditions, or the way the standard is specified (some standards are only described as an algorithmic description, while others are defined by bit-exact code) [COX199601].

however, the goal would be to make the vocoder variable-rate. For simultaneous voice and data applications, a compromise is to create a silence compression mechanism (see Table 2.7) as part of the coding standard. A common solution is to use a fixed rate for active speech and a low rate for background noise [COX199601]. The performance of the silence compression mechanism is critical to speech quality; if speech is declared too often, the gains of silence compression are not realized. The challenge is that for (loud) background noises, it may be difficult to distinguish between speech and noise. Another problem is that if the silence compression mechanism fails to recognize the onset of speech, the beginning of the speech will be cut off; this front-end clipping significantly impairs the intelligibility of the coded speech.

Table 2.7: Silence compression algorithms.

Voice Activity Detector (VAD)	Determines if the input signal is speech or background noise. If the signal is declared speech, it is coded at the full fixed bit rate; if the signal ids declared noise, it is coded at a lower bit rate. As appropriate, no bits are transmitted.
Comfort Noise Generation (CNG)	Mechanism invoked at the receiver end to reconstruct the main characteristic of the background noise.

The comfort noise generation mechanism must be designed in such a way that the encoder and decoder remain synchronized, even when there are no bits transmitted during some interval. This allows for smooth transitions between active and nonactive speech segments.

Delay

The delay of a speech coding system usually consists of three major components:

1. Frame delay
2. Speech processing delay
3. Bridging delay

Typically. low bit rate vocoders process a frame of speech data at a time, so that the speech parameters can be updated and transmitted for every frame. Hence, before the speech can be analyzed it is necessary to buffer a frame's worth of speech samples. The resulting delay is called *algorithmic delay*. It is sometimes necessary to analyze the signal beyond the frame boundary (this is referred to as *look-ahead*); here, additional speech samples need to be buffered, with additional concomitant delay. Note that this is the only implementation-independent delay (other delay components depend on the specific implementation; for example, how powerful is the processor used to run the algorithm, the kind of RAM used, etc.). Algorithmic delays are unavoidable, therefore, they need to be considered as part of the delay budget by the planner.

The second major component of the delay originates from the processing time it takes the encoder to analyze the speech and the processing time required by the decoder to reconstruct the speech. This processing delay depends on the speed of the hardware used to implement the vocoder. The combined algorithmic and processing delays is called the *one-way system delay*. The maximum tolerable value for the one-way system delay is 400 ms, if there are no echoes, but for ease and efficiency of communication it is preferable to have the one-way delay below 200 ms. If there are echoes, the tolerable one-way delay is 20–25 ms; therefore, the use of echo cancellation is often necessary.

In applications such as teleconferencing, it may be necessary to bridge several callers using a Multipoint Control Unit (MCU) to allow each person to communicate with the others. This requires decoding each bitstream, summing the decoded signals, and then re-encoding the combined signal. This process doubles the delay and at the same time it reduces the speech quality because of the multiple (tandem) encodings. Given the previous observation, a bridged system can tolerate a maximum one-way delay of 100 ms because the bridging will result in the doubling of the one-way system delay to 200 ms.

Algorithm's Complexity

As noted earlier, vocoders are often implemented on DSP hardware. Complexity can be measured in terms of computing speed in MIPS, Random Access Memory (RAM), and Read-Only Memory (ROM). Complexity determines cost; hence, in selecting a vocoder for an application, the developer must make an appropriate choice. When the vocoder shares a processor (with other applications) the developer must decide how much of these resources to allocate to the vocoder. Vocoders utilizing less than 15 MIPS are considered as having low-complexity; those using 30 MIPS or more are considered high-complexity. As discussed, increased complexity results in higher costs and greater power usage. Power usage is an important consideration in portable applications, since greater power usage implies reduced time between battery recharges or using larger batteries, which in turn means more expense and weight. Figure 2.9 depicts some additional complexity measures of these algorithms [RAD200401].

For the corporate/institutional planner, the vocoder technology is generally defined by the type of equipment and/or vendor that is selected for the system. In some instances (for example, softphones), it may be possible to set the vocoder methods via a menu choice.

Algorithm	Data Rate	Program Memory (Kbytes)	Data Memory (Kbytes)	Processor Loading (MCPS)
G.711	64 Kbps	0.5	1	0.07
G.722	48/56/64 Kbps	9.36	Tables: 1.61 Variables: 0.19	4.1
G.723.1	6.3 Kbps (high rate) 5.3 Kbps (low rate)	132	Tables: 19.4 Variables: 2.3	8.5
G.726	16/24/32/40 Kbps	7.87	Tables: 0.14 Variables: 0.512	7.5 (for 2 channels)
G.729 with Annex B	8 Kbps	115.9	Tables: 4.2 Variables: 8	13.6
G.729 A	8 Kbps	127	Tables: 2.71 Variables: 7.41	7.3
GSM-FR	13 Kbps	32	2	2.8

Courtesy: Radisys Corporation

Figure 2.9: Measures of complexity for various vocoders.

Quality

In terms of quality, the measure used in comparisons is how well the speech sounds for ideal conditions, namely clean speech, no transmission errors, and only one encoding (note, however, that in the real world these ideal conditions are often not met because there could be large amounts of background noise such as street noise, office noise, air conditioning noise, etc.). Table 2.8 shows the quality for the major coding schemes being utilized in voice over data networks.

Table 2.8: Quality of coding schemes.

Algorithm	G.723.1	G.729 G.729A	G.728	G.726 G.727	G.711
Rate (bps)	5.3–6.3	8	16	32	64
Quality	Good	Good	Good	Good	Good
Complexity	Highest	High	Lower	Low	Lowest

How well the vocoder performs under adverse conditions (e.g., what happens when there are channel errors or the loss of entire frames; how good does the vocoder sound when the speech is encoded and decoded in tandem, as is the case in a bridging application; how well does it sound when transcoding with another standard vocoder; how does it sound for a variety of languages) is the question that the Standards Bodies (e.g., ITU) try to answer during the testing phase of the standards drafting and generation process. The accepted measure of quality is MOS, which we introduced earlier in the chapter. With MOS, the score of multiple listeners are averaged to obtain a single figure-of-merit. Table 2.9 summarizes some of the key parameters that have been discussed in this section.

Table 2.9: Vocoder details.

Algorithm	Technology	Bit Rate (kbps)	MIPS	Compression delay (ms)	Framing Size	MOS
G.711	PCM	64	.34	0.75	0.125	4.1
G.726	ADPCM	32	13	1	0.125	3.85
G.728	LD-CELP	16	33	3–5	0.625	3.61
G.729	CS-ACELP	8	20	10	10	3.92
G.729a	CS-ACELP	8	10.5	10	10	3.9
G.723.1	MPMLQ	6.3	16	30	30	3.9
G.723.1	ACELP	5.3	16	30	30	3.8

Linear Prediction Analysis By Synthesis Coding

Basic Mechanisms

The ITU-T Recommendations G.723.1, G.728, and G.729 belong to a class of Linear Prediction Analysis-by-Synthesis (LPAS) vocoders. Code-excited linear predictive vocoders are the most common realization of the LPAS technique.

The decoded speech is produced by filtering the signal produced by the excitation generator through both a Long-Term (LT) predictor synthesis filter and a Short-Term (ST) predictor synthesis filter. The excitation signal is found by minimizing the mean-squared error signal (the difference between the original and decoded signal) over a block of samples[17]. It is weighted by filtering it through an appropriate filter. Both ST

---

[17] That is, the vocoder parameters are selected in such a manner that the error energy between the reference and reconstructed signal is minimized.

and LT predictors are adapted over time. Since the encoder analysis procedure includes the decoder synthesis procedure, the description of the encoder also defines the decoder.

The ST synthesis filter models the short-term correlations in the speech signal. This is an all-pole filter with an order between 8 and 16. The predictor coefficients of the short-term predictor are adapted in time, with rates varying from 30 to as high as 400 times per second. The LT predictor filter models the long-term correlations in the speech signal. Its parameters are a delay and a gain coefficient. For periodic signals, the delay corresponds to the pitch period (or possibly an integral number of pitch periods); for nonperiodic signals, the delay is random. Typically, the long-term predictor coefficients are adapted at rates varying from 100 to 200 times/s [COX199601].

A frequently-used alternative for the pitch filter is the adaptive codebook (this was briefly described in the earlier section). The LT synthesis filter is replaced by a codebook that contains the previous excitation at different delays. These vectors are searched, and the one that provides the best match is selected. To simplify the determination of the excitation for delays smaller than the length of the excitation frames, an optimal scaling factor can be determined for the selected vector. To achieve a low bit rate, the average number of bits per sample for each frame of excitation samples must be kept small.

The *multipulse excitation vocoder* represents the excitation as a sequence of pulses located at nonuniformly spaced intervals. The excitation analysis procedure determines both amplitudes and positions of the pulses. Finding these parameters all at once is a difficult problem and simpler procedures, such as determining locations and amplitudes one pulse at a time are typically used. The number of pulses required for an acceptable speech quality varies from four to six pulses every 5 milliseconds. For each pulse, both amplitude and location have to be transmitted, requiring about 7 or 8 bits per pulse [COX199601].

CELP vocoders approach the issue of reducing the number of bits per sample as follows: both encoder and decoder store the same collection of C possible sequences of length L in a codebook, and the excitation for each frame is described by the index to an appropriate vector in the codebook. This index is typically found by an exhaustive search of the codebook vectors and identifying the one that produces the smallest error between the original and decoded signals. To simplify the search procedure, many implementations use a gain-shape codebook where the gain is searched and quantized separately. The index requires $(\log_2 C)/L$ bits/sample, typically 0.2–2 bits/sample, and the gain requires 2 to 5 bits for each codebook vector.

ACELP introduces further simplifications by populating the codebook vectors with a multipulse structure: by using only a few nonzero unit pulses in each codebook vector, the search procedure can be sped up. The partitioning of the excitation space is known as an algebraic codebook; hence, the name of the vocoder.

Error Weighting Filter

The approach described above of minimizing a mean-squared error results in a quantization noise that has equal energy across the spectrum of the input signal. However, by making use of properties of the human auditory system, the vocoder designer can focus on reducing the *perceived* amount of noise. It has been found that greater amounts of quantization noise are undetectable in the frequency bands where the speech signal has high energy. Namely, the designer wants to shape the noise as a function of the spectral peaks in the speech signal. To put this masking effect to work in the vocoder design, the quantization noise has to be properly distributed among different frequency bands. This can be achieved by minimizing a weighted error from the short-term predictor filter.

Adaptive Postfilter

The noise in speech caused by the quantization of the excitation signal remains an area of vocoder design improvement (in particular, in the low-energy frequency regions, the noise can dominate the speech signal).

The perceived noise can be further reduced using a post-processing technique called *postfiltering* after reconstruction by the decoder. This operation trades-off spectral distortion in the speech versus suppression of the quantization noise, by emphasizing the spectral peaks and attenuating the spectral valleys. The postfilter is generally implemented as a combination ST/LT filter. The ST postfilter modifies the spectral envelope, it being based on the transmitted ST predictor coefficients (it can also be derived from the reconstructed signal). The parameters for the LT postfilter are either derived from the transmitted LT predictor coefficients or computed from the reconstructed speech [COX199601].

2.2.2.2 Key Coders

ITU G.729 CS-ACELP (Conjugate-Structure Algebraic-Code-Excited Linear-Prediction) standard (published in 1995) is a speech coding and decoding standard that provides 4 kHz speech bandwidth (telephone bandwidth) at a bit rate of 8 kbps. This coder is well-suited for telecommunications networks in which toll-quality speech is a requirement, and where total communications link delay and the ability to operate in noisy environments (possibly through several tandem encode/decode combinations) are important factors [ATL200401]. G.729 encodes 80 sample frames (10 ms) of 16-bit linear PCM data into ten 8-bit code words. G.729 provides near toll quality performance under clean channel conditions. The G.729 codec operates on 10 ms frames with 5 ms of look-ahead, allowing low transmission delays. The coder offers good speech quality in network impairments such as frame loss and bit errors, and is suitable for applications such as voice over frame relay, teleconferencing or visual telephony, wireless telephony and voice logging. Additional bandwidth savings are possible via voice activity detection (G.729 Annex B). The full CS-ACELP algorithm implementation runs on a single DSP and has quality similar to the ITU G.728 16 kbps coding standard. A good implementation is expected to pass all the floating-point tests provided by the ITU for algorithm verification [ASP200401]. The standard specifies a CELP that uses an algebraic codebook to code the excitation signal. The coder operates on speech frames of 10 msec (80 samples at an 8 kHz sample rate), computes the long-term predictor coefficients, and operates in an analysis-by-synthesis loop to find the excitation vector that minimizes the perceptually weighted error signal.

ITU G.728 low-delay code-excited linear prediction vocoder standard (published in 1992) is a 16 kbps algorithm for coding telephone-bandwidth speech for universal applications using low-delay code-excited linear prediction. This coder is well-suited to a wide range of applications, including both voice storage and voice communications. It is ideally suited for telecommunications networks in which toll-quality speech is a requirement and total communications link delay is an important factor [ASP200401]. In order to attain high speech quality at medium rates, it is necessary to increase coding gain by making some use of model-based coding. This generally involves relatively high delays of the order of 40–100 ms due to the block-based operation of most model-based coders. The G.728 coding algorithm is based on a standard analysis-by-synthesis CELP coding technique. However, several modifications are incorporated by vendors to meet the needs of low-delay high-quality speech coding. G.728 uses short excitation vectors (five samples, or 0.625 ms) and backward-adaptive linear predictors. The algorithmic delay of the resulting coder is 0.625 ms, resulting in an achievable end-to-end delay of less than 2 ms. The G.728 standard was designed to provide speech quality equivalent to or better than that of the G.721 32 kbps ADPCM international standard, even after three tandem connections. The G.728 coder was also designed to behave well in the presence of multiple speakers and background noise, and to be capable of handling nonspeech signals such as DTMF tones and voice-band modem signals at rates of up to 2,400 bps (if perceptual weighting and postfiltering are disabled). Techniques such as bandwidth expansion of the LPC filter coefficients and codebook structuring have been incorporated into the standard to improve resistance to moderate channel error conditions. The G.728 coder achieved a MOS score of 4.0.

ITU G.726 is a speech coding and decoding standard that provides 4 kHz speech bandwidth switchable at 40-, 32-, 24-, or 16-kbps. We discussed this standard earlier in the chapter in terms of its components and operation. The full algorithm implementation runs on a single DSP. Again, a good implementation is expected to pass all digital test sequences provided by the ITU for algorithm verification [ASP200401]. The G.726 standard specifies an ADPCM system for coding and decoding samples. The overall compression ratio is 2.8:1, 3.5:1, 4.67:1, or 7:1 for 14-bit linear data sampled at 8 kHz and 1.6:1, 2:1, 2.67:1, or 4:1 for 8-bit companded data sampled at 8 kHz. G.726, originally designed as a half-rate alternative to 64 kbps PCM coding (companding), is used in some digital network equipment for transmission of speech and voiceband data. This algorithm is ideal for any application that requires speech compression, or encoding for noise immunity, efficient regeneration, easy and effective encryption, and uniformity in transmitting voice and data signals. The 3, 4, and 5 bit versions of G.726 were once known as CCITT G.723. This standard was a subset of the G.726 algorithm. The 4-bit version of G.726 (32 kbps) was also known as CCITT standard G.721. Note that ITU G.723.1 now refers to a standard for a 5.3 kbps and 6.3 kbps dual-rate speech coder and should not be confused with ITU G.726 or the old CCITT G.723 ADPCM standard.

The Mixed-Excitation Linear Predictive (MELP) vocoder will be the new 2,400 bps Federal Standard speech coder. It was selected by the United States Department of Defense Digital Voice Processing Consortium (DDVPC) after a multiyear extensive testing program. The selection test concentrated on four areas: intelligibility, voice quality, talker recognizability, and communicability. The selection criteria also included hardware parameters such as processing power, memory usage, and delay. MELP was selected as the best of the seven candidates and even beat the FS1016 4,800 bps vocoder, a vocoder with twice the bit-rate [ASP200401]. MELP is robust in difficult background noise environments such as those frequently encountered in commercial and military communication systems. It is very efficient in its computational requirements. This translates into relatively low power consumption, an important consideration for portable systems. The MELP vocoder was developed by teams from Texas Instruments Corporate Research in Dallas and ASPI Digital. The MELP vocoder is based on technology developed at the Center for Signal and Image Processing at the Georgia Institute of Technology in Atlanta.

The ITU Rec. G.723.1 is a 6.3- and 5.3-kbps vocoder for multimedia communications that was designed originally for low bit rate videophones. The algorithm's frame size is 30 ms, and the one-way codec delay is 37.5 ms. In applications where low delay is important, the delay G.723.1 may not be tolerable; however, if the delay is tolerable, G.723.1 provides a lower-complexity lower-bandwidth alternative to G.729, at the expense of a small degradation in speech quality. Each of these three ITU recommendations (G.723.1, G.728, and G.729) has the potential to become a key commercial mechanism for voice over IP on the Internet and other networks, since all three are low-bandwidth and are simple enough in complexity to be executed on the host processor, such as a PC, or implemented on a modem chip. Hence, this chapter examines these standards in some level of detail.

The GSM-FR speech coder (ITU-T RPE/LTP) provides encoding 8 kHz sampled speech signals for transmission at a rate of 13 kbps. The encoder compresses linear-PCM narrow band speech input data, and uses Regular Pulse Excitation with Long-Term Prediction (RPE-LTP) algorithm. The GSM-FR coder has been optimized to compress speech to the highest quality. The primary applications of this coder are in digital circuit multiplex equipment, satellite telephony and wireless standards such as PACS, DECT and PHP (Japan).

At the other end of the quality spectrum, ITU G.722 is an audio encoding/decoding standard that provides 7 kHz audio bandwidth at 64 kbps. It is intended for conferencing applications. ITU G.722 has been fully implemented on a single DSP. The coding system uses Sub-Band Adaptive Differential Pulse Code Modulation (SB-ADPCM). The input signal to the coder is digitized using a 16-bit A-D sampled at 16 kHz. Output from the encoder is 8 bits at an 8 kHz sample rate for 64 kbps, which can be stored to disk for later playback. The decoder operates in exactly the opposite fashion. An 8-bit coded input signal is decoded by the SB-ADPCM decoders. The result is a 16 kHz sampled output. The overall compression ratio of the G.722 audio coder is 4 to 1 [ASP200401]. G.722, designed for high quality speech applications in the telecommunications market, can be used in a variety of applications, including audio coding for medium quality audio systems.

2.3 Signaling

This section provides a short description of call control signaling protocols that are applicable to VoP, including VoIP, VoMPLS, VoWi-Fi, and VoIPv6. Signaling is a critical mechanism to enable call setup as well as to enable delivery of advanced "supplementary" services. Signaling is needed for both on-net call establishment, as well as, and even more so, for interworking with the PSTN. Three approaches have arisen in the past few years regarding signaling in VoP applications (see Figure 2.10 [VOV200101]):

- All elements (NEs and CPEs) have intelligence; in this case, one would employ ITU-T H.323.
- The network is intelligent, but the end-nodes are "dumb." In this case, one would employ MGCP (Media Gateway Control Protocol), MEGACO/H.248 (Media Gateway Controller[18]), CCSS7 (Common Channel Signaling System 7), and BICC (Bearer-Independent Call Control).
- The end-nodes are intelligent, but the network is "dumb." In this case, one would employ Session Initiation Protocol (SIP).

Traditional carriers tend to subscribe to the first two models, while enterprise-oriented folks tend to subscribe to the last model. H.323 (various versions) has the largest market share to date, but it is expected that there will be an inversion at some point in the future. It is derived from ISDN signaling protocols and, therefore, has an affinity for PSTN-like and PSTN-interworking environments. It is a kind of ITU-T Q.931 on TCP/IP.

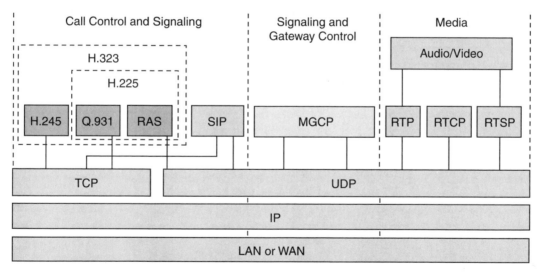

H.323 Version 1 and 2 supports H.245 over TCP, Q.931 over TCP, and RAS over UDP.
H.323 Version 3 and 4 supports H.245 over UDP/TCP, Q.931 over UDP/TCP, and RAS over UDP.
SIP supports TCP and UDP.

Figure 2.10: Comparison of signaling protocols.

---

[18] The MEGACO initiative has a genesis in IPDC (proposed by Level 3, 3Com, Alcatel, Cisco and others) and SGCP (Telcordia). These protocols were brought together by the IETF to form MGCP (Media Gateway Control Protocol); work continues under the responsibility of the MEGACO Working Group.

The chronology of the standardization efforts is as shown in Table 2.10.

Table 2.10: Chronology of the major signaling protocols.

Standard	Original Date	Proponents	Comments
ITU-T H.323 v1:	May 1996	Carriers	Complex
ITU-T H323 v2:	January 1998	Carriers	Complex
ITU-T H323 v3:	September 1999	Carriers	Complex
ITU-T H323 v4:	November 2000	Carriers	Complex
IETF SGCP (Simple Gateway Control Protocol)	July 1998	Telcordia, Cisco	Superseded
IPCD (IP Device Control)	August 1998	Level 3	Folded into MGCP 0.1
IETF MGCP 0.1	October 1998	IETF	
MDCP (Media Device Control Protocol)	December 1998	Lucent	Folded into MEGACO
MEGACO (MGCP+)	April 1999	IETF	
IETF SIP	March 1998	IETF, Industry Newcomers	Experiencing major deployment

Signaling is critical and is fundamental to the development of carrier-grade feature-rich telephony services. True end-to-end functionality is needed (spanning intraenterprise, intracarrier, and intercarrier domains) for VoIP to fulfill the service substitution role. Until VoIP signaling is taken more seriously by developers and until such time that these developers bring out truly standardized interoperable products, particularly in a carrier-to-carrier environment, the penetration of VoIP will remain at the margins, namely, a few percentage points globally in terms of actual paid-for call volume carried.

The different signaling protocols have been developed in different camps to address the need for real-time session signaling over packet-based networks. Each of these protocols has different origins and different supporters with differing priorities. H.323 was originally developed in the enterprise LAN community as a video-conferencing technique and has much in common with ISDN signaling protocols such as Q.931. MGCP/MEGACO comes from the carrier world and is closely associated with intradomain control of softswitches and media gateways, and so on. The IETF developed SIP, reusing many familiar Internet elements: SMTP, HTTP, URLs, MIME, and DNS. Despite all being signaling protocols, they are not equals and peers—they can and will coexist; however, there is some debate as to what extent [SIP200301]. Table 2.11 provides a basic comparison between the protocols [SIP200301].

Table 2.11: Comparison between three major signaling protocols.

	SIP	H.323	MGCP/MEGACO
Philosophy	Horizontal	Vertical	Vertical
Complexity	Medium-Low	High	High
Scope	Simple	Full	Partial
Scalability	Good	Reasonable	Moderate
New Service Revenues	Yes	No	No
Internet Cohesion	Yes	No	No
SS7 Compatibility	Poor	Reasonable	Good
Cost	Low	High	Moderate

2.3.1 H.323 Standards

ITU-T H.323 was initially developed for video and multimedia. It was later adopted by the VoIP community for direct voice applications; hence, we cover this protocol here in the context of voice and in the context of alternative solutions. According to ITU-T Recommendation H.323 Version 4, H.323: "Describes terminals and other entities that provide multimedia communications services over Packet-Cased Networks (PBN), which may not provide a guaranteed quality of service. H.323 entities may provide real-time audio, video and/or data communications." H.323 is an umbrella standard covering multimedia communications over LANs. H.323 defines: (1) Call establishment and teardown and, (2) Audio visual or multimedia conferencing. H.323 defines sophisticated multimedia conferencing supporting applications such as whiteboarding, data collaboration, or video conferencing. Basic call features include: call hold, call waiting, call transfer, call forwarding, caller identification, and call park. Figure 2.11 depicts the protocol model.

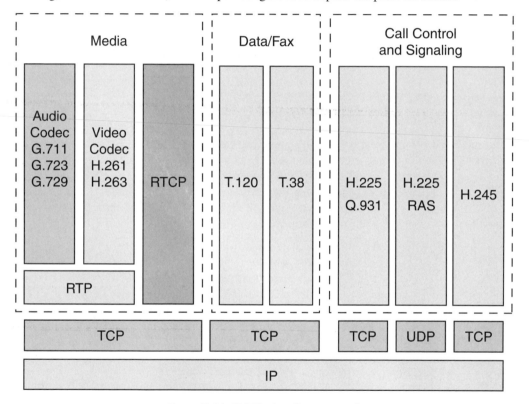

Figure 2.11: H.323 signaling protocol.

H.323 entities consist of (see Figure 2.12):

- Terminals
- Gateways
- Gatekeepers
- MCUs

The H.323-umbrella of protocols consists of:

- Parts of H.225.0 (registration, admission, status [RAS]), Q.931
- H.245
- RTP/RTCP
- Audio/video codec standards

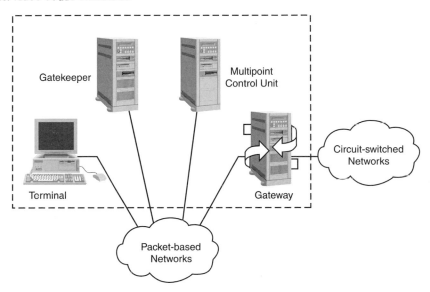

Figure 2.12: H.323 environment.

H.323 Entities: Terminals. Terminals are endsystems (or endpoints) on a LAN (see Figure 2.13). The terminal embodies capabilities to supports real-time, two-way communications with another H.323 entity. The terminal must support (1) Voice – audio codecs; and, (2) Signaling and setup – Q.931, H.245, RAS. Optional support includes video coders and data (whiteboarding). Audio codecs (G.711, G.723.1, G.728, etc.) and video codecs (H.261, H.263) compress and decompress media streams. Media streams transported on RTP/RTCP (RTP carries actual media while RTCP carries status and control information). RTP/RTCP is carried in User Datagram Protocol (UDP) datagrams. Signaling is transported reliably over TCP. RAS supports registration, admission, status; Q.931 handles call setup and termination, and H.245 provides capabilities exchange.

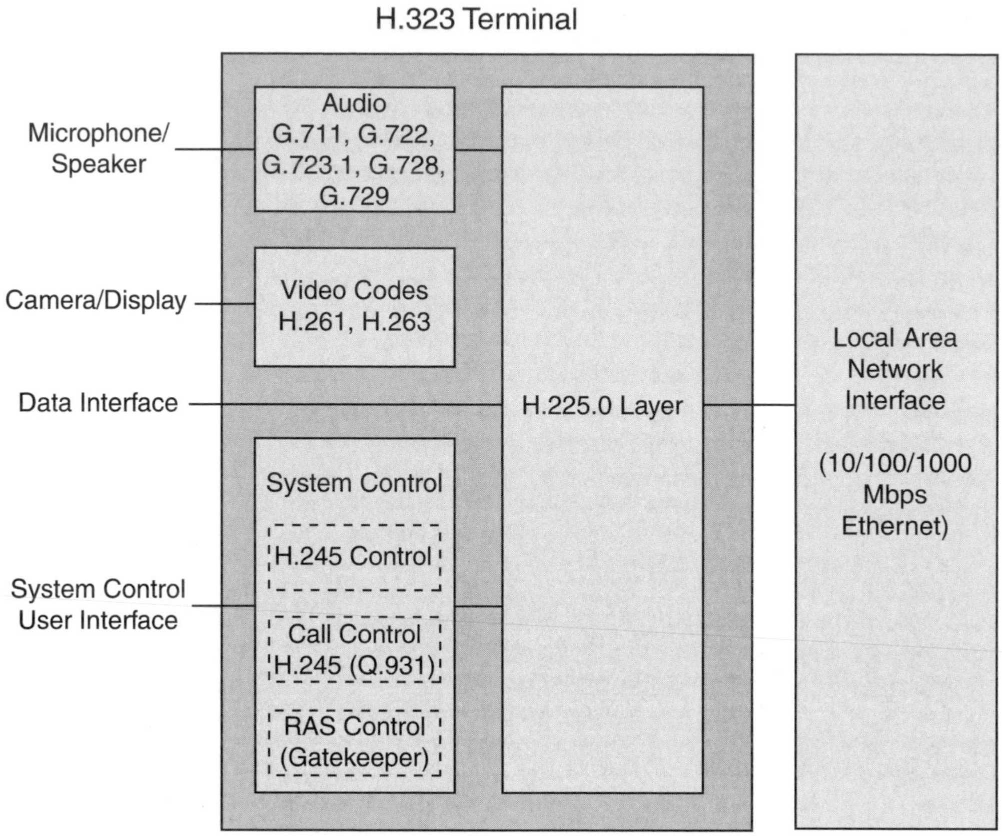

H.323 Terminal

Microphone/Speaker

Audio
G.711, G.722,
G.723.1, G.728,
G.729

Camera/Display

Video Codes
H.261, H.263

Data Interface

H.225.0 Layer

System Control User Interface

System Control

H.245 Control

Call Control
H.245 (Q.931)

RAS Control
(Gatekeeper)

Local Area Network Interface

(10/100/1000 Mbps Ethernet)

Figure 2.13: H.323 terminal.

H.323 Entities: Gateways. Gateways provide interfaces between the LAN and the switched circuit network. Gateway provides translation between entities in a packet switched network (example, IP/MPLS network) and circuit switched network (example, PSTN network). They can also provide transmission formats translation, communication procedures translation, H.323 and non-H.323 endpoints translations or codec translation. Gateways translates communication procedures and formats between networks, and handle call setup and clearing and compression and packetization of voice. Various types of gateways exist, however, the most common example is a IP/PSTN gateway. Naturally, the gateway must support the same protocol stack described above on the local side.

H.323 Entities: Gatekeeper. Gatekeepers are optional elements but must perform certain functions if present. Gatekeepers manage a zone (a collection of H.323 devices). Usually there is one gatekeeper per zone; alternate gatekeeper might exist for backup and load balancing. Typically gatekeepers are a software application, implemented on a PC, but can be integrated in a gateway or terminal. Some protocol messages pass through the gatekeeper while others pass directly between the two endpoints. The more messages that are routed between the gatekeeper, the more the load and responsibility (more information and more control). Notice that media streams never passes through the gatekeeper function. Mandatory gatekeeper functions include:

- Address translation (routing)
- Admission control
- Minimal bandwidth control – request processing
- Zone management

Optional gatekeeper functions include:

- Call control signaling – direct handling of Q.931 signaling between endpoints
- Call authorization, bandwidth management, and call management using some policy
- Gatekeeper Management Information (MIB)
- Directory services

H.323 Entities: Multipoint Control Unit (MCU). MCUs are endsystems that supports conferences between three or more endpoints. The MCU can be a stand-alone device (e.g., PC) or integrated into a gateway, gatekeeper, or terminal. Typically, the MCU consists of a Multipoint Controller (MC) and a Multipoint Processor (MP):

- MC – handles control and signaling for conference support;
- MP – receives streams from endpoints, processes them, and returns them to the endpoints in the conference.

MCUs can be of the centralized kind or of the decentralized kind.

The rest of this section briefly illustrates signaling interactions [RAD199801]. Figure 2.14 shows a gatekeeper-routed call signaling process. This discussion expands on the interaction shown in Figure 2.14 for a gatekeeper-routed call signaling (ITU-T Q.931/H.245) interaction between client A and client B [RAD199801]. This interaction supports the establishing a call between client A and client B.

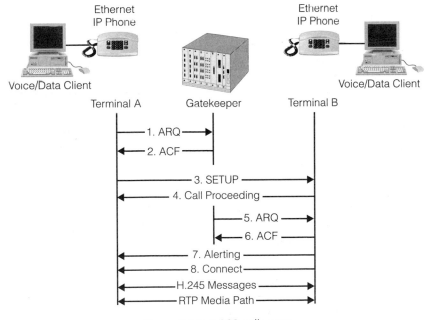

Figure 2.14: H.323 call setup.

83

The steps are:

1. Discover and register with the gatekeeper – RAS channel
    Discover Gatekeeper (RAS) works as follows:

    > Client transmits a Multicast Gatekeeper Request packet (who is my gatekeeper?)

    Gatekeeper responds with a Gatekeeper Confirmation packet or Gatekeeper Reject packet

    Registering with Gatekeeper (RAS) works as follows:

    > Client notifies gatekeeper of its address and aliases

    > Client transmits Gatekeeper Registration Request

    > Gatekeeper responds with either Registration Confirmation or Registration Rejection

    > In network deployment in diagram, both client A and client B register with gatekeeper A

2. Routed call setup between the endpoints through the gatekeeper – Q.931 call signaling
    Call Admission (RAS) handled as follows:

    > Client A initiates Admission Request (can I make this call?);
    > the packet includes a maximum bandwidth requirement for the call

    > Gatekeeper responds with Admission Confirmation

    > - Bandwidth for call is either confirmed or reduced
    > - Call signaling channel address of gatekeeper is provided

    Call Setup Through Gatekeeper (Q.931), as follows:

    > Client A sends call setup message to gatekeeper

    > Gatekeeper routes message to client B

    > If client B accepts, admission request with gatekeeper is initiated

    > If call accepted by gatekeeper, client B sends a connect message to client A specifying the H.245 call control channel for capabilities exchange

3. Initial communications and capability exchange – H.245 call control
    Capabilities Exchange (H.245):

    > Clients exchange call capabilities with Terminal Capability Set message that describes each client's ability to transmit media streams, i.e., audio/video codec capabilities of each client

    > If conferencing, determination of MCU is negotiated during this phase

    > After capabilities exchange, clients have a compatible method for transmitting media streams; multimedia communication channels can be opened

4. Establish multimedia communication/call services – H.245 call control

Establish Multimedia Communication:

> To open a logical channel for transmitting media streams, the calling client transmits an Open Logical Channel message (H.245)

> Receiving client responds with Open Logical Channel Acknowledgment message (H.245)

> Media streams are transmitted over an unreliable channel; control messages are transmitted over a reliable channel

> Once channels established, either client or gatekeeper can request call services, i.e., client or gatekeeper can initiate increase or decrease of call bandwidth

5. Call termination – H.245 call control and Q.931 call signaling

Call Termination:

> Either party can terminate the call

> Assume client A terminates call

> Client A completes transmission of media and closes logical channels used to transmit media

> - Client A transmits End Session Command (H.245)
> - Client B closes media logical channels and transmits End Session Command
> - Client A closes H.245 control channel
> - If call signaling channel is still open, a Release Complete message (Q.931) is sent between clients to close this channel

2.3.2 Introduction to Session Initiation Protocol (SIP)

According to IETF RFC 2543, Session Initiation Protocol (SIP) (now obsoleted by RFC 3261, RFC 3262, RFC 3263, RFC 3264, RFC 3265) is an application layer signaling protocol that defines initiation, modification and termination of interactive, multimedia communication sessions between users. SIP was designed for: (1) integration with existing IETF protocols; (2) scalability and simplicity; (3) mobility (including presence/proximity, multimodal and collaborative communications); and, (4) easy feature- and service-creation. SIP is designed to be fast and simple in the (enterprise) core of the network. SIP can support these features and applications, among others: basic call features (call waiting, call forwarding, call blocking, etc.); unified messaging, call forking, click-to-talk, instant messaging, and find me/follow me. SIP is a peer-to-peer protocol (as are other Internet protocols) where a client can establish a session with another client; by contrast MEGACO is a master-slave protocol. SIP can use any network transport protocol, for example, UDP (RFC 768), TCP (RFC 761), and Stream Control Transmission Protocol (SCTP) (RFC 2960). It appears that over time, SIP will achieve significant market penetration. SIP is examined in detail in Chapter 3. This section provides a quick summary with the goal of a comparison with the other signaling protocols.

SIP is the basis of the industry-standard, IP-centric converged communications architecture; it does for real-time interhuman communications what HTML did for browsing. SIP serves as a signaling mechanism to establish a wide variety of sessions, interactive communication that takes place between two or more entities over an IP network, from a simple two-way telephone call or an instant message exchange, to a collaborative multimedia conferencing session. SIP does not dictate the details within a session but instead negotiates interaction based on the capabilities of participants. Features and applications are integrated at the session and service layers, independent of access constraints and the processes of message transport. In SIP networks, voice is just another media. SIP is perceived as being capable of enabling innovative communications capabilities, where it is easy to introduce a new services. SIP supports a distributed architecture. Many telecommunications vendors are SIP-enabling their PBX and customer contact centers to enrich these environments and provide converged desktop functionality. The converged desktop tightly couples the telephone (whether digital, analog, or IP) and the PC for a richer experience, drives service ubiquity, and allows employees to have desktop functionality anywhere, anytime, using any device. SIP offers the mechanism for creating real-time integrated communications [NOR200501].

Figure 2.15 identifies key SIP components. User Agent is an application that initiates, receives, and terminates calls. There are two types:

- User Agent Clients (UAC) – An entity that initiates a call.
- User Agent Server (UAS) – An entity that receives a call.

Both UAC and UAS can terminate a call.

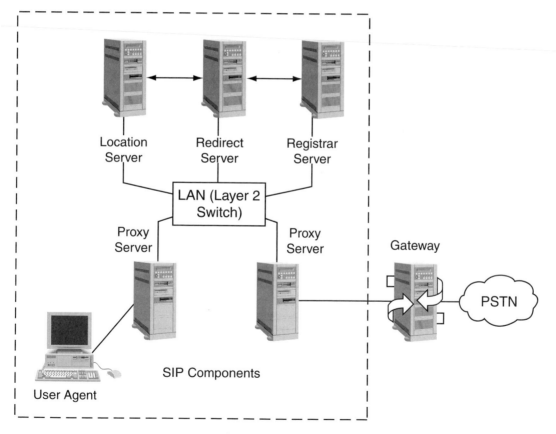

Figure 2.15: An SIP environment.

The *Proxy Server* is an intermediary program that acts as both a server and a client to make requests on behalf of other clients. Requests are serviced internally or by passing them on, possibly after translation, to other servers. The server interprets, rewrites or translates a request message before forwarding it.

The *Location Server* is utilized by an SIP redirect or proxy server to obtain information about a called party's possible location(s).

The *Redirect Server* is a server that accepts an SIP request, maps the address into zero or more new addresses and returns these addresses to the client. Unlike a proxy server, the redirect server does not initiate its own SIP request. Unlike a user agent server, the redirect server does not accept or terminate calls [VOV200101].

The *Registrar Server* is a server that accepts REGISTER requests. The register server may support authentication. A registrar server is typically co-located with a proxy or redirect server and may offer location services.

SIP components communicate by exchanging SIP messages (see Table 2.12). SIP borrows much of the syntax and semantics from HTTP. An SIP message looks like an HTTP message—message formatting, header and MIME (Multipurpose Internet Mail Extension) support. The SIP address is identified by an SIP URL; the URL has the format: user@host.

Table 2.12: SIP messages at a snapshot.

= == = = = = = = = = = = = = = = = =

SIP Messages

INVITE: Initiates a call by inviting the user to participate in the session

ACK: Confirms that the client has received a final response to an Invite request

BYE: Indicates a termination of the call

CANCEL: Cancels a pending request

REGISTER: Registers the user agent

OPTIONS: Used to query the capabilities of a server

INFO: Used to carry out-of-bound information, such as dual-tone multiple frequency (DTMF) digits

SIP Responses

1xx: Informational messages

2xx: Successful responses

3xx: Redirection responses

4xx: Request failure responses

5xx: Server failure responses

6xx: Global failure responses

= == = = = = = = = = = = = = = =

Establishing communication using SIP usually takes place in six steps:

1. Register, initiate, and locate the user.
2. Determine which media to use—involves delivering a description of the session to which the user is invited.
3. Determine the willingness of the called party to communicate. The called party must send a response message to indicate a willingness to communicate: accept or reject.
4. Set-up the call.
5. Modify or handle the call.
6. Terminate the call.

As noted above, SIP was designed for integration with IETF environments. Existing IETF protocol standards can be used to build an SIP application. The protocol can work with existing IETF protocols such as: RSVP (Reserve Network Sesources); real-time protocol (transport real-time data and provide QoS feedback); real-time streaming protocol (controls delivery of streaming media); session advertisement protocol (advertising multimedia session via multicast); session description protocol (describing multimedia sessions); multipurpose Internet mail extension (content description); hypertext transfer protocol (web pages delivery); and other IETF protocols. SIP supports flexible and intuitive feature creation using SIP-CGI (SIP-Common Gateway Interface) and CPL (Call Processing Language).

Facts about SIP noteworthy of observation include the following [CIS200501]:

SIP is a signaling protocol that is independent of transport protocol; it can run on top of several transport protocols, including UDP, TCP, and Stream Control Transmission Protocol (SCTP).

SIP does not mandate or include specific QoS capabilities; it works with other protocols that perform this function.

SIP is independent of any security protocol and may be used with several security protocols, such as Transport Layer Security (TLS) and IP Security (IPSec). SIP takes advantage of existing IP security standards to help ensure the integrity of communications sessions. SIP supports TLS, the successor to Secure Sockets Layer (SSL), to secure the signaling channel while Secure RTP (SRTP) encrypts the media to ensure voice privacy. Together, they represent a strong security system based on established standards including Advanced Encryption Standard (AES), the U.S. government encryption standard.

SIP is a peer-to-peer protocol, not an IP-to-PSTN gateway control protocol such as MGCP or H.248.

SIP provides methods to control sessions, but does not specify the applications and services that will use those sessions; as a result, SIP does not guarantee application behavior.

SIP is independent of the media used, allowing the flexibility to initiate sessions for different media types.

Interdomain Operation. One of the advantages of SIP is the ability to communicate between domains directly via the Internet. Consider arranging a multimedia conference with a business partner company. Today, one would likely do this through a third-party conferencing provider. If both the user and the partner company employed SIP-enabled multimedia infrastructures, the conference could be conducted directly via the Internet, with no need to reserve conference time or pay for the service. It would be as easy as putting multiple recipients on the "To:" line of an e-mail message. While this type of capability is eminently feasible using an all-IP infrastructure, it is far more difficult in an environment that involves TDM switching, which requires numerous PSTN-to-IP gateways. In either case, it requires compatibility between the applications on both ends.

Feature Support. SIP is intended to provide interoperability (enabling customers to use IP phones from one vendor, while employing SIP proxies from another, for example). Because many aspects of SIP-enabled applications are not defined in the standards, much of the implementation work is left up to the vendors.

Customers may find that not all features supported by one vendor's SIP endpoint will work with another vendor's call-control servers and proxies; they need to specify which features they will need in their implementations.

Challenges: NAT Traversal. As noted in Chapter 1, NAT is a common way to hide a private IP address from the public Internet and to extend the number of IP addresses that an enterprise can employ. NAT gateways translate a private address coming from inside an organization to a different address that is conveyed to outside IP devices. That translation can be difficult in an SIP session, since SIP needs to know the IP address of each endpoint device involved in a session. Many security component vendors address the NAT issue by examining Session Description Protocol (SDP) information, and may try to resolve addresses that were changed by NATs. This approach can cause problems in certain scenarios, particularly if the signaling information is encrypted between the client and the server. Another solution is a protocol called *STUN (simple traversal of UDP through network address translators)*. When a user sends a message to a server from inside a NAT, the server will reflect back whatever address the NAT gives it. STUN allows this reflected address to be used to establish an RTP session with the user inside the NAT, without involving any of the SIP proxies in the middle. Solutions to the NAT issue will vary depending on the exact scenario and environment; the industry has not settled on universally-accepted solutions. Both of these topics are revisited in Chapter 5.

However, by itself, SIP is not a communications panacea—it works with many other standards to foster open, reliable, rich multimedia communications.

Functionally, SIP and H.323 are similar ([VOV200101], [DAL199901] and [COL200201]). Both protocols provide for call control, call setup, and call teardown (with capabilities exchange). Both protocols provide basic call features such as call waiting, call hold, call transfer, call forwarding, call return, call identification, or call park.

"High power/early adopter users" are moving toward a mode of communication that has been called *always-on integrated communications*. Integrated communications are comprised up of asynchronous communications (e-mail, voicemail, short message services) and synchronous communications (IM, voice, video, and application sharing), along with presence and location intelligence. While each of these communication instances can be deployed on a discrete basis, "integrated" carries the implication of a seamless user experience across all these media. "Always-on" implies uninterrupted access (connectivity) any time, any place, in the sense of Pervasive computing—also called *invisible computing* or *ubiquitous computing*. Vendors are taking different approaches to meeting the advanced needs of the increasingly distributed and mobile users for secure and reliable real-time collaborative tools. For example, video conferencing and data web portal vendors are expanding into data and web conferencing, respectively. While IM and presence are often a common element of these systems, telephony, if offered at all, is very basic. In contrast, VoIP vendors seek to offer always-on interperson communications solutions in the form of business telephony (both desktop and mobile) and customer contact solutions, and presence-based, rich media capabilities in the form of video, IM, and application sharing. SIP is being embraced either as a gateway function or at the heart of their emerging architectures [NOR200501].

Central to the SIP-based architecture is the notion of presence. Real-time presence information is captured across a broad range of activities including being active on a device (telephone, PC, PDA, BlackBerry), having a session in progress (whether synchronous or asynchronous), or being at a location (office, functional area like a conference room, or surgery). This information can be combined with location intelligence and selectively made available to clients and to any SIP-enabled application, such as CRM, document handling, workflow, and customer service. SIP can be used to extend sessions and the notion of presence; for example, into vertically-targeted IM systems [NOR200501]. The topic of presence is revisited in Chapter 4.

2.3.3 MEGACO

As noted, MEGACO is a protocol that is evolving from MGCP and developed jointly by ITU and IETF: it is known as MEGACO in the IETF and H.248/H.GCP in the ITU-T. MEGACO has developed by the carrier community to address the issue of CCSS7/VoIP integration. The H.323 initiative had grown out of the LAN and had difficulties scaling to public network proportions; the architecture that it created was incompatible with the world of public telephony services, struggling with multiple gateways and the CCSS7. To address this problem, the new initiative exploded the gatekeeper model and removed the signaling control from the gateway, putting it in a "media gateway controller" or "softswitch." This device would control multiple "media gateways." This is effectively a decomposition of the gatekeeper to its CCSS7 equivalents. MGCP/MEGACO is the protocol used to communicate between the softswitch and the media gateways [SIP200301]. MEGACO brings a performance enhancement compared to MGCP: it can support thousands of ports on a gateway, multiple gateways, and can accommodation for connection-oriented media such as TDM and ATM.

MGCP/MEGACO "exploded" the gatekeeper model of the H.323 model and removed the signaling control from the gateway, putting it in a "media gateway controller" or "softswitch." This device controls multiple "media gateways." In the MGCP/MEGACO architecture, the intelligence (control) is unbundled from the media (data). It is a master-slave protocol where the master has absolute control and the slave simply executes commands. The master is the media gateway controller, or softswitch (or call agent) and the slave is the media gateway (this can be a VoIP gateway, an MPLS router, IP phone, etc.) [SIP200301].

MGCP/MEGACO is used for communication to the media gateways. MGCP/MEGACO instructs the media gateway to connect streams coming from outside a packet network on to a packet stream such as RTP. The softswitch issues commands to send and receive media from addresses, to generate tones, and to modify configuration. The architecture, however, requires a session initiation protocol for communication between gateway controllers.

When a gateway detects an off-hook condition, the softwsitch instructs the gateway controller via MEGACO commands, to put dial tone on the line and collect DTMF tones. After detecting the number, the gateway controller determines how to route the call and, using an intergateway signaling protocol such as SIP, H.323, or Q.BICC, contacts the terminating controller. The terminating controller could instruct the appropriate gateway to ring the dialed line. When the gateway detects the dialed line is off hook, both gateways could be instructed by their respective gateway controllers to establish two-way voice across the data network. Thus, these protocols have ways to detect conditions on endpoints and notify the gateway controller of their occurrence; place signals (such as dial tone) on the line; and create media streams between endpoints on the gateway and the data network, such as RTP streams [SIP200301].

There are two basic constructs in MGCP/MEGACO: terminations and contexts. Terminations represent streams entering or leaving the gateway (for example, analogue telephone lines, RTP streams, or MP3 streams). Terminations have properties, such as the maximum size of a jitter buffer, which can be inspected and modified by the gateway controller. A termination is given a name, or TerminationID, by the gateway. Some terminations, which typically represent ports on the gateway, such as analog loops or DS0s, are instantiated by the gateway when it boots and remain active all the time. Other terminations are created when they are needed, get used, and then are released. Such terminations are called *ephemerals* and are used to represent flows on the packet network, such as an RTP stream. Terminations may be placed into contexts, which are defined as when two or more termination streams are mixed and connected together. The normal, "active" context might have a physical termination (say, one DS0 in an E3) and one ephemeral one (the RTP stream connecting the gateway to the network). Contexts are created and released by the gateway under command of the gateway controller. Once created, a context is given a name (ContextID), and can have terminations added and removed from it. A context is created by adding the first termination, and it is released by removing the last termination. MGCP/MEGACO uses a series of commands to manipulate terminations, contexts, events, and signals [SIP200301]:

- Add – adds a termination to a context and may be used to create a new context at the same time
- Subtract – removes a termination from a context and may result in the context being released if no terminations remain
- Move – moves a termination from one context to another
- Modify – changes the state of the termination
- AuditValue and AuditCapabilities – return information about the terminations, contexts, and general gateway state and capabilities
- ServiceChange – creates a control association between a gateway and a gateway controller and also deals with some failover situations.

2.4 Numbering

This section looks at one facet of what is needed to affect a packet/circuit integration in the PSTN both for basic services as well as for advanced new services, namely, a workable cross-addressing mechanism. The section briefly looks at ENUM (RFC 2916, original issue: September 2000) proposal. ENUM defines a Domain Name System (DNS)-based architecture and protocols for mapping a telephone number to a set of attributes (e.g., URLs) that can be used to contact a resource associated with that number. This IETF protocol is designed to assist in the convergence of the PSTN and the IP network, since supports is the mapping of a telephone number from the PSTN to Internet services.

Directory Services are part of an overall networking functionality (e.g., X.500 Directory Services). VoIP must support effective addressing and address translation. These functions can be supported in a customer-resident gateway, but for full scalability, the directory function is best located in the network at large. A protocol is then needed to support various interactions with the Directory.

ENUM was developed as a (potential) solution to the question of how network elements can find services on the Internet using only a telephone number, and how telephones, having an input mechanism limited to twelve keys on a keypad, can be used to access Internet services [ENU200301]. ENUM at its most basic level aims at facilitating the convergence of PSTN and IP networks; it is the mapping of a telephone number from the public switched telephone network to Internet functionalities. The convergence can be facilitated, that is the PSTN can be organically linked to the Internet by making the telephone number part of an Internet address. ENUM is a proposed approach to supporting a number of directory-related functions.

With ENUM, the telephone number can also serve as basis for a person's e-mail address. This agreement allows a person to reach multiple services by knowing a single contact address (number). ENUM supports a capability that takes "a telephone number in, and gives a URL out." The protocol takes a complete, international telephone number and resolves it to a series of URLs using a DNS-based architecture.

ENUM was developed as a solution to the question of how to find services on the Internet using only a telephone number, and how telephones, which have an input mechanism limited to twelve keys on a keypad, can be used to access Internet services [ENU200301]. Because ENUM puts telephone numbers into the DNS, it allows for a gamut of applications based only on a telephone number. Proponents see the most promising application as being an improvement in VoIP for telephone calls made over the Internet; additional applications include addressing for fax machines, e-mail, instant messaging, and websites.

Although the technical issue is somewhat straightforward, the "politics" are sensitive. The issue has generated controversy, because "number administration" has intrinsic prestige, influence, and power. The issue is particularly thorny on the international arena. People that control addressing control certain aspects of network operations, design, and even ownership. There is a competition that positions the heavily-regulated telephone industry against Internet entities that are uninterested in government regulation. Some nations are involved in the debate out of concern that a merged network could undermine state-owned telephone networks [CHA200101].

"ENUM" has a number of meanings. It is the name of a protocol that resolves fully qualified telephone numbers to fully qualified domain name addresses using a DNS-based architecture. It is the name of a chartered working group of the IETF chartered to develop protocols that map telephone numbers to resources found on the Internet using the DNS. It is also the title of RFC 2916, the approved protocol document that discusses the use of DNS for the storage of ITU-T E.164 numbers and the available services connected to an E.164 number. It should be noted that ENUM does not change the Numbering Plan and does not change telephony numbering or its administration in any way and it will not drain already-scarce numbering resources given that it uses existing numbers.

E.164 is the specification of the international telephone numbering plan administered by the ITU that specifies the format, structure, and administrative hierarchy of telephone numbers. Specifically, "E.164" refers to the ITU document that describes the structure of telephone numbers. The ITU issues country codes to each nation; the administration of telephone numbers within each country is governed by that country's telecommunications regulatory agency. A fully qualified E.164 number is designated by a country code, an area or city code, and a phone number. For example, a fully qualified E.164 number for the phone number 555-1234 in New York City (area code 212) in the United States (country code 1) is +1-212-555-1234. E.164 numbers are appropriate for use in ENUM because they are an existing system for global traceability.

Under this proposal, the number 1-212-555-1234 would become 4.3.2.1.5.5.5.2.1.2.1.e164.arpa as an Internet address (that is, the telephone number backward—separated by periods—with the extension ".e164. arpa" added in). The system would recognize both addresses as belonging to the same individual or entity[19]. The extension ".e164.arpa" was picked to appease both camps: as noted, E.164 is the specification for the carriers' numbering scheme, while arpa refers to the Advanced Research Projects Agency, the U.S. agency that funded much of the Internet work in the 1970s and 1980s.

Telephone numbers currently identify many different types of end terminals, supporting different services and protocols. Telephone numbers are used to identify telephones stations, fax machines, pagers, data modems, email clients, text terminals for the hearing impaired, and so on. A prospective caller may wish to discover which services and protocols are supported by the terminal named by a given telephone number. The caller may also require more information beyond simply the telephone number to communicate with the terminal. As an example, certain telephones can receive short e-mail messages. The telephone number does not embody sufficient information to be able to send email; the sender must have more information (equivalent to the information in a mailto: URL). From the callee's perspective, the owner of the telephone number or device may wish to control the information that prospective callers may receive. The architecture must allow for different service providers competing openly to furnish the directory information required by clients to reach the desired telephone numbers. To address these issues, the IETF Working Group specified a while back a DNS-based architecture and supportive protocols that fulfill the following requirements:

1. The system must enable resolving input telephone numbers into a set of URLs which represent different ways to start communication with a device associated to the input phone number.

---

[19] User's are not required to use the common number, and they could still keep separate telephone and e-mail addresses. But if the use it, they gain the advantages intrinsic with the system.

2. The system must scale to handle quantities of telephone numbers and queries comparable to current PSTN usage. It is highly desirable that the system respond to queries with speed comparable to current PSTN queries, including in the case of a query failure.

3. The system must have some means to insert the information needed to answer queries into the servers via the Internet. The source of this information may be individual owners of telephone numbers (or the devices associated to the number), or it may be service providers that own servers that can answer service-specific queries. The system is designed not preclude the insertion of information by competing service providers (in such a manner that allows for the source of the information to be authenticated).

4. The system must enable the authorization of requests and of updates.

5. The effort must carefully consider and document the security and performance requirements for the proposed system and its use.

6. The effort must take into account the impact of developments in the area of local number portability on the proposed system.

Naturally, the protocol put forth needs to take into consideration how number resolution using the ENUM system is affected by the PSTN infrastructure for telephone numbering plans, such as the ITU-T E.164 standard.

The documentation developed in the recent past (Internet-Drafts: *Number Portability in the GSTN: An Overview* and Request for Comments (RFC) 2916: *E.164 number and DNS*) specifies the architecture and protocols (query, update) of the ENUM system.

Proponents argue that a government-sanctioned standard (e.g., through adoption via the ITU) for a centralized directory system is needed to avoid the consumer confusion that arises as the plethora of devices people use to communicate becomes more pronounced [CHA200101].

As inferred from the material presented this far, the IETF has looked at the issue and generated RFC(s) and Drafts. Liaisons with the ITU have occurred. Lately, a number of U.S. governmental and nongovernmental groups have been studying where to asset their power over the telephone-Internet database:

- The Federal Communication Commission (FCC), which oversees the U.S. telephone system;
- The Commerce Department, which has taken charge of the Internet's addressing system;
- The State Department, which has responsibility for cross-border issues and is the department that has representation at the ITU;
- The Federal Trade Commission, which is responsible with protecting consumers (e.g., privacy issues); and,
- The Internet Corporation for Assigned Names and Numbers (ICANN), which has some jurisdiction over the Internet addressing system.

Because the ENUM plan sits at the point-of-interaction of the telephone and Internet worlds, it presents a nontrivial policy issue in the U.S. and elsewhere. One of the issues the regulators could decide is who will take charge of the database that will map the telephone numbers to the Internet addresses. Some have lobbied the government to approve the proposal quickly (e.g., NeuStar); others want the regulators to steer clear of the debate (e.g., VeriSign).

Proponents, in the meantime continue to play advocacy rolls. The ENUM Alliance has been formed to assist ENUM implementers in the commercial marketplace. More work remains to be done in this arena.

2.5 VoIP and Wireless Networks

2.5.1 Approaches

To date, VoIP has seen a variety of applications as follows:

- Within an enterprise setting, utilizing intersite intranet trunking in conjunction with PSTN gateways for off-net access;
- Consumer VoIP services obtained over cable TV or Digital Subscriber Line (DSL) infrastructures, also, sometimes, in conjunction with pure-play VoIP long-haul providers; and,
- (Of late a resurgence of) Internet-based telephony with industry moves in this space by several search-engine and related companies (voice transmission over the Internet goes back about a decade, but not as a full-fledged provider-offered service).

In addition to these wireline applications, there is a symbiotic relationship, and emerging service support, between VoIP and wireless networks. VoIP can be carried by a number of wireless technologies such as 3G, WLAN, and Broadband Wireless Access (BWA). We refer to all of these as *VoIP over Wireless* (VoW) (the terminology Wireless VoIP—wVoIP—is used by some).

In the past ten years cellular/wireless players have pressured the landline voice market. Separately, VoIP is now having a measurable impact on fixed wireline networks; some forecasts claim up to 10% of the wireline voice traffic will be handled via VoIP within five years. But with the integrated use of VoIP over wireless technologies, specifically with VoW, the table could be turned upside down once again: VoW may bring new players to the mobile voice market thus further pressuring the voice revenues of existing mobile network operators. With VoW enables fixed operators, Mobile Virtual Network Operators (MVNOs), and VoIP service providers to bypass existing cellular voice services [3GS200501].

This section briefly looks at this topic. Our discussion is fairly terse since an entire textbook could be dedicated to this topic. We look at three VoW areas: (2) VoIP in enterprise WLAN (VoWi-Fi) environments; (2) VoIP over hotspot/WiMAX, and (3) VoIP in 3G cellular networks.

(i) Enterprise WLAN Applications (VoWi-Fi)

As VoIP-based technology begins to replace the traditional corporate PBX, there is interest in using a VoIP-ready phone in a mobility mode. With this deployment enterprise associates can make use of a wireless H.323/SIP handset anywhere on site where the intranet provides WLAN connectivity. There is also interest in being able to roam the VoWi-Fi connection to a cellular service when the user leaves the building and enters the metro and/or national footprint.

VoWi-Fi requires a number of enhancements to traditional WLANs, including:

- Higher WLAN speeds to support an adequate number of VoIP users. The transition to an IEEE 802.11g and/or 802.11n environment is a basic necessity in this context;
- Support of QoS over the wireless (and also core intranet) infrastructure. The deployment of IEEE 802.11e QoS-supporting technology is another basic necessity;
- Secure (voice) communications is highly desirable. The deployment of IEEE 802.11i security capabilities is yet another requirement;
- Roaming between Access Points, floors, and subnets is needed, as is a handoff to a cellular service when the corporate WLAN service is no longer available. The deployment of IEEE 802.11r roaming capabilities addresses this requirement (this capability, however, is expected to be available and/or implemented further into the future). Roaming also brings up the question of whether a traditional IP solution is adequate or if one needs to utilize Mobile IP (MIP) (IETF RFC 3344) [PER200201]; this is a fairly complex issue.

It should be noted that enterprise users that have WLAN-ready laptops can, in theory, use softphones (software clients) to support VoIP connectivity with other corporate VoIP users. We do not consider this a full VoW environment because this arrangement has severe limitations as follows: (1) this application only works where there are a small handful of users off any one Access Point; (2) the softphone is generally proprietary; and, (3) this solution does not readily interwork with non-VoIP and/or off-net voice uses.

(ii) VoIP over Hotspot/WiMax

Service providers are also considering offering VoIP services in IEEE 802.11b/11g hotspot environments. Furthermore there is interest in delivering metro-wide VoIP services using Wi-MAX (IEEE 802.16-based) connectivity. Since WiMAX is newer, we focus here on this technology (see Table 2.13 for a technical comparison of WiMAX to Wi-Fi [WIM200501]).

The IEEE 802.16 Working Group has developed a Point-to-Multipoint (PMP) broadband wireless access standard for systems in the frequency ranges 10–66 GHz and sub 11 GHz. This technology is targeted at metropolitan area environments. The IEEE 802.16 standard covers both the Media Access Control (MAC) and the physical (PHY) layers. A number of PHY considerations were taken into account for the target environment. At higher frequencies, Line-Of-Sight (LOS) is a must. This requirement eases the effect of multipath, allowing for wide channels, typically greater than 10 MHz in bandwidth. This gives the IEEE 802.16 protocol the ability to provide very high capacity links on both the uplink and the downlink. For sub 11 GHz Nonline-Of-Sight (NLOS) capability is a requirement. The original IEEE 802.16 MAC was enhanced to accommodate different PHYs and services, which address the needs of different environments. The standard is designed to accommodate either Time Division Duplexing (TDD) or Frequency Division Duplexing (FDD) deployments, allowing for both full and half-duplex terminals in the FDD case. [WIM200501]. 802.16a has a LOS radius of 50 km and a NLOS of 10 km or thereabouts, depending on the type of obstacles in the topography. WiMAX is the marketing name of the IEEE 802.16 standard.

The MAC was designed specifically for the PMP wireless access environment. It supports higher layer or transport protocols such as ATM, Ethernet, or IP, and is designed to easily accommodate future protocols that have not yet been developed. The MAC is designed for very high bit rates (up to 268 Mbps each way) of the truly broadband physical layer, while delivering ATM compatible QoS, UGS (Unsolicited Grant Service), rtPS (real-time Polling Service), nrtPS (nonreal-time Polling Service), and Best Effort services. The frame structure allows terminals to be dynamically assigned uplink and downlink burst profiles according to their link conditions. This allows a trade-off between capacity and robustness in real-time, and provides roughly a two times increase in capacity on average when compared to nonadaptive systems, while maintaining appropriate link availability. The 802.16 MAC uses a variable length Protocol Data Unit (PDU) along with a number of other concepts that greatly increase the efficiency of the standard. Multiple MAC PDUs may be concatenated into a single burst to save PHY overhead. Additionally, multiple Service Data Units (SDU) for the same service may be concatenated into a single MAC PDU, saving on MAC header overhead. Fragmentation allows very large SDUs to be sent across frame boundaries to guarantee the QoS of competing services. Payload header suppression can be used to reduce the overhead caused by the redundant portions of SDU headers. The MAC uses a self-correcting bandwidth request/grant scheme that eliminates the overhead and delay of acknowledgments, while simultaneously allowing better QoS handling than traditional acknowledgment schemes. Terminals have a variety of options available to them for requesting bandwidth depending upon the QoS and traffic parameters of their services. Terminals can be polled individually or in groups; they can steal bandwidth already allocated to make requests for more; they can signal the need to be polled, and they can piggyback requests for bandwidth. [WIM200501].

A typical WiMAX network consists of a base station supported by a tower-mounted or on building-mounted antenna. The base station connects to the appropriate terrestrial network (PSTN, Internet, etc.) Applications include but are not limited to point-to-point communication between stations; point-to-multipoint communication between the base station and clients; backhaul services for Wi-Fi (802.11) hotspots; broadband Internet services to home users; and, private line services for users in remote locations.

Table 2.13: Comparison between 802.11 and 802.16.

	802.11 (Wi-Fi)	**802.16 (WiMAX)**	**Technical Differences**
Range	Maximum 100 meters. Access points needs to be added for greater coverage.	Line-Of-Sight (LOS): Up to 50 km. (NLOS): between 6 to 10 km. Nonline-Of-Sight.	802.16 PHY tolerates greater multi-path, delay spread (reflections) via implementation of a 256 FFT vs. 64 FFT for 802.11.
Spectrum	Uses unlicensed spectrum only.	Uses licensed and unlicensed spectrum.	802.16 ranges from 2 to 66 GHz and 802.11 is restricted to 2.4 and 5.8 GHz.
Coverage	Suitable for indoor use only.	Designed for outdoor LOS and NLOS services.	802.16 has superior system gain enabling it to penetrate through obstacles at longer distances.
Scalability	Number of users range from one to tens. Intended for LAN use. Channel sizes are fixed at 20 MHz.	Number of users can be thousands. Channel sizes are flexible (1.75 ~ 20 MHz).	The MAC protocol of 802.16 uses Dynamic TDMA, but 802.11 uses CSMA/CA.
Data Rate	2.7 bps/Hz, maximum of 54 Mbps in a 20 MHz channel.	5 bps/Hz, maximum of 100 Mbps in a 20 MHz channel.	802.16 can maintain ATM compatible QoS; UGS, rtPS, nrtPS and Best Effort; it can also support bit rates as high as 268 Mbps each way.
QoS	Can only support best offer service.	Supports multiple QoS and it is built into MAC.	802.11: Uses CSMA/CA (wireless Ethernet). 802.16: Dynamic TDMA-based MAC with no-demand bandwidth allocation.

Courtesy: WiMAX Forum

(iii) VoIP in 3G Cellular Networks

Over the past decade, mobile communications technology has evolved from First-Generation (1G) analog voice-only communications, to Second Generation (2G) digital, voice and data communications; the demand for more cost-effective and feature-enhanced mobile applications has led to the development of new-generation generation wireless systems (or simply 3G). State-of-the-art 3G handsets are designed to provide multimegabit Internet access with an "always on" feature and data rates of up to 2.048 Mbps [MAV200201].

In reference to cellular applications, the core network of traditional cellular systems is typically based on a circuit-switched architecture similar to that utilized in wireline networks. The wireless service providers are now in the process of evolving their core networks to IP technology.

Wireless telecommunications started as a subdiscipline of wireline telephony, and the absence of global standards resulted in regional standardization. Two major mobile telecommunications standards have emerged: (1) Time Division Multiple Access/Code Division Multiple Access (TDMA/CDMA) developed by the Telecommunications Industry Association (TIA) in North America, and (2) Global System for Mobile

Communications (GSM) developed by European Telecommunications Standards Institute (ETSI) in Europe. As one moves toward Third-Generation (3G) wireless services, there is a need to develop standards that are more global in scope [PAT200001].

In the late 1990s, there were discussions for the development of standards for a 3G mobile system with a core network based on evolutions of the GSM and an access network based on all the radio access technologies (i.e., both frequency- and time-division duplex modes) supported by the plethora of different carriers (in different countries). This project was called the *Third Generation Partnership Project (3GPP)* [3GP200501]. Around the turn of this decade, the American National Standards Institute (ANSI) decided to establish the Third-Generation Partnership Project 2 (3GPP2), a 3G partnership initiative for evolved ANSI/TIA/Electronics Industry Association (EIA) networks [3GP200502]. In addition, there also was the establishment of a strategic group called *International Mobile Telecommunications-2000 (IMT-2000)* within the International Telecommunication Union (ITU) [IMT200001], that focused its work on defining interfaces between 3G networks evolved from GSM on one hand and ANSI on the other, with the goal to enable seamless roaming between 3GPP and 3GPP2 networks. Because of the worldwide ("universal") roaming characteristic, 3GPP started referring to 3G mobile systems as the Universal Mobile Telecommunication System (UMTS) [BOS200101]. Since then, there has been advocacy for and progress toward an all-IP UMTS network architecture. The "all-IP UMTS" specifications replaced the earlier circuit-switched transport technologies by utilizing packet-switched transport technologies and introduce multimedia support in the UMTS core network [BOS200101].

Figure 2.16 depicts some basic industry transition paths to 3G wireless. As implied in the previous paragraph, currently, the 3G world is split into two camps: (1) the cdma2000 which is an evolution of the IS-95 standard, and (2) the Wideband Code Division Multiple Access (W-CDMA)/Time Division Synchronous CDMA (TD-SCDMA)/Enhanced Data Rates for GSM evolution (EDGE) camp whose standards are all improvements of GSM, IS-136 and Packet Data Cellular (PDC)—these are all second-generation standards. In the U.S., Verizon Wireless and Sprint PCS were the first two carriers to develop 3G networks. The other major carriers have already advanced to the 2.5G technology with the vision to soon join the 3G community [MAV200201].

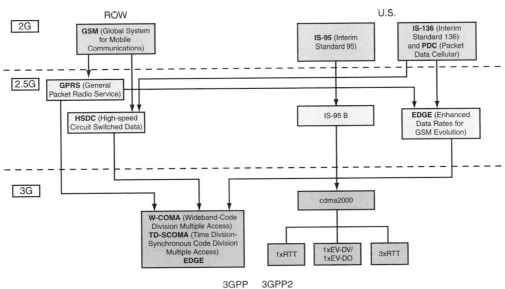

Figure 2.16: Migration path(s) to 3G wireless networks.

The original scope of 3GPP was to produce globally-applicable Technical Specifications and Technical Reports for a 3G Mobile System based on evolved GSM core networks and the radio access technologies that they support (i.e., Universal Terrestrial Radio Access (UTRA) both FDD and TDD modes). The scope was subsequently amended to include the maintenance and development of the GSM Technical Specifications and Technical Reports including evolved radio access technologies (e.g., General Packet Radio Service (GPRS) and EDGE) [TDD200501].

3GPP and 3GPP2 also address the issue of the limited data throughput capabilities of 2G/2.5G systems, motivating providers to start work on 3G wideband radio technologies that can provide higher data rates e.g., for Internet access, messaging, location-based services, etc. This work resulted in 3G wireless radio technologies that provide data rates of 144 kbps for vehicular, 384 kbps for pedestrian, and 2 Mbps for indoor environments, and meet the ITU IMT-2000 requirements. Now that the radio technology standards to support higher data rates have been developed, the providers are focusing on development of standards for all-IP networks [PAT200001].

3GPP

The basic characteristics of an all-IP network are: end-to-end IP connectivity, distributed control and services, and gateways to legacy networks [PAT200001]. As noted earlier in this chapter, there are two major protocol suites for supporting VoIP: SIP, standardized by the IETF, and H.323, standardized by the ITU. It was decided in 3GPP to use only SIP as the call control protocol between terminals and the mobile network. Interworking with other H.323 terminals (e.g., fixed H.323 hosts) is performed by a dedicated server in the network. New elements in this architecture, compared to a traditional 2G cellular network, are as follows (also see Figure 2.17) [BOS200101]:

Mobile Switching Center (MSC) Server: The MSC server controls all calls coming from circuit-switched mobile terminals and mobile terminated calls from a PSTN/GSM network to a circuit-switched terminal. The MSC server interacts with the Media Gateway Control Function (MGCF) for calls to/from the PSTN. There is a functional split of the MSC, where the call control and services part is maintained in the MSC server, and the switch is replaced by an IP router (Media Gateway (MG)). This functional split reduces the deployment cost and guarantees the support of all existing services.

Call State Control Function (CSCF): The CSCF is a SIP server that provides/controls multimedia services for packet-switched (IP) terminals, both mobile and fixed.

MG at the Universal Terrestrial Access Network (UTRAN) side: The MG transforms VoIP packets into UMTS radio frames. The MG is controlled by the MGCF by means of Media Gateway Control Protocol ITU H.248. The media gateway is added to fulfill requirement two. In Figure 2.17, the MG is drawn at the UTRAN side of the Iu interface, so the Iu interface, between the core network and UTRAN, is IP-based. The MG can also be located at the core network side of the Iu interface (without impact on the UTRAN.)

MG at the PSTN Side: All calls coming from the PSTN are translated to VoIP calls for transport in the UMTS core network. This MG is controlled by the MGCF using the ITU H.248 protocol.

Signaling Gateway (SG): An SG relays all call-related signaling to/from the PSTN/UTRAN on an IP bearer and sends the signaling data to the MGCF. The SG does not perform any translation at the signaling level.

MGCF: The first task of the MGCF is to control the MGs via H.248. Also, the MGCF performs translation at the call control signaling level between ISDN User Part (ISUP) signaling, used in the PSTN, and SIP signaling, used in the UMTS multimedia domain.

Home Subscriber Server (HSS): The HSS is the extension of the Home Location Register (HLR) database with the subscribers' multimedia profile data.

For the transport of data traffic, UMTS uses the General-Packet Radio Service (GPRS) network. For voice calls, there are two options: for packet switched mobile terminals, voice data is transported over the GPRS network using the GPRS Tunneling Protocol (GTP) on top of IP; all mobility is solved by the GPRS protocols. For circuit-switched mobile terminals, voice samples are transported over IP between the MGs using the Iu Frame Protocol; in the latter case there is no tunneling, hence, mobility has to be solved in a different way, namely by media gateway handovers.

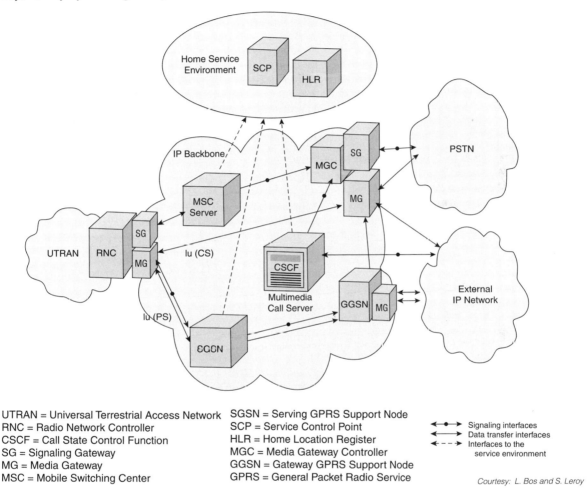

UTRAN = Universal Terrestrial Access Network
RNC = Radio Network Controller
CSCF = Call State Control Function
SG = Signaling Gateway
MG = Media Gateway
MSC = Mobile Switching Center

SGSN = Serving GPRS Support Node
SCP = Service Control Point
HLR = Home Location Register
MGC = Media Gateway Controller
GGSN = Gateway GPRS Support Node
GPRS = General Packet Radio Service

Signaling interfaces
Data transfer interfaces
Interfaces to the
  service environment

Courtesy: L. Bos and S. Leroy

Figure 2.17: An all-IP 3G cellular service.

An essential architectural principle of the 3GPP framework is to provide separation of service control from connection control. 3GPP started with GPRS as the core packet network, and overlaid it with call control and gateway functions required for supporting VoIP and other multimedia services. The functions are provided via IETF-developed protocols to maintain compatibility with the industry direction in all-IP networks. To support VoIP, call control functions (analogous to call control in a circuit-switched environment) are provided by the Call State Control Function (CSCF) (refer back to Figure 2.17). The mobile terminal communicates with the CSCF via SIP protocols. The CSCF performs call control functions, service switching functions, address translation functions, and vocoder negotiation functions. For communication to the Public

Switched Telephone Network (PSTN) and legacy networks, PSTN gateways are utilized. To support roaming to 2G wireless networks, roaming gateway functions are also provided. The Serving GPRS Support Node (SGSN) uses existing GSM registration and authentication schemes to verify the identity of the data user. This makes the SGSN access-technology-dependent. The GPRS HLR is enhanced for services that use IP protocols. The data terminal makes itself known to the packet network by doing a "GPRS-attach." The IP address is anchored in the GPRS gateway node, GGSN, during the entire data session. This limits the mobility of the data terminal to within GPRS-based networks. To provide mobility with other networks a MIP Foreign Agent (FA) can be incorporated in the GGSN [PAT200001].

3G Release 1999 was the first release of the 3GPP specifications; it was essentially a consolidation of the underlying GSM specifications and the development of the new UTRAN radio access network. The foundations were laid for future high-speed traffic transfer in both circuit-switched and packet-switched modes. That release was followed over the years by Release 4, 5, and 6 [TDD200501]. Release 1999 was an introductory specification on the architecture of the UMTS network. According to Release 1999, UMTS is composed by a UTRAN and two core networks (circuit-switched core-network [CS-CN] and packet-switched core-network [PS-CN]), which link up to services networks such as the PSTN and the Internet. Thus, using both traditional circuit-switched and modern packet-switched networks, UMTS Release 1999 supports various services including voice, data (fax, SMS), and Internet access. Later on, Release 4 adapted to the same architecture adds more services to the UMTS network. The co-existence of two core networks however, signified many limitations compared to competitive 3G systems, especially in video/multimedia services. Release 5 was a solution to these limitations that came along to modernize the UMTS architecture currently employed in 3G networks around the world. In this final phase, the Packet-Switched Core-Network (PS-CN) dominates over the Circuit-Switched Core-Network (CS-CN) and takes the responsibility of telephony services. Systems based on UMTS Release 5 have much lower infrastructure and maintenance costs and provide enhanced services. Release 6 added additional capabilities [MAV200201].

As seen at the macro level in Figure 2.18, a new component is added to the basic UMTS architecture: the supplementary IP Multimedia Subsystem (IMS) will support both telephony and multimedia services. The IMS' role in the UMTS architecture is to interact both with the PSTN and the Internet to provide all types of multimedia services to users. The CSCF element in the IMS infrastructure is responsible for signaling messages between all IMS components in order to control multimedia sessions originated by the user. Consequently, there is a Proxy-CSCF (P-CSCF), an Interrogating-CSCF (I-CSCF) and a Serving-CSCF (S-CSCF), all responsible for particular signaling functions using SIP. The P-CSCF's responsibility is to act as the QoS enforcement point and to provide local control for emergency services. I-CSCF is an optional component that interacts with the HSS to find the location of the S-CSCF (it is optional because the P-CSCF can be set up to negotiate directly with the S-CSCF). The S-CSCF controls all the session management functions for the IMS. Depending on the capabilities of the IMS and the capacity requirements, there may be more than one S-CSCF node while others can be eventually added to the system. The function of HSS is to handle all user information such as subscription and location queries. HSS communicates with the CSCFs via an IP-based protocol called *Cx interface*; all other IMS components interact with each other via SIP. The Media Gateway Control Function (MGCF) is in charge of controlling one or more MGs; the MGCF interacts with the S-CSCF and the Transport Signaling Gateway (T-SGW). MGs are bit processors for end to end users; their function is to convert PCM in the PSTN to IP-based formats and vice versa. Finally, the T-SGW is included in the IMS because of the need to convert Signaling System Number 7 (SS7) to IP since the PSTN is only SS7-compatible [MAV200201].

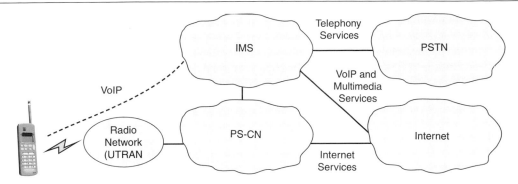

UTRAN = UMTS Terrestrial Access Network
PS-CN = Packet Switched Core Network
IMS = IP Multimedia Subsystem

Figure 2.18: UMTS release 5 basic architecture.

3GPP2

3GPP2 has also undertaken the work to enhance the IP architecture for multimedia services (including voice). The approach here is to capitalize on the synergies of Internet technologies and use a single network for all services. 3GPP2 has created a new packet data architecture building on the CDMA 2G and 3G air interface data services. 3GPP2 has taken advantage of 3G high data rates and existing work in IETF on MIP to enhance the network architecture to provide IP capabilities. One advantage of using globally accepted-IETF protocols is ease in interworking and roaming with other IP networks. The other major advantage is that it can provide private network access (virtual private networking) via a MIP tunnel with IP security [PAT200001].

In the 3GPP2 architecture, IP connectivity reaches all the way to the Base Station Transceiver (BTS). Both the Base Station Controller (BSC) and BTS are contained in the IP-based radio access network node. This means that the BSC will be a router-based IP node containing some critical radio control functions (e.g., power control, soft handoff frame selection). The remaining control functions, such as call/session control, mobility management, and gateway functions, are moved out to the managed IP network. This allows for a distributed and modular control architecture. Since much of the communication will be between wireless and legacy terminals, gateway functions are provided for roaming to 2G wireless networks and interworking with the PSTN. In the 3GPP2 architecture, the mobile terminal uses Mobile-IP-based protocols to identify itself. The Packet Data Serving Node (PDSN) contains a MIP Foreign Agent (FA) functionality. When the mobile terminal attaches to the FA, the FA establishes a Mobile IP tunnel to the Home Agent (HA) and sends a registration message to the HA. The HA accesses the Authorization, Authentication, and Accounting (AAA) server to authenticate the mobile terminal. The IP address of the mobile terminal is now anchored in the HA for the duration of the data session. The data device connected to the mobile terminal can be handed over to any other access device that supports Mobile IP. Thus, this approach can provide mobility across different access networks (wireless, wireline, etc.). However, since it essentially uses address translation to provide mobility, it cannot do fast handoff due to the latency of address updates from distant agents [PAT200001].

Comparison

In provision of mobility for IP sessions, the 3GPP and 3GPP2 architectures are different because of the underlying base networks and evolution strategies. In 3GPP, GPRS-based mobility was already defined, so the IP network enhancements were considered on top of GPRS. On the other hand, 3GPP2 needed to develop a mobility mechanism for packet data since one did not exist previously. As noted, 3GPP2 has decided to use MIP as the basis for packet data mobility [PAT200001].

To illustrate the similarities and differences of the two approaches, mobility needs to be separated into three levels: air interface mobility, link-level mobility, and network-level mobility. Air interface mobility supports cell-to-cell handoff within a radio access network. Link-level mobility maintains a Point-to-Point Protocol (PPP) context across multiple radio access networks. Network-level mobility provides mobility across networks. In both approaches, air interface mobility is handled in the radio access network. Air interface mobility is specific to the radio technology, so harmonization of the two depends on the harmonization efforts underway for global CDMA. In 3GPP, link-level mobility is handled by GTP; this protocol is used to provide mobility to other 3GPP-defined networks. The 3GPP architecture also provides an option in which an FA may be located in the GGSN. This allows roaming from GPRS-based networks to other IP access networks. In 3GPP2, link-level mobility is provided by defining a tunneling protocol as an extension of MIP. The MIP architecture allows the mobile device to have a point of presence and roam across any IP network. Registration and authentication in the 3GPP architecture for access and data networks are integrated and utilize the schemes used for wireless. In the 3GPP2 architecture, the registration and authentication for access and data networks are performed separately. For a data network, authentication and registration as defined in MIP is used; hence, the data architecture is access-independent [PAT200001].

2.5.2 Wireless VoIP Service Offering Dynamics

Market research has shown that mobile consumers are interested in being able to hold conversations at reduced fees, conduct web conferences, receive live music and/or video clips from radio/TV stations, and interact with each other using Multimedia Messages (MMS) [MAV200201]. This section looks at some of the possible strategies and/or approaches planned by mobile operators to supports these evolving needs.

3G Operators

Industry observers have suggested that VoW using technologies such as Wi-Fi and WiMAX could shift customers and revenue away from cellular carriers; making VoIP available over the cellular network is way of addressing this potential revenue loss. After many delays, 3G networks are now being rolled out. Any technology hoping to position itself as the next generation of networking needs to support voice services in general and VoIP in particular; after all, voice services continue to represent the largest portion of any carrier's income (in the range of 80% of their revenue, as an average—naturally, new multimedia-oriented services are also emerging[20]). This is motivating 3G cellular operators to look in new directions.

3G wireless networks offer all the normal mobile telephony services plus "high speed" data access. 3G operators may initially limit data access to their own branded data services or at least price open Internet access significantly higher than access to their own traditional data services. The mobile market, however, is very competitive, and there consumer and business requirements for access to the open Internet. In fact, flat rate bundles for data access services are already available in some markets. This data-channel access can be used to support VoIP services [TUR200501]. Wireless operators that are looking to continue to displace wireline voice revenues as their business posture need to reduce their overall delivery costs as users move from 2G TDM to 3G VoIP [WIW200501].

For example, equipment upgrades can introduce high-speed data capabilities to UMTS networks. Specifically, new technologies becoming available at press time enables carriers to provide new "blended lifestyle services" via any wireline, wireless, or Wi-Fi/WiMax endpoint by providing a variety of 3GPP IMS functional elements (as discussed in the previous subsections), including the call session control functions, the media resource function controller, the policy decision function, and the breakout gateway control function.

---

[20] Some refer to the cell phone screen as the 3rd screen; the first screen being the TV set, and the second screen being the PC. This third screen opens up the possibility for new multimedia services.

Because this equipment expands the data channel on 3G cellular networks, these upgrades also lays the foundation for operators to introduce VoIP and more advanced multimedia services on their mobile networks (here one can transmit the IP-voice datagrams over the data channel.) VoIP over 3G gives operators the ability to support a greater number of voice users at a lower cost, in turn helping to ensure that voice services can continue to be delivered profitably. Some researchers estimate that 3G wireless can deliver voice by way of VoIP for a quarter of the cost per minute compared to 2G TDM methods [WIW200501].

For mobile operators that have invested heavily in 2G and 3G cellular networks there may be relatively little incentive to offer VoIP services according to observers (their existing networks already deliver better-quality voice services at lower cost than VoIP can achieve today.) However, VoIP may look more attractive to those service providers seeking to bypass mobile operators' traditional voice tariffs, particularly if an opportunity to undercut those tariffs using VoIP arises due to significant drops in 3G data pricing. A number of mobile operators have launched unlimited-use data tariffs that could make them vulnerable to customers using VoIP to cut their spend [3GS200501]. 3G service-provider VoIP offerings could appear in the U.S. in the 2008 or 2009 timeframe. That would come after the operators upgrades their 2.5G/3G networks. For example, upgrades to 1xEV-DO provide peak data rates of about 1.8 Mbps, compared to typical rates of 300–400 Kbps for the current generation of 1xEV-DO [MOB200501], [MOB200502].

Calculations of the threat to 3G revenues from broadband wireless (WiMAX) have focused mainly on data, but as some 3G carriers start to put VoIP in a more central position in their strategies, they could find that this service segment is also impacted. The 3G UMTS and CDMA technologies may have been the first to promise both voice and broadband-class data on one network and device, but the emergence of usable VoW has moved formerly data-only approaches into this space too. A potential early limit on VoIP over 3G data access could be the limited upstream capability of the initial 3G services. W-CDMA can deliver up to 384 kbps downstream but only 64 kbps upstream; it is preferable to have data rates exceeding 64 kbps, but if that is all that is available one can make-do for most VoIP services [TUR200501]. Roadmaps for data networks such as CDMA EVDO (Evolution–Data Only) and UMTS' data only strand, TDD[21] now include VoIP [WIW200501].

The shift is already visible in the CDMA market, even without taking challenges from broadband wireless into account. The next upgrade of EV-DO (Evolution—Data Only) equipment, called *Rev A*, which promises peak data rates of 3.1 Mbps, would also carry VoIP, and so could perhaps make a further upgrade to the next CDMA generation, EV-DV (Evolution—Data and Voice) unnecessary. Rev A equipment was expected to start shipping by press time and, although EV-DO with VoIP will take advantage of the spectral efficiencies of CDMA less well than EV-DV, this will be outweighed by early availability and lower prices [WIW200501].

In the UMTS space, manufacturers have already developed TDD mobile handset offering VoIP as well as the usual broadband packet based services and providers have completed the first successful transmission of a call from a mobile VoIP handset over UMTS TDD, and claim the network is ideal for voice because it features high capacity, low latency, and low power requirements. Their services will be more compelling if they can offer voice and, therefore, they will be less likely to opt for a pure IP solution such as 802.16 instead of TDD. TDD-ready handsets were becoming commercially available as of press time [WIW200501].

---

[21] UMTS TDD Mobile Broadband technology is a packet data implementation of the international 3GPP UMTS standard. Unlike W-CDMA, which uses Frequency Division Duplex (FDD), UMTS TDD is designed to work in a single unpaired frequency band. One of the largest benefits of using TDD is that TDD supports variable asymmetry, meaning an operator can dictate how much capacity is allocated to downlink versus uplink. As the traffic patterns for data typically heavily favor the downlink, this results in better use of spectrum assets and higher efficiency [TDD200501].

Hotspot/WiMAX Operators

For operators considering deployment of broadband wireless access technologies (e.g., WiMAX), being able to offer VoIP could strengthen the business case for investing in such networks by moving operators beyond a focus on low-margin Internet access. Fixed/wireline operators have shown interest in use of wireless VoIP in trying to defend against fixed mobile substitution by developing services that combine VoIP over WLAN/hotspot/WiMAX with cellular voice elsewhere [3GS200501], [TUR200501].

Even when offering VoIP, the 3G industry can no longer expect to take all the voice revenue for itself. The promise of this business model could be severely disrupted by VoIP over WiMAX, especially for operators that are now relying primarily on low-cost-voice-minutes for growth. In the IP world, users will become increasingly accustomed to having inexpensive voice bundled into an overall flat rate package, and operator delivery costs will be even lower than for 3G. Their current hopes related to ARPU (Average Revenue Per Unit) from voice could be severely threatened by broadband wireless options as these become more seamless and quality assured than the current voice over Wi-Fi options [WIW200501].

Fixed-Mobile Convergence (FMC) Operators

Recently, there has been interest in Fixed-Mobile Convergence (FMC). Mobile network operators plan to leverage emerging IMS service platforms to deliver "one phone, one number" telephony over both fixed and mobile infrastructure. This means a mobile handset will use 2G/3G mobile infrastructure when the user is outdoors and VoIP over Wi-Fi when the user is at work or at home. Mobile operators see IMS and FMC as an opportunity to take additional market share from traditional fixed line operators. However, once high speed Internet access becomes available on mobile phones, a plethora of VoIP services will follow [TUR200501].

Most telephone calls originate from inside buildings, where cellular mobile coverage is poorest. As such, residential users are often forced to keep their fixed-line services for use when they are at home; the same applies in office buildings, with the added problem that wireless operators have not been in a position to offer the Centrex or PBX features that enterprises require. In theory, however, that could change with the advent of IMS and FMC [TUR200501].

In Japan, for example, NTT DoCoMo is trying to address the coverage problem in major office buildings with nano-cells and in-building repeaters, but the more widely applicable approach is to introduce FMC services for business and residential users based on IMS. Fixed-mobile convergence is attractive to operators because IMS is a logical extension of their existing networks and the resulting services make the most of operators' installed base. In addition, FMC represents an opportunity for the mobile operator to sell new services directly to enterprises: currently most mobile services are sold to consumers, even though the bills are often paid by enterprises; establishing a direct relationship with the enterprise opens new service and new revenue possibilities for mobile operators [TUR200501].

To enable converged handsets FMC relies on broadband Internet access for the "fixed" portion and WLANs now and WiMAX in the future for the "mobile" portion. WLANs are deployed at a large percentage of enterprises and home-based Wi-Fi setups are spreading rapidly. Broadband Internet access is also available in thousands of public hotspots. The first round of convergence depends upon handsets that support 2G, 3G and Wi-Fi connections on the same phone. Mobile operators then use an IMS platform to transparently combine regular mobile service on their 2G or 3G mobile network with VoIP services over Wi-Fi and/or fixed broadband access. Because of the fact that the mobile portion of FMC uses the existing mobile number and the existing mobile switching network elements, mobile operators have an advantage [TUR200501].

Without broadband Internet access, the VoIP service providers are less of a threat to mobile operators' FMC services. The business proposition of fixed-mobile convergence is to hit the sweet spot of high convenience and low cost [TUR200501]. VoIP vendors will be in a better position to provide their own FMC if WiMAX delivers on its promise of wireless broadband Internet access; however, widespread WiMAX deployments is expected to take a number of years. Instead, the VoIP competitive threat may be enabled by the mobile operators' own data services [TUR200501].

Providers of Internet VoIP Services

At press time there were numerous VoIP services available over the public Internet for wireline environments and, more recently, for wireless environments. We call these providers "independent VoIP providers". Skype was one of the most prominent providers in the wireline space (by press time it had grown to be the world's largest VoIP provider with 38 million registered users—now part of Ebay). To illustrate the trend toward wireless VoIP services by the independent providers we make note that Skype has worked with Motorola to embed the Skype client in Motorola's Wi-Fi-enabled mobile phones; this development means that Internet VoIP service providers could be cutting into mobile operators' consumer voice revenues—especially roaming charges—in the immediate future [TUR200501]. Independent VoIP services are a looming threat for 3G mobile operators. European providers are at greater risk than their U.S. counterparts because European per-minute rates are higher, roaming is more frequent, and flat-rate bundles of minutes are still a fairly new concept. FMC will afford VoIP service providers a viable competitive advantage for the foreseeable future. But as 3G data services get better, Wi-Fi continues to spread, and WiMAX emerges, the VoIP service providers will begin to compete head-to-head with the mobile operators offering their own converged services [TUR200501].

Speculations About the Future: Mobile Operators of the 2010 Decade

There is a realization that there is an advantage for mobile operators to roll-out FMC services as quickly as possible to build a strong base and a strong brand while their competitive advantage survives. Further out, however, there will be some business decisions to be made. The following speculations on what future (industry) positioning may occur are based on [TUR200501].

The services layer (first content delivery and then converged telephony) will become independent of the underlying network (broadband Internet access). Ultimately, a voice connection between two people on the open Internet may not incur any extra charge beyond that for Internet access. Users will pay for mobile broadband Internet access, and then acquire additional content, products and services on the open market just as they do today for web-based services: they get broadband access in their home but acquire most of their content and services from other brands over the Internet (including but not limited to Google, Yahoo, Amazon, eBay, and Vonage) and not from their broadband access provider.

Eventually, mobile operators will need to split their integrated mobile telephony business into a mobile access business and one or more Internet brands. A FMC service will take them beyond their own networks. Can this be built into an independent VoIP service that works anywhere? And, can the walled garden content services be built into Internet brands? Operators need to consider their brand development, so that when the broadband Internet is truly mobile-accessible, they have built Internet brands in addition to (and separable from) their telephony brand. At this juncture, mobile operators have an opportunity to profit from an aggressive rollout of FMC services, with their inherent but short-term advantages, to build a global branded VoIP service that can survive when competing VoIP providers start leveraging mobile broadband Internet access.

2.5.3 Wireless Summary

This brief foray into the wireless arena shows that there is a lot of interest and opportunity for VoIP. In particular, VoIP needs to be commercially robust and it needs to be a 3G service as we defined the term in Chapter 1. IPv6 is an enabling technology in support of that 3G VoIP goal.

2.6 Conclusion

This chapter presented a short overview of some key concepts in VoIP. The chapter is not intended to be exhaustive and the interested reader is encouraged to consult other references, as needed.

Basic VoIP Signaling and SIP Concepts

3.1 Introduction

This chapter describes the Session Initiation Protocol (SIP) in some detail. The anticipation is that an SIP will play a key role in 3G VoIP networks based on IPv6; hence, the coverage we allocate to this topic. An SIP is an application-layer control (signaling) protocol for creating, modifying, and terminating sessions with one or more participants. These sessions include Internet telephone calls, multimedia distribution, and multimedia conferences. SIP invitations used to create sessions carry session descriptions that allow participants to agree on a set of compatible media types. SIP makes use of elements called *proxy servers* to help route requests to the user's current location, authenticate and authorize users for services, implement provider call-routing policies, and provide features to users. SIP also provides a registration function that allows users to upload their current locations for use by proxy servers. SIP runs on top of several different transport protocols. Besides basic VoIP telephony, SIP can support presence/proximity, multimodal and collaborative communications.

We stressed in Chapter 2 the importance of signaling. Signaling is a critical constituent element of a truly-global VoIP construct that is able to support public telephony. SIP can in principle be leveraged to this end. This (relatively) new protocol is described in a series of IETF RFCs, as shown in Table 3.1:

Table 3.1: IETF SIP RFCs.

RFC 3261	SIP: Session Initiation Protocol, *J. Rosenberg, H. Schulzrinne, G. Camarillo, A. Johnston, J. Peterson, R. Sparks, M. Handley, E. Schooler (June 2002) (Obsoletes RFC 2543) (Updated by RFC 3265, RFC 3853)*
RFC 3262	Reliability of Provisional Responses in Session Initiation Protocol (SIP), *J. Rosenberg, H. Schulzrinne (June 2002) (Obsoletes RFC 2543)*
RFC 3263	Session Initiation Protocol (SIP): Locating SIP Servers, *J. Rosenberg, H. Schulzrinne (June 2002) (Obsoletes RFC 2543)*
RFC 3264	An Offer/Answer Model with Session Description Protocol (SDP), *J. Rosenberg, H. Schulzrinne (June 2002) (Obsoletes RFC 2543)*
RFC 3265	Session Initiation Protocol (SIP)-Specific Event Notification, *A. B. Roach (June 2002) (Obsoletes RFC 2543) (Updates RFC 3261)*

While the protocol linkage between SIP and IPv6 is a pragmatic one at this juncture, what makes the two have a symbiotic affinity, at least in the view of the author, is that IPv6 is a "replacement" (and not a trivial one at that) of the current global IP infrastructure; similarly, SIP may well be a replacement for the current ISDN-era signaling, ITU-T's H.323 being a derivative of ISDN signaling [MIN199401]. One perspective on this is that SIP could help make VoIP ubiquitous, but in fact, for VoIP to be able to support a global user population in a seamless manner, the larger address space and the flow mechanisms of IPv6 are needed (note that the traditional E.164 ISDN-era addressing scheme allows for 15 digits or 1015 (a quadrillion) unique combinations (at least mathematically)—that's a million times larger than what IPv4 allows.

This treatment is based on IETF RFC 3261 [SCH200201]. This discussion is strictly for pedagogical purposes. All normative and/or development work should make direct and explicit reference to the latest IETF/RFC documentation. SIP is a sophisticated multipurpose protocol; hence, it is fairly extensive as seen anecdotally by the size of this chapter (the chapter actually only provides a summary of RFC 3261, and, furthermore, does not cover the extensive normative apparatus listed in Appendix B of Chapter 1 or even the short list of the table above). The reader focused completely and exclusively on IPv6 issues may choose to skip this chapter on first reading without a loss in continuity.

3.2   Overview

There are many applications of the Internet that require the creation and management of a session, where a session is considered an exchange of data between an association of participants. The implementation of these applications is complicated by the practices of participants: users may move between endpoints, they may be addressable by multiple names, and they may communicate in several different media—sometimes simultaneously. Numerous protocols have been developed that carry various forms of real-time multimedia session data such as voice, video, or text messages. SIP works in concert with these protocols by enabling Internet endpoints (called *user agents*) to discover one another and to agree on a characterization of a session they would like to share. For locating prospective session participants, and for other functions, SIP enables the creation of an infrastructure of network hosts (called *proxy servers*) to which user agents can send registrations, invitations to sessions, and other requests. SIP is an agile, general-purpose tool for creating, modifying, and terminating sessions that works independently of underlying transport protocols and without dependency on the type of session that is being established.

3.3   Fundamental SIP Functionality

SIP is an application-layer control protocol that enables one to establish, modify, and terminate multimedia sessions (conferences) such as Internet telephony calls. SIP can also invite participants to already existing sessions, such as multicast conferences. Media can be added to (and removed from) an existing session. SIP transparently supports name mapping and redirection services, which supports personal mobility—users can maintain a single externally-visible identifier regardless of their network location. SIP works with both IPv4 and IPv6.

SIP supports five facets related to establishing and terminating multimedia communications:

- User location—determination of the end system to be used for communication;
- User availability—determination of the willingness of the called party to engage in communications;
- User capabilities—determination of the media and media parameters to be used;
- Session setup—"ringing," establishment of session parameters at both called and calling party; and,
- Session management—including transfer and termination of sessions, modifying session parameters, and invoking services.

SIP is not a vertically-integrated communications system. Rather, SIP is a component that can be used with other IETF protocols to build a complete multimedia architecture. Typically, these architectures include protocols such as the Real-Time Transport Protocol (RTP) (RFC 1889) for transporting real-time data and providing QoS feedback, the Real-Time Streaming Protocol (RTSP) (RFC 2326) for controlling delivery of streaming media, the Media Gateway Control Protocol (MEGACO) (RFC 3015) for controlling gateways to the Public Switched Telephone Network (PSTN), and, the Session Description Protocol (SDP) (RFC 2327) for describing multimedia sessions. Therefore, SIP should be used in conjunction with other protocols in order to provide complete services to the users; however, the basic functionality and operation of SIP does not depend on any of these protocols.

It should be noted that SIP does not provide services. Rather, SIP provides primitives that can be used to implement different services. For example, SIP can locate a user and deliver an opaque object to his/her current location. If this primitive is used to deliver a session description written in SDP, for instance, the endpoints can agree on the parameters of a session. If the same primitive is used to deliver a photo of the caller as well as the session description, a "caller ID" service can be easily implemented. As this example shows, a single primitive is typically used to provide several different services.

Furthermore, SIP does not offer conference control services such as floor control or voting, and does not prescribe how a conference is to be managed. SIP can be used to initiate a session that uses some other conference control protocol. Since SIP messages and the sessions they establish can pass through entirely different networks, SIP cannot, and does not, provide any kind of network resource reservation capabilities.

The nature of the services provided make security considerations particularly important. To that end, SIP provides a suite of security services, that include denial-of-service prevention, authentication (both user-to-user and proxy-to-user), integrity protection, and encryption/privacy services.

3.4 Overview of Operation

This section introduces the basic operations of SIP using simple examples (this section does not contain any normative statements). Some of the key protocol details are contained in the appendix to this chapter; additional low-level details are found in the RFC itself.

The first example shows the basic functions of SIP: location of an end point, signal of a desire to communicate, negotiation of session parameters to establish the session, and teardown of the session once established.

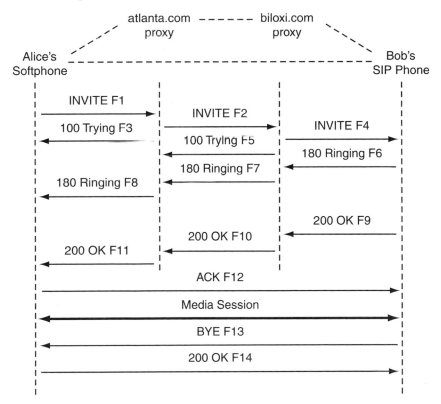

Figure 3.1: SIP session setup example with SIP trapezoid.

Alice "calls" Bob using his SIP identity, a type of Uniform Resource Identifier (URI) called an *SIP URI*. It has a similar form to an email address, typically containing a username and a host name. In this case, it is sip: bob@biloxi.com, where biloxi.com is the domain of Bob's SIP service provider. Alice has a SIP URI of sip: alice@atlanta.com. Alice might have typed in Bob's URI or perhaps clicked on a hyperlink or an entry in an address book. SIP also provides a secure URI, called a *SIPS URI*. An example would be sips:bob@biloxi.com. A call made to a SIPS URI guarantees that secure, encrypted transport (namely, TLS-based communication) is used to carry all SIP messages from the caller to the domain of the callee. From there, the request is sent securely to the callee, but with security mechanisms that depend on the policy of the domain of the callee.

SIP is based on an HTTP-like request/response transaction model. Each transaction consists of a request that invokes a particular method, or function, on the server and at least one response. In this example, the transaction begins with Alice's softphone sending an INVITE request addressed to Bob's SIP URI. INVITE is an example of a SIP method that specifies the action that the requestor (Alice) wants the server (Bob) to take. The INVITE request contains a number of header fields. Header fields are named attributes that provide additional information about a message. The ones present in an INVITE include a unique identifier for the call, the destination address, Alice's address, and information about the type of session that Alice wishes to establish with Bob. The INVITE (message F1 in Figure 3.1) might look like this:

```
INVITE sip:bob@biloxi.com SIP/2.0
Via: SIP/2.0/UDP pc33.atlanta.com;branch=z9hG4bK776asdhds
Max-Forwards: 70
To: Bob <sip:bob@biloxi.com>
From: Alice <sip:alice@atlanta.com>;tag=1928301774
Call-ID: a84b4c76e66710@pc33.atlanta.com
CSeq: 314159 INVITE
Contact: <sip:alice@pc33.atlanta.com>
Content-Type: application/sdp
Content-Length: 142
```

(Alice's SDP not shown)

The first line of the text-encoded message contains the method name (INVITE). The lines that follow are a list of header fields. This example contains a minimum required set. The header fields are briefly described below.

Via contains the address (pc33.atlanta.com) at which Alice is expecting to receive responses to this request. It also contains a branch parameter that identifies this transaction.

To contains a display name (Bob) and a SIP or SIPS URI (sip:bob@biloxi.com) towards which the request was originally directed. Display names are described in RFC 2822.

From also contains a display name (Alice) and a SIP or SIPS URI (sip:alice@atlanta.com) that indicate the originator of the request. This header field also has a tag parameter containing a random string (1928301774) that was added to the URI by the softphone. It is used for identification purposes.

Call-ID contains a globally unique identifier for this call, generated by the combination of a random string and the softphone's host name or IP address. The combination of the To tag, From tag, and Call-ID completely defines a peer-to-peer SIP relationship between Alice and Bob and is referred to as a dialog.

CSeq (Command Sequence) contains an integer and a method name. The CSeq number is incremented for each new request within a dialog and is a traditional sequence number.

Contact contains a SIP or SIPS URI that represents a direct route to contact Alice, usually composed of a username at a Fully Qualified Domain Name (FQDN). While an FQDN is preferred, many end systems do not have registered domain names, so IP addresses are permitted. While the Via header field tells other elements where to send the response, the Contact header field tells other elements where to send future requests.

Max-Forwards (not shown) serves to limit the number of hops a request can make on the way to its destination. It consists of an integer that is decremented by one at each hop.

Content-Type contains a description of the message body (not shown).

Content-Length contains an octet (byte) count of the message body.

The complete set of SIP header fields is defined in the RFC.

The details of the session, such as the type of media, codec, or sampling rate (e.g., as discussed in Chapter 2), are not described using SIP. Rather, the body of a SIP message contains a description of the session, encoded in some other protocol format; one such format is the Session Description Protocol (SDP) (RFC 2327). The SDP message (not shown in the example) is carried by the SIP message in a way that is analogous to a document attachment being carried by an email message, or a web page being carried in an HTTP message.

Since the softphone does not know the location of Bob or the SIP server in the biloxi.com domain, the softphone sends the INVITE to the SIP server that serves Alice's domain, atlanta.com. The address of the atlanta.com SIP server could have been configured in Alice's softphone, or it could have been discovered by DHCP, for example.

The atlanta.com SIP server is a type of SIP server known as a proxy server. A proxy server receives SIP requests and forwards them on behalf of the requestor. In this example, the proxy server receives the INVITE request and sends a 100 (Trying) response back to Alice's softphone. The 100 (Trying) response indicates that the INVITE has been received and that the proxy is working on her behalf to route the INVITE to the destination. Responses in SIP use a three-digit code followed by a descriptive phrase. This response contains the same To, From, Call-ID, CSeq and branch parameter in the Via as the INVITE, which allows Alice's softphone to correlate this response to the sent INVITE. The atlanta.com proxy server locates the proxy server at biloxi.com, possibly by performing a particular type of DNS (Domain Name Service) lookup to find the SIP server that serves the biloxi.com domain. As a result, it obtains the IP address of the biloxi. com proxy server and forwards, or proxies, the INVITE request there. Before forwarding the request, the atlanta.com proxy server adds an additional Via header field value that contains its own address (the INVITE already contains Alice's address in the first Via). The biloxi.com proxy server receives the INVITE and responds with a 100 (Trying) response back to the atlanta.com proxy server to indicate that it has received the INVITE and is processing the request. The proxy server consults a database, generically called a *location service*, that contains the current IP address of Bob. (We shall see in the next section how this database can be populated.) The biloxi.com proxy server adds another Via header field value with its own address to the INVITE and proxies it to Bob's SIP phone.

Bob's SIP phone receives the INVITE and alerts Bob to the incoming call from Alice so that Bob can decide whether to answer the call, that is, Bob's phone rings. Bob's SIP phone indicates this in a 180 (Ringing) response, which is routed back through the two proxies in the reverse direction. Each proxy uses the Via header field to determine where to send the response and removes its own address from the top. As a result, although DNS and location service lookups were required to route the initial INVITE, the 180 (Ringing) response can be returned to the caller without lookups or without state being maintained in the proxies. This also has the desirable property that each proxy that sees the INVITE will also see all responses to the INVITE.

When Alice's softphone receives the 180 (Ringing) response, it passes this information to Alice, perhaps using an audio ringback tone or by displaying a message on Alice's screen.

In this example, Bob decides to answer the call. When he picks up the handset, his SIP phone sends a 200 (OK) response to indicate that the call has been answered. The 200 (OK) contains a message body with the SDP media description of the type of session that Bob is willing to establish with Alice. As a result, there is a two-phase exchange of SDP messages: Alice sent one to Bob, and Bob sent one back to Alice. This two-phase exchange provides basic negotiation capabilities and is based on a simple offer/answer model of SDP exchange. If Bob did not wish to answer the call or was busy on another call, an error response would have been sent instead of the 200 (OK), which would have resulted in no media session being established. The 200 (OK) (message F9 in Figure 3.1) might look like this as Bob sends it out:

```
SIP/2.0 200 OK
Via: SIP/2.0/UDP server10.biloxi.com
  ;branch=z9hG4bKnashds8;received=192.0.2.3
Via: SIP/2.0/UDP bigbox3.site3.atlanta.com
  ;branch=z9hG4bK77ef4c2312983.1;received=192.0.2.2
Via: SIP/2.0/UDP pc33.atlanta.com
  ;branch=z9hG4bK776asdhds ;received=192.0.2.1
To: Bob <sip:bob@biloxi.com>;tag=a6c85cf
From: Alice <sip:alice@atlanta.com>;tag=1928301774
Call-ID: a84b4c76e66710@pc33.atlanta.com
CSeq: 314159 INVITE
Contact: <sip:bob@192.0.2.4>
Content-Type: application/sdp
Content-Length: 131
```

(Bob's SDP not shown)

The first line of the response contains the response code (200) and the reason phrase (OK). The remaining lines contain header fields. The Via, To, From, Call-ID, and CSeq header fields are copied from the INVITE request. (There are three Via header field values—one added by Alice's SIP phone, one added by the atlanta.com proxy, and one added by the biloxi.com proxy.) Bob's SIP phone has added a tag parameter to the To header field. This tag will be incorporated by both endpoints into the dialog and will be included in all future requests and responses in this call. The Contact header field contains a URI with which Bob can be directly reached at his SIP phone. The Content-Type and Content-Length refer to the message body (not shown) that contains Bob's SDP media information.

In addition to DNS and location service lookups shown in this example, proxy servers can make flexible "routing decisions" to decide where to send a request. For example, if Bob's SIP phone returned a 486 (Busy Here) response, the biloxi.com proxy server could proxy the INVITE to Bob's voicemail server. A proxy server can also send an INVITE to a number of locations at the same time. This type of parallel search is known as forking.

In this case, the 200 (OK) is routed back through the two proxies and is received by Alice's softphone, which then stops the ringback tone and indicates that the call has been answered. Finally, Alice's softphone sends an acknowledgment message, ACK, to Bob's SIP phone to confirm the reception of the final response (200 (OK)). In this example, the ACK is sent directly from Alice's softphone to Bob's SIP phone, bypassing the two proxies. This occurs because the endpoints have learned each other's address from the Contact header fields through the INVITE/200 (OK) exchange, which was not known when the initial INVITE was sent. The lookups performed by the two proxies are no longer needed, so the proxies drop out of the call flow. This completes the INVITE/200/ACK three-way handshake used to establish SIP sessions.

Alice and Bob's media session has now begun, and they send media packets using the format to which they agreed in the exchange of SDP. In general, the end-to-end media packets take a different path from the SIP signaling messages.

During the session, either Alice or Bob may decide to change the characteristics of the media session. This is accomplished by sending a re-INVITE containing a new media description. This re-INVITE references the existing dialog so that the other party knows that it is to modify an existing session instead of establishing a new session. The other party sends a 200 (OK) to accept the change. The requestor responds to the 200 (OK) with an ACK. If the other party does not accept the change, he sends an error response such as 488 (Not Acceptable Here), which also receives an ACK. However, the failure of the re-INVITE does not cause the existing call to fail—the session continues using the previously negotiated characteristics.

At the end of the call, Bob disconnects (hangs up) first and generates a BYE message. This BYE is routed directly to Alice's softphone, again bypassing the proxies. Alice confirms receipt of the BYE with a 200 (OK) response, which terminates the session and the BYE transaction. No ACK is sent—an ACK is only sent in response to a response to an INVITE request. The reasons for this special handling for INVITE will be discussed later, but relate to the reliability mechanisms in SIP, the length of time it can take for a ringing phone to be answered, and forking. For this reason, request handling in SIP is often classified as either INVITE or non-INVITE, referring to all other methods besides INVITE.

In some cases, it may be useful for proxies in the SIP signaling path to see all the messaging between the endpoints for the duration of the session. For example, if the biloxi.com proxy server wished to remain in the SIP messaging path beyond the initial INVITE, it would add to the INVITE a required routing header field known as Record-Route that contained a URI resolving to the hostname or IP address of the proxy. This information would be received by both Bob's SIP phone and (due to the Record-Route header field being passed back in the 200 (OK)) Alice's softphone and stored for the duration of the dialog. The biloxi.com proxy server would then receive and proxy the ACK, BYE, and 200 (OK) to the BYE. Each proxy can independently decide to receive subsequent messages, and those messages will pass through all proxies that elect to receive it. This capability is frequently used for proxies that are providing mid-call features.

Registration is another common operation in SIP. Registration is one way that the biloxi.com server can learn the current location of Bob. Upon initialization, and at periodic intervals, Bob's SIP phone sends REGISTER messages to a server in the biloxi.com domain known as an SIP registrar. The REGISTER messages associate Bob's SIP or SIPS URI (sip:bob@biloxi.com) with the machine into which he is currently logged (conveyed as a SIP or SIPS URI in the Contact header field). The registrar writes this association, also called a *binding*, to a database, called the *location service*, where it can be used by the proxy in the biloxi.com domain. Often, a registrar server for a domain is co-located with the proxy for that domain. It is an important concept that the distinction between types of SIP servers is logical, not physical.

Bob is not limited to registering from a single device. For example, both his SIP phone at home and the one in the office could send registrations. This information is stored together in the location service and allows a proxy to perform various types of searches to locate Bob. Similarly, more than one user can be registered on a single device at the same time.

The location service is just an abstract concept. It generally contains information that allows a proxy to input a URI and receive a set of zero or more URIs that tell the proxy where to send the request. Registrations are one way to create this information, but not the only way. Arbitrary mapping functions can be configured at the discretion of the administrator.

Finally, it is important to note that in SIP, registration is used for routing incoming SIP requests and has no role in authorizing outgoing requests. Authorization and authentication are handled in SIP either on a request-by-request basis with a challenge/response mechanism, or by using a lower-layer scheme.

Additional operations in SIP, such as querying for the capabilities of a SIP server or client using OPTIONS, or canceling a pending request using CANCEL, will be introduced in later sections.

3.5 Structure of the Protocol

SIP is structured as a layered protocol, which means that its behavior is described in terms of a set of fairly independent processing stages with only a loose coupling between each stage. The protocol behavior is described as layers for the purpose of presentation, allowing the description of functions common across elements in a single section. It does not dictate an implementation in any way. When one states that an element "contains" a layer, one means it is compliant to the set of rules defined by that layer.

Not every element specified by the protocol contains every layer. Furthermore, the elements specified by SIP are logical elements, not physical ones. A physical realization can choose to act as different logical elements, perhaps even on a transaction-by-transaction basis.

- The lowest layer of SIP is its syntax and encoding. Its encoding is specified using an augmented Backus-Naur Form (BNF) grammar.
- The second layer is the transport layer. It defines how a client sends requests and receives responses and how a server receives requests and sends responses over the network. All SIP elements contain a transport layer.
- The third layer is the transaction layer. Transactions are a fundamental component of SIP. A transaction is a request sent by a client transaction (using the transport layer) to a server transaction, along with all responses to that request sent from the server transaction back to the client. The transaction layer handles application-layer retransmissions, matching of responses to requests, and application-layer timeouts. Any task that a User Agent Client (UAC) accomplishes takes place using a series of transactions. User agents contain a transaction layer, as do stateful proxies. Stateless proxies do not contain a transaction layer. The transaction layer has a client component (referred to as a client transaction) and a server component (referred to as a server transaction), each of which are represented by a finite state machine that is constructed to process a particular request.
- The layer above the transaction layer is called the *Transaction User (TU)*. Each of the SIP entities, except the stateless proxy, is a transaction user. When a TU wishes to send a request, it creates a client transaction instance and passes it the request along with the destination IP address, port, and transport to which to send the request. A TU that creates a client transaction can also cancel it. When a client cancels a transaction, it requests that the server stop further processing, revert to the state that existed before the transaction was initiated, and generate a specific error response to that transaction. This is done with a CANCEL request, which constitutes its own transaction, but references the transaction to be cancelled.

The SIP elements, that is, user agent clients and servers, stateless and stateful proxies and registrars, contain a core that distinguishes them from each other. Cores, except for the stateless proxy, are transaction users. While the behavior of the UAC and UAS cores depends on the method, there are some common rules for all methods. For a UAC, these rules govern the construction of a request; for a UAS, they govern the processing of a request and generating a response. Since registrations play an important role in SIP, a UAS that handles a REGISTER is given the special name registrar. Section A.5 (in this chapter's appendix) describes UAC and UAS core behavior for the REGISTER method; Section A.6 describes UAC and UAS core behavior for the OPTIONS method, used for determining the capabilities of a UA.

Certain other requests are sent within a dialog. A dialog is a peer-to-peer SIP relationship between two user agents that persists for some time. The dialog facilitates sequencing of messages and proper routing of requests between the user agents. The INVITE method is the only way defined in this specification to establish a dialog. When a UAC sends a request that is within the context of a dialog, it follows the common UAC rules as discussed in Section A.3 but also the rules for mid-dialog requests. Section A.7 discusses dialogs and presents the procedures for their construction and maintenance, in addition to construction of requests within a dialog.

The most important method in SIP is the INVITE method, which is used to establish a session between participants. A session is a collection of participants, and streams of media between them, for the purposes of communication. Section A.8 discusses how sessions are initiated, resulting in one or more SIP dialogs. Section A.9 discusses how characteristics of that session are modified through the use of an INVITE request within a dialog. Section A.10 discusses how a session is terminated. Section A.12 covers transactions, while Section A.13 covers the topic of the transport of SIP messages.

The procedures of Sections A.3, A.5, A.6, A.7, A.8, A.9, and A.10 deal entirely with the UA core (Section A.4 describes cancellation, which applies to both UA core and proxy core). Section A.11 discusses the proxy element, which facilitates routing of messages between user agents.

3.6 SIP Details

Some, but for sure not all, details of the SIP protocol are discussed in Appendix A. All normative and/or development work should make direct and explicit reference to the latest IETF/RFC documentation.

Because the theme of this book is end-to-end VoIP service mechanisms, we provide a view into some of the SIP details in the material that follows. We see SIP and IPv6 as two key enabling capabilities for 3G VoIP being introduced in the next 2–3 years. In fact, we see the issue around IPv6 as to "when" rather than "if." However, nothing discussed in this book precludes the continued use of IPv4, anymore than digital PSTN switching/transmission (including ISDN and DSL) has completely eliminated analog telephone systems still in use is some locations in the U.S. and in the rest of the world.

Appendix A

A.1 Definitions

This section identifies nomenclature used in describing the SIP protocol.

Address-of-Record: An Address-Of-Record (AOR) is a SIP or SIPS URI that points to a domain with a location service that can map the URI to another URI where the user might be available. Typically, the location service is populated through registrations. An AOR is frequently thought of as the "public address" of the user.

Back-to-Back User Agent: A Back-to-Back User Agent (B2BUA) is a logical entity that receives a request and processes it as a User Agent Server (UAS). In order to determine how the request should be answered, it acts as a User Agent Client (UAC) and generates requests. Unlike a proxy server, it maintains dialog state and must participate in all requests sent on the dialogs it has established. Since it is a concatenation of a UAC and UAS, no explicit definitions are needed for its behavior.

Call: A call is an informal term that refers to some communication between peers, generally set up for the purposes of a multimedia conversation.

Call Leg: Another name for a dialog; no longer used in this specification.

Call Stateful: A proxy is call stateful if it retains state for a dialog from the initiating INVITE to the terminating BYE request. A call stateful proxy is always transaction stateful, but the converse is not necessarily true.

Client: A client is any network element that sends SIP requests and receives SIP responses. Clients may or may not interact directly with a human user. User agent clients and proxies are clients.

Conference: A multimedia session that contains multiple participants.

Core: Core designates the functions specific to a particular type of SIP entity, i.e., specific to either a stateful or stateless proxy, a user agent or registrar. All cores, except those for the stateless proxy, are transaction users.

Dialog: A dialog is a peer-to-peer SIP relationship between two UAs that persists for some time. A dialog is established by SIP messages, such as a 2xx response to an INVITE request. A dialog is identified by a call identifier, local tag, and a remote tag. A dialog was formerly known as a call leg in RFC 2543.

Downstream: A direction of message forwarding within a transaction that refers to the direction that requests flow from the user agent client to user agent server.

Final Response: A response that terminates a SIP transaction, as opposed to a provisional response that does not. All 2xx, 3xx, 4xx, 5xx and 6xx responses are final.

Header: A header is a component of a SIP message that conveys information about the message. It is structured as a sequence of header fields.

Header Field: A header field is a component of the SIP message header. A header field can appear as one or more header field rows. Header field rows consist of a header field name and zero or more header field values. Multiple header field values on a given header field row are separated by commas. Some header fields can only have a single header field value, and as a result, always appear as a single header field row.

Header Field Value: A header field value is a single value; a header field consists of zero or more header field values.

Home Domain: The domain providing service to a SIP user. Typically, this is the domain present in the URI in the address-of-record of a registration.

Informational Response: Same as a provisional response.

Initiator, Calling Party, Caller: The party initiating a session (and dialog) with an INVITE request. A caller retains this role from the time it sends the initial INVITE that established a dialog until the termination of that dialog.

Invitation: An INVITE request.

Invitee, Invited User, Called Party, Callee: The party that receives an INVITE request for the purpose of establishing a new session. A callee retains this role from the time it receives the INVITE until the termination of the dialog established by that INVITE.

Location Service: A location service is used by a SIP redirect or proxy server to obtain information about a callee's possible location(s). It contains a list of bindings of address-of-record keys to zero or more contact addresses. The bindings can be created and removed in many ways; this specification defines a REGISTER method that updates the bindings.

Loop: A request that arrives at a proxy, is forwarded, and later arrives back at the same proxy. When it arrives the second time, its Request-URI is identical to the first time, and other header fields that affect proxy operation are unchanged, so that the proxy would make the same processing decision on the request it made the first time. Looped requests are errors, and the procedures for detecting them and handling them are described by the protocol.

Loose Routing: A proxy is said to be loose routing if it follows the procedures defined in this specification for processing of the Route header field. These procedures separate the destination of the request (present in the Request-URI) from the set of proxies that need to be visited along the way (present in the Route header field). A proxy compliant to these mechanisms is also known as a loose router.

Message: Data sent between SIP elements as part of the protocol. SIP messages are either requests or responses.

Method: The method is the primary function that a request is meant to invoke on a server. The method is carried in the request message itself. Example methods are INVITE and BYE.

Outbound Proxy: A proxy that receives requests from a client, even though it may not be the server resolved by the Request-URI. Typically, a UA is manually configured with an outbound proxy, or can learn about one through auto-configuration protocols.

Parallel Search: In a parallel search, a proxy issues several requests to possible user locations upon receiving an incoming request. Rather than issuing one request and then waiting for the final response before issuing the next request as in a sequential search, a parallel search issues requests without waiting for the result of previous requests.

Provisional Response: A response used by the server to indicate progress, but that does not terminate a SIP transaction. 1xx responses are provisional, other responses are considered final.

Proxy, Proxy Server: An intermediary entity that acts as both a server and a client for the purpose of making requests on behalf of other clients. A proxy server primarily plays the role of routing, which means its job is to ensure that a request is sent to another entity "closer" to the targeted user. Proxies are also useful for enforcing policy (for example, making sure a user is allowed to make a call). A proxy interprets, and, if necessary, rewrites specific parts of a request message before forwarding it.

Recursion: A client recurses on a 3xx response when it generates a new request to one or more of the URIs in the Contact header field in the response.

Redirect Server: A redirect server is a user agent server that generates 3xx responses to requests it receives, directing the client to contact an alternate set of URIs.

Registrar: A registrar is a server that accepts REGISTER requests and places the information it receives in those requests into the location service for the domain it handles.

Regular Transaction: A regular transaction is any transaction with a method other than INVITE, ACK, or CANCEL.

Request: A SIP message sent from a client to a server, for the purpose of invoking a particular operation.

Response: A SIP message sent from a server to a client, for indicating the status of a request sent from the client to the server.

Ringback: Ringback is the signaling tone produced by the calling party's application indicating that a called party is being alerted (ringing).

Route Set: A route set is a collection of ordered SIP or SIPS URI which represent a list of proxies that must be traversed when sending a particular request. A route set can be learned, through headers like Record-Route, or it can be configured.

Server: A server is a network element that receives requests in order to service them and sends back responses to those requests. Examples of servers are proxies, user agent servers, redirect servers, and registrars.

Sequential Search: In a sequential search, a proxy server attempts each contact address in sequence, proceeding to the next one only after the previous has generated a final response. A 2xx or 6xx class final response always terminates a sequential search.

Session: From the SDP specification: "A multimedia session is a set of multimedia senders and receivers and the data streams flowing from senders to receivers. A multimedia conference is an example of a multimedia session." (RFC 2327 [1]) (A session as defined for SDP can comprise one or more RTP sessions.) As defined, a callee can be invited several times, by different calls, to the same session. If SDP is used, a session is defined by the concatenation of the SDP user name, session id, network type, address type, and address elements in the origin field.

SIP Transaction: A SIP transaction occurs between a client and a server and comprises all messages from the first request sent from the client to the server up to a final (non-1xx) response sent from the server to the client. If the request is INVITE and the final response is a non-2xx, the transaction also includes an ACK to the response. The ACK for a 2xx response to an INVITE request is a separate transaction.

Spiral: A spiral is a SIP request that is routed to a proxy, forwarded onwards, and arrives once again at that proxy, but this time differs in a way that will result in a different processing decision than the original request. Typically, this means that the request's Request-URI differs from its previous arrival. A spiral is not an error condition, unlike a loop. A typical cause for this is call forwarding. A user calls joe@example.com. The example.com proxy forwards it to Joe's PC, which in turn, forwards it to bob@example.com. This request is proxied back to the example.com proxy. However, this is not a loop. Since the request is targeted at a different user, it is considered a spiral, and is a valid condition.

Stateful Proxy: A logical entity that maintains the client and server transaction state machines defined by this specification during the processing of a request, also known as a transaction stateful proxy. The behavior of a stateful proxy is further defined in Section A.11. A (transaction) stateful proxy is not the same as a call stateful proxy.

Stateless Proxy: A logical entity that does not maintain the client or server transaction state machines defined in this specification when it processes requests. A stateless proxy forwards every request it receives downstream and every response it receives upstream.

Strict Routing: A proxy is said to be strict routing if it follows the Route processing rules of RFC 2543 and many prior work in progress versions of this RFC. That rule caused proxies to destroy the contents of the Request-URI when a Route header field was present. Strict routing behavior is not used in this specification, in favor of a loose routing behavior. Proxies that perform strict routing are also known as strict routers.

Target Refresh Request: A target refresh request sent within a dialog is defined as a request that can modify the remote target of the dialog.

Transaction User (TU): The layer of protocol processing that resides above the transaction layer. Transaction users include the UAC core, UAS core, and proxy core.

Upstream: A direction of message forwarding within a transaction that refers to the direction that responses flow from the user agent server back to the user agent client.

URL-encoded: A character string encoded according to RFC 2396, Section 2.4.

User Agent Client (UAC): A user agent client is a logical entity that creates a new request, and then uses the client transaction state machinery to send it. The role of UAC lasts only for the duration of that transaction. In other words, if a piece of software initiates a request, it acts as a UAC for the duration of that transaction. If it receives a request later, it assumes the role of a user agent server for the processing of that transaction.

UAC Core: The set of processing functions required of a UAC that reside above the transaction and transport layers.

User Agent Server (UAS): A user agent server is a logical entity that generates a response to a SIP request. The response accepts, rejects, or redirects the request. This role lasts only for the duration of that transaction. In other words, if a piece of software responds to a request, it acts as a UAS for the duration of that transaction. If it generates a request later, it assumes the role of a user agent client for the processing of that transaction.

UAS Core: The set of processing functions required at a UAS that resides above the transaction and transport layers.

User Agent (UA): A logical entity that can act as both a user agent client and user agent server.

The role of UAC and UAS, as well as proxy and redirect servers, are defined on a transaction-by-transaction basis. For example, the user agent initiating a call acts as a UAC when sending the initial INVITE request and as a UAS when receiving a BYE request from the callee. Similarly, the same software can act as a proxy server for one request and as a redirect server for the next request.

Proxy, location, and registrar servers defined above are logical entities; implementations may combine them into a single application.

A.2 SIP Messages

SIP is a text-based protocol and uses the UTF-8 charset (RFC 2279). A SIP message is either a request from a client to a server, or a response from a server to a client. Both Request and Response messages use the basic format of RFC 2822, even though the syntax differs in character set and syntax specifics. (SIP allows header fields that would not be valid RFC 2822 header fields, for example.) Both types of messages consist of a start-line, one or more header fields, an empty line indicating the end of the header fields, and an optional message-body.

```
generic-message = start-line
            *message-header
            CRLF
            [ message-body ]
start-line    = Request-Line / Status-Line
```

The start-line, each message-header line, and the empty line must be terminated by a Carriage-Return Line-Feed sequence (CRLF). Note that the empty line must be present even if the message-body is not. Except for the above difference in character sets, much of SIP's message and header field syntax is identical to HTTP/1.1. Rather than repeating the syntax and semantics here, we use [HX.Y] to refer to Section X.Y of the HTTP/1.1 specification (RFC 2616). However, it should be noted that SIP is not an extension of HTTP.

A.2.1 Requests

SIP requests are distinguished by having a Request-Line for a start-line. A Request-Line contains a method name, a Request-URI, and the protocol version separated by a Single Space (SP) character. The Request-Line ends with CRLF. No CR or LF are allowed except in the end-of-line CRLF sequence. No Linear Whitespace (LWS) is allowed in any of the elements.

```
Request-Line = Method SP Request-URI SP SIP-Version CRLF
```

Method: This specification defines six methods: REGISTER for registering contact information, INVITE, ACK, and CANCEL for setting up sessions, BYE for terminating sessions, and OPTIONS for querying servers about their capabilities. SIP extensions, documented in standards track RFCs, may define additional methods.

Request-URI: The Request-URI is a SIP or SIPS URI or a general URI (RFC 2396). It indicates the user or service to which this request is being addressed. The Request-URI must not contain unescaped spaces or control characters and must not be enclosed in "<>." SIP elements may support Request-URIs with schemes other than "sip" and "sips," for example the "tel" URI scheme of RFC 2806. SIP elements may translate non-SIP URIs using any mechanism at their disposal, resulting in SIP URI, SIPS URI, or some other scheme.

SIP-Version: Both request and response messages include the version of SIP in use, and follow [H3.1] (with HTTP replaced by SIP, and HTTP/1.1 replaced by SIP/2.0) regarding version ordering, compliance requirements, and upgrading of version numbers. To be compliant with this specification, applications sending SIP messages must include a SIP-Version of "SIP/2.0." The SIP-Version string is case-insensitive, but implementations must send upper-case. Unlike HTTP/1.1, SIP treats the version number as a literal string. In practice, this should make no difference.

A.2.2 Responses

SIP responses are distinguished from requests by having a Status-Line as their start-line. A Status-Line consists of the protocol version followed by a numeric Status-Code and its associated textual phrase, with each element separated by a single SP character. No CR or LF is allowed except in the final CRLF sequence.

```
Status-Line = SIP-Version SP Status-Code SP Reason-Phrase CRLF
```

The Status-Code is a 3-digit integer result code that indicates the outcome of an attempt to understand and satisfy a request. The Reason-Phrase is intended to give a short textual description of the Status-Code. The Status-Code is intended for use by automata, whereas the Reason-Phrase is intended for the human user. A client is not required to examine or display the Reason-Phrase.

While RFC 3261 suggests specific wording for the reason phrase, implementations may choose other text, for example, in the language indicated in the Accept-Language header field of the request.

The first digit of the Status-Code defines the class of response. The last two digits do not have any categorization role. For this reason, any response with a status code between 100 and 199 is referred to as a "1xx response," any response with a status code between 200 and 299 as a "2xx response," and so on. SIP/2.0 allows six values for the first digit:

1xx: Provisional—request received, continuing to process the request;

2xx: Success—the action was successfully received, understood, and accepted;

3xx: Redirection—further action needs to be taken in order to complete the request;

4xx: Client Error—the request contains bad syntax or cannot be fulfilled at this server;

5xx: Server Error—the server failed to fulfill an apparently valid request;

6xx: Global Failure—the request cannot be fulfilled at any server.

A.2.3 Header Fields

SIP header fields are similar to HTTP header fields in both syntax and semantics. In particular, SIP header fields follow the [H4.2] definitions of syntax for the message-header and the rules for extending header fields over multiple lines. However, the latter is specified in HTTP with implicit whitespace and folding. This specification conforms to RFC 2234 and uses only explicit whitespace and folding as an integral part of the grammar. [H4.2] also specifies that multiple header fields of the same field name whose value is a comma-separated list can be combined into one header field. That applies to SIP as well, but the specific rule is different because of the different grammars. Specifically, any SIP header whose grammar is of the form

```
header = "header-name" HCOLON header-value *(COMMA header-value)
```

allows for combining header fields of the same name into a comma-separated list. The Contact header field allows a comma-separated list unless the header field value is "*."

A.2.3.1   Header Field Format

Header fields follow the same generic header format as that given in RFC 2822. Each header field consists of a field name followed by a colon (":") and the field value.

```
field-name: field-value
```

The formal grammar for a message-header allows for an arbitrary amount of whitespace on either side of the colon; however, implementations should avoid spaces between the field name and the colon and use a Single Space (SP) between the colon and the field-value.

```
Subject:        lunch
Subject    :    lunch
Subject         :lunch
Subject: lunch
```

Thus, the above are all valid and equivalent, but the last is the preferred form.

Header fields can be extended over multiple lines by preceding each extra line with at least one SP or Horizontal Tab (IIT). The line break and the whitespace at the beginning of the next line are treated as a single SP character. Thus, the following are equivalent:

```
Subject: I know you're there, pick up the phone and talk to me!
Subject: I know you're there,
      pick up the phone
      and talk to me!
```

The relative order of header fields with different field names is not significant. However, it is recommended that header fields which are needed for proxy processing (Via, Route, Record-Route, Proxy-Require, Max-Forwards, and Proxy-Authorization, for example) appear towards the top of the message to facilitate rapid parsing. The relative order of header field rows with the same field name is important. Multiple header field rows with the same field-name may be present in a message if and only if the entire field-value for that header field is defined as a comma-separated list (that is, if it follows the grammar defined in Section A.2.3). It must be possible to combine the multiple header field rows into one "field-name: field-value" pair, without changing the semantics of the message, by appending each subsequent field-value to the first, each separated by a comma. The exceptions to this rule are the WWW-Authenticate, Authorization, Proxy-Authenticate, and Proxy-Authorization header fields. Multiple header field rows with these names may be present in a message, but since their grammar does not follow the general form listed in Section A.2.3, they must not be combined into a single header field row.

Implementations must be able to process multiple header field rows with the same name in any combination of the single-value-per-line or comma-separated value forms.

The following groups of header field rows are valid and equivalent:

```
Route: <sip:alice@atlanta.com>
Subject: Lunch
Route: <sip:bob@biloxi.com>
Route: <sip:carol@chicago.com>

Route: <sip:alice@atlanta.com>, <sip:bob@biloxi.com>
Route: <sip:carol@chicago.com>
Subject: Lunch

Subject: Lunch
Route: <sip:alice@atlanta.com>, <sip:bob@biloxi.com>,
    <sip:carol@chicago.com>
```

Each of the following blocks is valid but not equivalent to the others:

```
Route: <sip:alice@atlanta.com>
Route: <sip:bob@biloxi.com>
Route: <sip:carol@chicago.com>

Route: <sip:bob@biloxi.com>
Route: <sip:alice@atlanta.com>
Route: <sip:carol@chicago.com>

Route: <sip:alice@atlanta.com>,<sip:carol@chicago.com>,<sip:bob@biloxi.com>
```

The format of a header field-value is defined per header-name. It will always be either an opaque sequence of TEXT-UTF8 octets, or a combination of whitespace, tokens, separators, and quoted strings. Many existing header fields will adhere to the general form of a value followed by a semicolon-separated sequence of parameter-name, parameter-value pairs:

```
field-name: field-value *(;parameter-name=parameter-value)
```

Even though an arbitrary number of parameter pairs may be attached to a header field value, any given parameter-name must not appear more than once.

When comparing header fields, field names are always case-insensitive. Unless otherwise stated in the definition of a particular header field, field values, parameter names, and parameter values are case-insensitive. Tokens are always case-insensitive. Unless specified otherwise, values expressed as quoted strings are case-sensitive. For example,

```
Contact: <sip:alice@atlanta.com>;expires=3600
```

is equivalent to

```
CONTACT: <sip:alice@atlanta.com>;ExPiReS=3600
```

Similarly,

```
Content-Disposition: session;handling=optional
```

is equivalent to

```
content-disposition: Session;HANDLING=OPTIONAL
```

The following two header fields are not equivalent:

```
Warning: 370 devnull "Choose a bigger pipe"
Warning: 370 devnull "CHOOSE A BIGGER PIPE"
```

A.2.3.2   Header Field Classification

Some header fields only make sense in requests or responses. These are called *request header fields* and *response header fields*, respectively. If a header field appears in a message not matching its category (such as a request header field in a response), it must be ignored. The RFC defines the classification of each header field.

A.2.3.3 Compact Form

SIP provides a mechanism to represent common header field names in an abbreviated form. This may be useful when messages would otherwise become too large to be carried on the transport available to it (exceeding the Maximum Transmission Unit (MTU) when using UDP, for example). A compact form may be substituted for the longer form of a header field name at any time without changing the semantics of the message. A header field name may appear in both long and short forms within the same message. Implementations must accept both the long and short forms of each header name.

A.2.4 Bodies

Requests, including new requests defined in extensions to this specification, may contain message bodies unless otherwise noted. The interpretation of the body depends on the request method. For response messages, the request method and the response status code determine the type and interpretation of any message body. All responses may include a body.

A.2.4.1 Message Body Type

The Internet media type of the message body must be given by the Content-Type header field. If the body has undergone any encoding such as compression, then this must be indicated by the Content-Encoding header field; otherwise, Content-Encoding must be omitted. If applicable, the character set of the message body is indicated as part of the Content-Type header-field value. The "multipart" MIME type defined in RFC 2046 may be used within the body of the message. Implementations that send requests containing multipart message bodies must send a session description as a nonmultipart message body if the remote implementation requests this through an Accept header field that does not contain multipart message bodies. SIP messages may contain binary bodies or body parts. When no explicit charset parameter is provided by the sender, media subtypes of the "text" type are defined to have a default charset value of "UTF-8."

A.2.4.2 Message Body Length

The body length in bytes is provided by the Content-Length header field. The "chunked" transfer encoding of HTTP/1.1 is not be used for SIP. (Note: The chunked encoding modifies the body of a message in order to transfer it as a series of chunks, each with its own size indicator.)

A.2.5 Framing SIP Messages

Unlike HTTP, SIP implementations can use UDP or other unreliable datagram protocols. Each such datagram carries one request or response. Implementations processing SIP messages over stream-oriented transports must ignore any CRLF appearing before the start-line. The Content-Length header field value is used to locate the end of each SIP message in a stream. It will always be present when SIP messages are sent over stream-oriented transports.

A.3 General User Agent Behavior

A user agent represents an end system. It contains a User Agent Client (UAC), which generates requests, and a User Agent Server (UAS), which responds to them. A UAC is capable of generating a request based on some external stimulus (the user clicking a button, or a signal on a PSTN line) and processing a response. A UAS is capable of receiving a request and generating a response based on user input, external stimulus, the result of a program execution, or some other mechanism. When a UAC sends a request, the request passes through some number of proxy servers, which forward the request towards the UAS. When the UAS generates a response, the response is forwarded towards the UAC.

UAC and UAS procedures depend strongly on two factors. First, based on whether the request or response is inside or outside of a dialog, and second, based on the method of a request. Dialogs are discussed thoroughly in Section A.7; they represent a peer-to-peer relationship between user agents and are established by specific SIP methods, such as INVITE.

This section discusses the method-independent rules for UAC and UAS behavior when processing requests that are outside of a dialog. This includes, of course, the requests which themselves establish a dialog.

A.3.1 UAC Behavior

This section covers UAC behavior outside of a dialog.

A.3.1.1   Generating the Request

A valid SIP request formulated by a UAC must, at a minimum, contain the following header fields: To, From, CSeq, Call-ID, Max-Forwards, and Via; all of these header fields are mandatory in all SIP requests. These six header fields are the fundamental building blocks of a SIP message, as they jointly provide for most of the critical message routing services including the addressing of messages, the routing of responses, limiting message propagation, ordering of messages, and the unique identification of transactions. These header fields are in addition to the mandatory request line, which contains the method, Request-URI, and SIP version. Examples of requests sent outside of a dialog include an INVITE to establish a session and an OPTIONS to query for capabilities.

Request-URI

The initial Request-URI of the message should be set to the value of the URI in the To field. One notable exception is the REGISTER method; behavior for setting the Request-URI of REGISTER is given in Section A.5. It may also be undesirable for privacy reasons or convenience to set these fields to the same value (especially if the originating UA expects that the Request-URI will be changed during transit).

In some special circumstances, the presence of a pre-existing route set can affect the Request-URI of the message. A pre-existing route set is an ordered set of URIs that identify a chain of servers, to which a UAC will send outgoing requests that are outside of a dialog. Commonly, they are configured on the UA by a user or service provider manually, or through some other non-SIP mechanism. When a provider wishes to configure a UA with an outbound proxy, it is recommended that this be done by providing it with a pre-existing route set with a single URI, that of the outbound proxy.

To

The To header field, first and foremost, specifies the desired "logical" recipient of the request, or the address-of-record of the user or resource that is the target of this request. This may or may not be the ultimate recipient of the request. The To header field may contain a SIP or SIPS URI, but it may also make use of other URI schemes (the tel URL (RFC 2806), for example) when appropriate. All SIP implementations must support the SIP URI scheme. Any implementation that supports TLS must support the SIPS URI scheme. The To header field allows for a display name.

A UAC may learn how to populate the To header field for a particular request in a number of ways. Usually the user will suggest the To header field through a human interface, perhaps inputting the URI manually or selecting it from some sort of address book. Frequently, the user will not enter a complete URI, but rather a string of digits or letters (for example, "bob"). It is at the discretion of the UA to choose how to interpret this input. Using the string to form the user part of a SIP URI implies that the UA wishes the name to be resolved in the domain to the right-hand side (RHS) of the at-sign in the SIP URI (for instance, sip:bob@example. com). Using the string to form the user part of a SIPS URI implies that the UA wishes to communicate securely, and that the name is to be resolved in the domain to the RHS of the at-sign. The RHS will frequently be the home domain of the requestor, which allows for the home domain to process the outgoing request. This is useful for features like "speed dial" that require interpretation of the user part in the home domain. The tel URL may be used when the UA does not wish to specify the domain that should interpret a telephone number that has been input by the user. Rather, each domain through which the request passes would be given that opportunity. As an example, a user in an airport might log in and send requests through an outbound proxy in the airport. If they enter "411" (this is the phone number for local directory assistance in the United States), that needs to be interpreted and processed by the outbound proxy in the airport, not the user's home domain. In this case, tel:411 would be the right choice.

A request outside of a dialog must not contain a To tag; the tag in the To field of a request identifies the peer of the dialog. Since no dialog is established, no tag is present.

The following is an example of a valid To header field:

```
To: Carol <sip:carol@chicago.com>
```

From

The From header field indicates the logical identity of the initiator of the request, possibly the user's address-of-record. Like the To header field, it contains a URI and optionally a display name. It is used by SIP elements to determine which processing rules to apply to a request (for example, automatic call rejection). As such, it is very important that the From URI not contain IP addresses or the FQDN of the host on which the UA is running, since these are not logical names.

The From header field allows for a display name. A UAC should use the display name "Anonymous," along with a syntactically correct, but otherwise meaningless URI (like sip:thisis@anonymous.invalid), if the identity of the client is to remain hidden.

Usually, the value that populates the From header field in requests generated by a particular UA is pre-provisioned by the user or by the administrators of the user's local domain. If a particular UA is used by multiple users, it might have switchable profiles that include a URI corresponding to the identity of the profiled user. Recipients of requests can authenticate the originator of a request in order to ascertain that they are who their From header field claims they are. The From field must contain a new "tag" parameter, chosen by the UAC.

Examples:

```
From: "Bob" <sips:bob@biloxi.com> ;tag=a48s
From: sip:+12125551212@phone2net.com;tag=887s
From: Anonymous <sip:c8oqz84zk7z@privacy.org>;tag=hyh8
```

Call-ID

The Call-ID header field acts as a unique identifier to group together a series of messages. It must be the same for all requests and responses sent by either UA in a dialog. It should be the same in each registration from a UA.

In a new request created by a UAC outside of any dialog, the Call-ID header field must be selected by the UAC as a globally unique identifier over space and time unless overridden by method-specific behavior. All SIP UAs must have a means to guarantee that the Call-ID header fields they produce will not be inadvertently generated by any other UA. Note that when requests are retried after certain failure responses that solicit an amendment to a request (for example, a challenge for authentication), these retried requests are not considered new requests and, therefore, do not need new Call-ID header fields.

Use of cryptographically random identifiers (RFC 1750) in the generation of Call-IDs is recommended. Implementations may use the form "localid@host." Call-IDs are case-sensitive and are simply compared byte-by-byte. Using cryptographically random identifiers provides some protection against session hijacking and reduces the likelihood of unintentional Call-ID collisions. No provisioning or human interface is required for the selection of the Call-ID header field value for a request.

Example:

```
Call-ID: f81d4fae-7dec-11d0-a765-00a0c91e6bf6@foo.bar.com
```

CSeq

The CSeq header field serves as a way to identify and order transactions. It consists of a sequence number and a method. The method must match that of the request. For non-REGISTER requests outside of a dialog, the sequence number value is arbitrary. The sequence number value must be expressible as a 32-bit unsigned integer and must be less than $2**31$. As long as it follows the above guidelines, a client may use any mechanism it would like to select CSeq header field values.

Example:

```
CSeq: 4711 INVITE
```

Max-Forwards

The Max-Forwards header field serves to limit the number of hops a request can transit on the way to its destination. It consists of an integer that is decremented by one at each hop. If the Max-Forwards value reaches 0 before the request reaches its destination, it will be rejected with a 483 (Too Many Hops) error response.

A UAC must insert a Max-Forwards header field into each request it originates with a value that should be 70. This number was chosen to be sufficiently large to guarantee that a request would not be dropped in any SIP network when there were no loops, but not so large as to consume proxy resources when a loop does occur. Lower values should be used with caution and only in networks where topologies are known by the UA.

Via

The Via header field indicates the transport used for the transaction and identifies the location where the response is to be sent. A Via header field value is added only after the transport that will be used to reach the next hop has been selected.

When the UAC creates a request, it must insert a Via into that request. The protocol name and protocol version in the header field must be SIP and 2.0, respectively. The Via header field value must contain a branch parameter. This parameter is used to identify the transaction created by that request. This parameter is used by both the client and the server.

The branch parameter value must be unique across space and time for all requests sent by the UA. The exceptions to this rule are CANCEL and ACK for non-2xx responses. As discussed below, a CANCEL request will have the same value of the branch parameter as the request it cancels. An ACK for a non-2xx response will also have the same branch ID as the INVITE whose response it acknowledges. The uniqueness property of the branch ID parameter, to facilitate its use as a transaction ID, was not part of RFC 2543.

The branch ID inserted by an element compliant with this specification must always begin with the characters "z9hG4bK." These seven characters are used as a magic cookie (seven is deemed sufficient to ensure that an older RFC 2543 implementation would not pick such a value), so that servers receiving the request can determine that the branch ID was constructed in the fashion described by this specification (that is, globally unique). Beyond this requirement, the precise format of the branch token is implementation-defined. The Via header maddr, ttl, and sent-by components will be set when the request is processed by the transport layer.

Contact

The Contact header field provides a SIP or SIPS URI that can be used to contact that specific instance of the UA for subsequent requests. The Contact header field must be present and contain exactly one SIP or SIPS URI in any request that can result in the establishment of a dialog. For the methods defined in this specification, that includes only the INVITE request. For these requests, the scope of the Contact is global. That is, the Contact header field value contains the URI at which the UA would like to receive requests, and this URI must be valid even if used in subsequent requests outside of any dialogs. If the Request-URI or top Route header field value contains a SIPS URI, the Contact header field must contain a SIPS URI as well.

Supported and Require

If the UAC supports extensions to SIP that can be applied by the server to the response, the UAC should include a Supported header field in the request listing the option tags for those extensions.

The option tags listed must only refer to extensions defined in standards-track RFCs. This is to prevent servers from insisting that clients implement nonstandard, vendor-defined features in order to receive service. Extensions defined by experimental and informational RFCs are explicitly excluded from usage with the Supported header field in a request, since they too are often used to document vendor-defined extensions.

If the UAC wishes to insist that a UAS understand an extension that the UAC will apply to the request in order to process the request, it must insert a Require header field into the request listing the option tag for that extension. If the UAC wishes to apply an extension to the request and insist that any proxies that are traversed understand that extension, it must insert a Proxy-Require header field into the request listing the option tag for that extension. As with the Supported header field, the option tags in the Require and Proxy-Require header fields must only refer to extensions defined in standards-track RFCs.

Additional Message Components

After a new request has been created, and the header fields described above have been properly constructed, any additional optional header fields are added, as are any header fields specific to the method. SIP requests may contain a MIME-encoded message-body. Regardless of the type of body that a request contains, certain header fields must be formulated to characterize the contents of the body.

A.3.1.2 Sending the Request

The destination for the request is then computed. Unless there is local policy specifying otherwise, the destination must be determined by applying the DNS procedures as follows. If the first element in the route set indicated a strict router (resulting in forming the request), the procedures must be applied to the Request-URI of the request. Otherwise, the procedures are applied to the first Route header field value in the request (if one exists), or to the request's Request-URI if there is no Route header field present. These procedures yield an ordered set of address, port, and transports to attempt.

Local policy may specify an alternate set of destinations to attempt. If the Request-URI contains a SIPS URI, any alternate destinations must be contacted with TLS. Beyond that, there are no restrictions on the alternate destinations if the request contains no Route header field. This provides a simple alternative to a pre-existing route set as a way to specify an outbound proxy. However, that approach for configuring an outbound proxy is not recommended; a pre-existing route set with a single URI should be used instead. If the request contains a Route header field, the request should be sent to the locations derived from its topmost value, but may be sent to any server that the UA is certain will honor the Route and Request-URI policies specified in this RFC (as opposed to those in RFC 2543). In particular, a UAC configured with an outbound proxy should attempt to send the request to the location indicated in the first Route header field value instead of adopting the policy of sending all messages to the outbound proxy.

This ensures that outbound proxies that do not add Record-Route header field values will drop out of the path of subsequent requests. It allows endpoints that cannot resolve the first Route URI to delegate that task to an outbound proxy.

The UAC should try each address until a server is contacted. Each try constitutes a new transaction, and therefore each carries a different topmost Via header field value with a new branch parameter. Furthermore, the transport value in the Via header field is set to whatever transport was determined for the target server.

A.3.1.3   Processing Responses

Responses are first processed by the transport layer and then passed up to the transaction layer. The transaction layer performs its processing and then passes the response up to the TU. The majority of response processing in the TU is method specific. However, there are some general behaviors independent of the method.

Transaction Layer Errors

In some cases, the response returned by the transaction layer will not be a SIP message, but rather a transaction layer error. When a timeout error is received from the transaction layer, it must be treated as if a 408 (Request Timeout) status code has been received. If a fatal transport error is reported by the transport layer (generally, due to fatal ICMP errors in UDP or connection failures in TCP), the condition must be treated as a 503 (Service Unavailable) status code.

Unrecognized Responses

A UAC must treat any final response it does not recognize as being equivalent to the x00 response code of that class, and must be able to process the x00 response code for all classes. For example, if a UAC receives an unrecognized response code of 431, it can safely assume that there was something wrong with its request and treat the response as if it had received a 400 (Bad Request) response code. A UAC must treat any provisional response different than 100 that it does not recognize as 183 (Session Progress). A UAC must be able to process 100 and 183 responses.

Vias

If more than one Via header field value is present in a response, the UAC should discard the message. The presence of additional Via header field values that precede the originator of the request suggests that the message was misrouted or possibly corrupted.

Processing 3xx Responses

Upon receipt of a redirection response (for example, a 301 response status code), clients should use the URI(s) in the Contact header field to formulate one or more new requests based on the redirected request. This process is similar to that of a proxy recursing on a 3xx class response as detailed in Section A.11. A client starts with an initial target set containing exactly one URI, the Request-URI of the original request. If a client wishes to formulate new requests based on a 3xx class response to that request, it places the URIs to try into the target set. Subject to the restrictions in this specification, a client can choose which Contact URIs it places into the target set. As with proxy recursion, a client processing 3xx class responses must not add any given URI to the target set more than once. If the original request had a SIPS URI in the Request-URI, the client may choose to recurse to a non-SIPS URI, but should inform the user of the redirection to an insecure URI.

Any new request may receive 3xx responses themselves containing the original URI as a contact. Two locations can be configured to redirect to each other. Placing any given URI in the target set only once prevents infinite redirection loops.

As the target set grows, the client may generate new requests to the URIs in any order. A common mechanism is to order the set by the "q" parameter value from the Contact header field value. Requests to the URIs may be generated serially or in parallel. One approach is to process groups of decreasing q-values serially and process the URIs in each q-value group in parallel. Another is to perform only serial processing in decreasing q-value order, arbitrarily choosing between contacts of equal q-value.

If contacting an address in the list results in a failure, as defined in the next paragraph, the element moves to the next address in the list, until the list is exhausted. If the list is exhausted, then the request has failed. Failures should be detected through failure response codes (codes greater than 399); for network errors the client transaction will report any transport layer failures to the transaction user. Note that some response codes indicate that the request can be retried; requests that are reattempted should not be considered failures. When a failure for a particular contact address is received, the client should try the next contact address. This will involve creating a new client transaction to deliver a new request.

In order to create a request based on a contact address in a 3xx response, a UAC must copy the entire URI from the target set into the Request-URI, except for the "method-param" and "header" URI parameters. It uses the "header" parameters to create header field values for the new request, overwriting header field values associated with the redirected request.

Note that in some instances, header fields that have been communicated in the contact address may instead append to existing request header fields in the original redirected request. As a general rule, if the header field can accept a comma-separated list of values, then the new header field value may be appended to any existing values in the original redirected request. If the header field does not accept multiple values, the value in the original redirected request may be overwritten by the header field value communicated in the contact address. For example, if a contact address is returned with the following value:

```
sip:user@host?Subject=foo&Call Info=<http://www.foo.com>
```

Then any Subject header field in the original redirected request is overwritten, but the HTTP URL is merely appended to any existing Call-Info header field values. It is recommended that the UAC reuse the same To, From, and Call-ID used in the original redirected request, but the UAC may also choose to update the Call-ID header field value, for example, for new requests.

Finally, once the new request has been constructed, it is sent using a new client transaction, and therefore must have a new branch ID in the top Via field. In all other respects, requests sent upon receipt of a redirect response should reuse the header fields and bodies of the original request.

In some instances, Contact header field values may be cached at UAC temporarily or permanently depending on the status code received and the presence of an expiration interval.

Processing 4xx Responses
- Certain 4xx response codes require specific UA processing, independent of the method.
- If a 401 (Unauthorized) or 407 (Proxy Authentication Required) response is received, the UAC should follow the authorization procedures of described in the RFC to retry the request with credentials.

- If a 413 (Request Entity Too Large) response is received, the request contained a body that was longer than the UAS was willing to accept. If possible, the UAC should retry the request, either omitting the body or using one of a smaller length.
- If a 415 (Unsupported Media Type) response is received, the request contained media types not supported by the UAS. The UAC should retry sending the request, this time only using content with types listed in the Accept header field in the response, with encodings listed in the Accept-Encoding header field in the response, and with languages listed in the Accept-Language in the response.
- If a 416 (Unsupported URI Scheme) response is received, the Request-URI used a URI scheme not supported by the server. The client should retry the request, this time, using a SIP URI.
- If a 420 (Bad Extension) response is received, the request contained a Require or Proxy-Require header field listing an option-tag for a feature not supported by a proxy or UAS. The UAC should retry the request, this time omitting any extensions listed in the Unsupported header field in the response.

In all of the above cases, the request is retried by creating a new request with the appropriate modifications. This new request constitutes a new transaction and should have the same value of the Call-ID, To, and From of the previous request, but the CSeq should contain a new sequence number that is one higher than the previous.

With other 4xx responses, including those yet to be defined, a retry may or may not be possible depending on the method and the use case.

A.3.2 UAS Behavior

When a request outside of a dialog is processed by a UAS, there is a set of processing rules that are followed, independent of the method. Section A.7 gives guidance on how a UAS can tell whether a request is inside or outside of a dialog. Note that request processing is atomic. If a request is accepted, all state changes associated with it must be performed. If it is rejected, all state changes must not be performed. UASs should process the requests in the order of the steps that follow in this section (that is, starting with authentication, then inspecting the method, the header fields, and so on throughout the remainder of this section).

A.3.2.1 Method Inspection

Once a request is authenticated (or authentication is skipped), the UAS must inspect the method of the request. If the UAS recognizes but does not support the method of a request, it must generate a 405 (Method Not Allowed) response. The UAS must also add an Allow header field to the 405 (Method Not Allowed) response. The Allow header field must list the set of methods supported by the UAS generating the message.

If the method is one supported by the server, processing continues.

A.3.2.2 Header Inspection

If a UAS does not understand a header field in a request (that is, the header field is not defined in this specification or in any supported extension), the server must ignore that header field and continue processing the message. A UAS should ignore any malformed header fields that are not necessary for processing requests.

To and Request-URI

The To header field identifies the original recipient of the request designated by the user identified in the From field. The original recipient may or may not be the UAS processing the request, due to call forwarding or other proxy operations. A UAS may apply any policy it wishes to determine whether to accept requests when the To header field is not the identity of the UAS. However, it is recommended that a UAS accept requests even if they do not recognize the URI scheme (for example, a tel: URI) in the To header field, or if the To header field does not address a known or current user of this UAS. If, on the other hand, the UAS decides to reject the request, it should generate a response with a 403 (Forbidden) status code and pass it to the server transaction for transmission.

However, the Request-URI identifies the UAS that is to process the request. If the Request-URI uses a scheme not supported by the UAS, it should reject the request with a 416 (Unsupported URI Scheme) response. If the Request-URI does not identify an address that the UAS is willing to accept requests for, it should reject the request with a 404 (Not Found) response. Typically, a UA that uses the REGISTER method to bind its address-of-record to a specific contact address will see requests whose Request-URI equals that contact address. Other potential sources of received Request-URIs include the Contact header fields of requests and responses sent by the UA that establish or refresh dialogs.

Merged Requests

If the request has no tag in the To header field, the UAS core must check the request against ongoing transactions. If the From tag, Call-ID, and CSeq exactly match those associated with an ongoing transaction, but the request does not match that transaction (based on the matching rules in the RFC), the UAS core should generate a 482 (Loop Detected) response and pass it to the server transaction.

The same request has arrived at the UAS more than once, following different paths, most likely due to forking. The UAS processes the first such request received and responds with a 482 (Loop Detected) to the rest of them.

Require

Assuming the UAS decides that it is the proper element to process the request, it examines the Require header field, if present. The Require header field is used by a UAC to tell a UAS about SIP extensions that the UAC expects the UAS to support in order to process the request properly. Its format is described in the RFC. If a UAS does not understand an option-tag listed in a Require header field, it must respond by generating a response with status code 420 (Bad Extension). The UAS must add an Unsupported header field, and list in it those options it does not understand amongst those in the Require header field of the request.

Note that Require and Proxy-Require must not be used in a SIP CANCEL request, or in an ACK request sent for a non-2xx response. These header fields must be ignored if they are present in these requests.

An ACK request for a 2xx response must contain only those Require and Proxy-Require values that were present in the initial request.

Example:

```
UAC->UAS:   INVITE sip:watson@bell-telephone.com SIP/2.0
      Require: 100rel

UAS->UAC:   SIP/2.0 420 Bad Extension
      Unsupported: 100rel
```

This behavior ensures that the client-server interaction will proceed without delay when all options are understood by both sides, and only slow down if options are not understood (as in the example above). For a well-matched client-server pair, the interaction proceeds quickly, saving a round-trip often required by negotiation mechanisms. In addition, it also removes ambiguity when the client requires features that the server does not understand. Some features, such as call handling fields, are only of interest to end systems.

A.3.2.3   Content Processing

Assuming the UAS understands any extensions required by the client, the UAS examines the body of the message, and the header fields that describe it. If there are any bodies whose type (indicated by the Content-Type), language (indicated by the Content-Language) or encoding (indicated by the Content-Encoding) are not understood, and that body part is not optional (as indicated by the Content-Disposition header field), the UAS must reject the request with a 415 (Unsupported Media Type) response. The response must contain an Accept header field listing the types of all bodies it understands in the event the request contained bodies of types not supported by the UAS. If the request contained content encodings not understood by the UAS, the response must contain an Accept-Encoding header field listing the encodings understood by the UAS. If the request contained content with languages not understood by the UAS, the response must contain an Accept-Language header field indicating the languages understood by the UAS. Beyond these checks, body handling depends on the method and type.

A.3.2.4   Applying Extensions

A UAS that wishes to apply some extension when generating the response must not do so unless support for that extension is indicated in the Supported header field in the request. If the desired extension is not supported, the server should rely only on baseline SIP and any other extensions supported by the client. In rare circumstances, where the server cannot process the request without the extension, the server may send a 421 (Extension Required) response. This response indicates that the proper response cannot be generated without support of a specific extension. The needed extension(s) must be included in a Require header field in the response. This behavior is not recommended, as it will generally break interoperability.

Any extensions applied to a non-421 response must be listed in a Require header field included in the response. Of course, the server must not apply extensions not listed in the Supported header field in the request. As a result of this, the Require header field in a response will only ever contain option tags defined in standards-track RFCs.

A.3.2.5   Processing the Request

Assuming all of the checks in the previous subsections are passed, the UAS processing becomes method-specific. Section A.5 covers the REGISTER request, Section A.6 covers the OPTIONS request, Section A.8 covers the INVITE request, and Section A.10 covers the BYE request.

A.3.2.6   Generating the Response

When a UAS wishes to construct a response to a request, it follows the general procedures detailed in the following subsections. Additional behaviors specific to the response code in question, which are not detailed in this section, may also be required. Once all procedures associated with the creation of a response have been completed, the UAS hands the response back to the server transaction from which it received the request.

Sending a Provisional Response

One largely nonmethod-specific guideline for the generation of responses is that UASs should not issue a provisional response for a non-INVITE request. Rather, UASs should generate a final response to a non-INVITE request as soon as possible. When a 100 (Trying) response is generated, any Timestamp header field present in the request must be copied into this 100 (Trying) response. If there is a delay in generating the response, the UAS should add a delay value into the Timestamp value in the response. This value must contain the difference between the time of sending of the response and receipt of the request, measured in seconds.

Headers and Tags

The From field of the response must equal the From header field of the request. The Call-ID header field of the response must equal the Call-ID header field of the request. The CSeq header field of the response must equal the CSeq field of the request. The Via header field values in the response must equal the Via header field values in the request and must maintain the same ordering.

If a request contained a To tag in the request, the To header field in the response must equal that of the request. However, if the To header field in the request did not contain a tag, the URI in the To header field in the response must equal the URI in the To header field; additionally, the UAS must add a tag to the To header field in the response (with the exception of the 100 (Trying) response, in which a tag may be present). This serves to identify the UAS that is responding, possibly resulting in a component of a dialog ID. The same tag must be used for all responses to that request, both final and provisional (again excepting the 100 (Trying)). Procedures for the generation of tags are defined in the RFC.

A.3.2.7 Stateless UAS Behavior

A stateless UAS is a UAS that does not maintain transaction state. It replies to requests normally, but discards any state that would ordinarily be retained by a UAS after a response has been sent. If a stateless UAS receives a retransmission of a request, it regenerates the response and resends it, just as if it were replying to the first instance of the request. A UAS cannot be stateless unless the request processing for that method would always result in the same response if the requests are identical. This rules out stateless registrars, for example. Stateless UASs do not use a transaction layer; they receive requests directly from the transport layer and send responses directly to the transport layer.

The stateless UAS role is needed primarily to handle unauthenticated requests for which a challenge response is issued. If unauthenticated requests were handled statefully, then malicious floods of unauthenticated requests could create massive amounts of transaction state that might slow or completely halt call processing in a UAS, effectively creating a denial of service condition; for more information see the RFC.

The most important behaviors of a stateless UAS are the following:

- A stateless UAS must not send provisional (1xx) responses;
- A stateless UAS must not retransmit responses;
- A stateless UAS must ignore ACK requests;
- A stateless UAS must ignore CANCEL requests;
- To header tags must be generated for responses in a stateless manner—in a manner that will generate the same tag for the same request consistently.

In all other respects, a stateless UAS behaves in the same manner as a stateful UAS. A UAS can operate in either a stateful or stateless mode for each new request.

A.3.3 Redirect Servers

In some architectures, it may be desirable to reduce the processing load on proxy servers that are responsible for routing requests, and improve signaling path robustness, by relying on redirection.

Redirection allows servers to push routing information for a request back in a response to the client, thereby taking themselves out of the loop of further messaging for this transaction while still aiding in locating the target of the request. When the originator of the request receives the redirection, it will send a new request based on the URI(s) it has received. By propagating URIs from the core of the network to its edges, redirection allows for considerable network scalability.

A redirect server is logically constituted of a server transaction layer and a transaction user that has access to a location service of some kind (see Section A.5 for more on registrars and location services). This location service is effectively a database containing mappings between a single URI and a set of one or more alternative locations at which the target of that URI can be found.

A redirect server does not issue any SIP requests of its own. After receiving a request other than CANCEL, the server either refuses the request or gathers the list of alternative locations from the location service and returns a final response of class 3xx. For well-formed CANCEL requests, it should return a 2xx response. This response ends the SIP transaction. The redirect server maintains transaction state for an entire SIP transaction. It is the responsibility of clients to detect forwarding loops between redirect servers.

When a redirect server returns a 3xx response to a request, it populates the list of (one or more) alternative locations into the Contact header field. An "expires" parameter to the Contact header field values may also be supplied to indicate the lifetime of the Contact data.

The Contact header field contains URIs giving the new locations or user names to try, or may simply specify additional transport parameters. A 301 (Moved Permanently) or 302 (Moved Temporarily) response may also give the same location and username that was targeted by the initial request but specify additional transport parameters such as a different server or multicast address to try, or a change of SIP transport from UDP to TCP or vice versa. However, redirect servers must not redirect a request to a URI equal to the one in the Request-URI; instead, provided that the URI does not point to itself, the server may proxy the request to the destination URI, or may reject it with a 404.

If a client is using an outbound proxy, and that proxy actually redirects requests, a potential arises for infinite redirection loops. Note that a Contact header field value may also refer to a different resource than the one originally called. For example, an SIP call connected to PSTN gateway may need to deliver a special informational announcement such as "The number you have dialed has been changed."

A Contact response header field can contain any suitable URI indicating where the called party can be reached, not limited to SIP URIs. For example, it could contain URIs for phones, fax, or irc (if they were defined) or a mailto: (RFC 2368) URL.

The "expires" parameter of a Contact header field value indicates how long the URI is valid. The value of the parameter is a number indicating seconds. If this parameter is not provided, the value of the Expires header field determines how long the URI is valid. Malformed values should be treated as equivalent to 3600.

This provides a modest level of backwards compatibility with RFC 2543, which allowed absolute times in this header field. If an absolute time is received, it will be treated as malformed, and then default to 3600.

Redirect servers must ignore features that are not understood (including unrecognized header fields, any unknown option tags in Require, or even method names) and proceed with the redirection of the request in question.

A.4 Canceling a Request

The previous section has discussed general UA behavior for generating requests and processing responses for requests of all methods. In this section, we discuss a general purpose method, called *CANCEL*.

The CANCEL request, as the name implies, is used to cancel a previous request sent by a client. Specifically, it asks the UAS to cease processing the request and to generate an error response to that request. CANCEL has no effect on a request to which a UAS has already given a final response. Because of this, it is most useful to CANCEL requests to which it can take a server a long time to respond. For this reason, CANCEL is best for INVITE requests, which can take a long time to generate a response. In that usage, a UAS that receives a CANCEL request for an INVITE, but has not yet sent a final response, would "stop ringing," and then respond to the INVITE with a specific error response (a 487).

CANCEL requests can be constructed and sent by both proxies and user agent clients. A stateful proxy responds to a CANCEL, rather than simply forwarding a response it would receive from a downstream element. For that reason, CANCEL is referred to as a "hop-by-hop" request since it is responded to at each stateful proxy hop.

A.4.1 Client Behavior

A CANCEL request should not be sent to cancel a request other than INVITE. Since requests other than INVITE are responded to immediately, sending a CANCEL for a non-INVITE request would always create a race condition.

The following procedures are used to construct a CANCEL request. The Request-URI, Call-ID, To, the numeric part of CSeq, and From header fields in the CANCEL request must be identical to those in the request being cancelled, including tags. A CANCEL constructed by a client must have only a single Via header field value matching the top Via value in the request being cancelled. Using the same values for these header fields allows the CANCEL to be matched with the request it cancels. However, the method part of the CSeq header field must have a value of CANCEL. This allows it to be identified and processed as a transaction in its own right.

If the request being cancelled contains a Route header field, the CANCEL request must include that Route header field's values. This is needed so that stateless proxies are able to route CANCEL requests properly. The CANCEL request must not contain any Require or Proxy-Require header fields. Once the CANCEL is constructed, the client should check whether it has received any response (provisional or final) for the request being cancelled (herein referred to as the "original request").

If no provisional response has been received, the CANCEL request must not be sent; rather, the client must wait for the arrival of a provisional response before sending the request. If the original request has generated a final response, the CANCEL should not be sent, as it is an effective no-op, since CANCEL has no effect on requests that have already generated a final response. When the client decides to send the CANCEL, it creates a client transaction for the CANCEL and passes it the CANCEL request along with the destination address, port, and transport. The destination address, port, and transport for the CANCEL must be identical to those used to send the original request.

If it was allowed to send the CANCEL before receiving a response for the previous request, the server could receive the CANCEL before the original request.

Note that both the transaction corresponding to the original request and the CANCEL transaction will complete independently. However, a UAC canceling a request cannot rely on receiving a 487 (Request Terminated) response for the original request, as an RFC 2543-compliant UAS will not generate such a response. If there is no final response for the original request in 64*T1 seconds (T1 is defined in the RFC), the client should then consider the original transaction cancelled and should destroy the client transaction handling the original request.

A.4.2 Server Behavior

The CANCEL method requests that the TU at the server side cancel a pending transaction. The TU determines the transaction to be cancelled by taking the CANCEL request, and then assuming that the request method is anything but CANCEL or ACK and applying the transaction matching procedures described in the RFC. The matching transaction is the one to be cancelled.

The processing of a CANCEL request at a server depends on the type of server. A stateless proxy will forward it, a stateful proxy might respond to it and generate some CANCEL requests of its own, and a UAS will respond to it.

A UAS first processes the CANCEL request according to the general UAS processing described in Section A.3.2. However, since CANCEL requests are hop-by-hop and cannot be resubmitted, they cannot be challenged by the server in order to get proper credentials in an Authorization header field. Note also that CANCEL requests do not contain a Require header field.

If the UAS did not find a matching transaction for the CANCEL according to the procedure above, it should respond to the CANCEL with a 481 (Call Leg/Transaction Does Not Exist). If the transaction for the original request still exists, the behavior of the UAS on receiving a CANCEL request depends on whether it has already sent a final response for the original request. If it has, the CANCEL request has no effect on the processing of the original request, no effect on any session state, and no effect on the responses generated for the original request. If the UAS has not issued a final response for the original request, its behavior depends on the method of the original request. If the original request was an INVITE, the UAS should immediately respond to the INVITE with a 487 (Request Terminated). A CANCEL request has no impact on the processing of transactions with any other method defined in this specification.

Regardless of the method of the original request, as long as the CANCEL matched an existing transaction, the UAS answers the CANCEL request itself with a 200 (OK) response. This response is constructed following the procedures described in Section A.3.2.6 noting that the To tag of the response to the CANCEL and the To tag in the response to the original request should be the same. The response to CANCEL is passed to the server transaction for transmission.

A.5 Registrations

A.5.1 Overview

SIP offers a discovery capability. If a user wants to initiate a session with another user, SIP must discover the current host(s) at which the destination user is reachable. This discovery process is frequently accomplished by SIP network elements such as proxy servers and redirect servers which are responsible for receiving a request, determining where to send it based on knowledge of the location of the user, and then sending it there. To do this, SIP network elements consult an abstract service known as a location service, which provides address bindings for a particular domain. These address bindings map an incoming SIP or SIPS URI—sip: bob@biloxi.com, for example—to one or more URIs that are somehow "closer" to the desired user, sip: bob@engineering.biloxi.com, for example. Ultimately, a proxy will consult a location service that maps a received URI to the user agent(s) at which the desired recipient is currently residing.

Registration creates bindings in a location service for a particular domain that associates an address-of-record URI with one or more contact addresses. Thus, when a proxy for that domain receives a request whose Request-URI matches the address-of-record, the proxy will forward the request to the contact addresses registered to that address-of-record. Generally, it only makes sense to register an address-of-record at a domain's location service when requests for that address-of-record would be routed to that domain. In most cases, this means that the domain of the registration will need to match the domain in the URI of the address-of-record.

There are many ways by which the contents of the location service can be established. One way is administratively. In the above example, Bob is known to be a member of the engineering department through access to a corporate database. However, SIP provides a mechanism for a UA to create a binding explicitly. This mechanism is known as registration. Registration entails sending a REGISTER request to a special type of UAS known as a registrar. A registrar acts as the front end to the location service for a domain, reading and writing mappings based on the contents of REGISTER requests. This location service is then typically consulted by a proxy server that is responsible for routing requests for that domain.

An illustration of the overall registration process is given in Figure 3.2. Note that the registrar and proxy server are logical roles that can be played by a single device in a network; for purposes of clarity, the two are separated in this illustration. Also note that UAs may send requests through a proxy server in order to reach a registrar if the two are separate elements.

SIP does not mandate a particular mechanism for implementing the location service. The only requirement is that a registrar for some domain must be able to read and write data to the location service, and a proxy or a redirect server for that domain must be capable of reading that same data. A registrar may be co-located with a particular SIP proxy server for the same domain.

A.5.2 Constructing the REGISTER Request

REGISTER requests add, remove, and query bindings. A REGISTER request can add a new binding between an address-of-record and one or more contact addresses. Registration on behalf of a particular address-of-record can be performed by a suitably authorized third party. A client can also remove previous bindings or query to determine which bindings are currently in place for an address-of-record. Except as noted, the construction of the REGISTER request and the behavior of clients sending a REGISTER request is identical to the general UAC behavior.

A REGISTER request does not establish a dialog. A UAC may include a Route header field in a REGISTER request based on a pre-existing route set. The Record-Route header field has no meaning in REGISTER requests or responses, and must be ignored if present. In particular, the UAC must not create a new route set based on the presence or absence of a Record-Route header field in any response to a REGISTER request.

The following header fields, except Contact, must be included in a REGISTER request. A Contact header field may be included.

> *Request-URI:* The Request-URI names the domain of the location service for which the registration is meant (for example, "sip:chicago.com"). The "userinfo" and "@" components of the SIP URI must not be present.

> *To:* The To header field contains the address of record whose registration is to be created, queried, or modified. The To header field and the Request-URI field typically differ, as the former contains a user name. This address-of-record must be a SIP URI or SIPS URI.

> *From:* The From header field contains the address-of-record of the person responsible for the registration. The value is the same as the To header field unless the request is a third-party registration.

> *Call-ID:* All registrations from a UAC should use the same Call-ID header field value for registrations sent to a particular registrar.

If the same client were to use different Call-ID values, a registrar could not detect whether a delayed REGISTER request might have arrived out of order.

> *CSeq:* The CSeq value guarantees proper ordering of REGISTER requests. A UA must increment the CSeq value by one for each REGISTER request with the same Call-ID.

> *Contact:* REGISTER requests may contain a Contact header field with zero or more values containing address bindings.

UAs must not send a new registration (that is, containing new Contact header field values, as opposed to a retransmission) until they have received a final response from the registrar for the previous one or the previous REGISTER request has timed out.

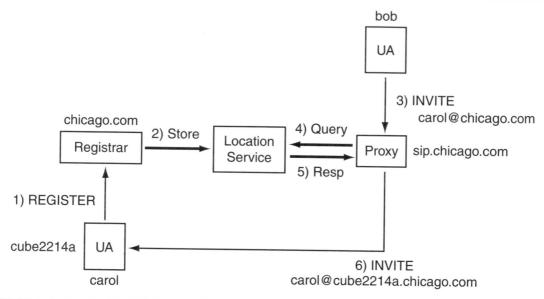

Figure 3.2: REGISTER example.

The following Contact header parameters have a special meaning in REGISTER requests:

action: The "action" parameter from RFC 2543 has been deprecated. UACs should not use the "action" parameter.

expires: The "expires" parameter indicates how long the UA would like the binding to be valid. The value is a number indicating seconds. If this parameter is not provided, the value of the Expires header field is used instead. Implementations may treat values larger than $2**32-1$ (4294967295 seconds or 136 years) as equivalent to $2**32-1$. Malformed values should be treated as equivalent to 3600.

Refer to the RFC for additional information on REGISTER usage.

A.5.3 Processing REGISTER Requests

A registrar is a UAS that responds to REGISTER requests and maintains a list of bindings that are accessible to proxy servers and redirect servers within its administrative domain. A registrar handles requests according to Section A.3.2 and Section A.12.2, but it accepts only REGISTER requests. A registrar must not generate 6xx responses.

A registrar may redirect REGISTER requests as appropriate. One common usage would be for a registrar listening on a multicast interface to redirect multicast REGISTER requests to its own unicast interface with a 302 (Moved Temporarily) response. Registrars must ignore the Record-Route header field if it is included in a REGISTER request. Registrars must not include a Record-Route header field in any response to a REGISTER request. A registrar might receive a request that traversed a proxy which treats REGISTER as an unknown request and which added a Record-Route header field value.

A registrar has to know (for example, through configuration) the set of domain(s) for which it maintains bindings. REGISTER requests must be processed by a registrar in the order that they are received. REGISTER requests must also be processed atomically, meaning that a particular REGISTER request is either processed completely or not at all. Each REGISTER message must be processed independently of any other registration or binding changes.

When receiving a REGISTER request, a registrar follows these steps.

1. The registrar inspects the Request-URI to determine whether it has access to bindings for the domain identified in the Request-URI. If not, and if the server also acts as a proxy server, the server should forward the request to the addressed domain, following the general behavior for proxying messages described in Section A.11.

2. To guarantee that the registrar supports any necessary extensions, the registrar must process the Require header field values as described for UASs in Section A.3.2.2.

3. A registrar should authenticate the UAC. Mechanisms for the authentication of SIP user agents are described in the RFC. Registration behavior in no way overrides the generic authentication framework for SIP. If no authentication mechanism is available, the registrar may take the From address as the asserted identity of the originator of the request.

4. The registrar should determine if the authenticated user is authorized to modify registrations for this address-of-record. For example, a registrar might consult an authorization database that maps user names to a list of addresses-of-record for which that user has authorization to modify bindings. If the authenticated user is not authorized to modify bindings, the registrar must return a 403 (Forbidden) and skip the remaining steps.

In architectures that support third-party registration, one entity may be responsible for updating the registrations associated with multiple addresses-of-record.

5. The registrar extracts the address-of-record from the To header field of the request. If the address-of-record is not valid for the domain in the Request-URI, the registrar must send a 404 (Not Found) response and skip the remaining steps. The URI must then be converted to a canonical form. To do that, all URI parameters must be removed (including the user-param), and any escaped characters must be converted to their unescaped form. The result serves as an index into the list of bindings.

6. The registrar checks whether the request contains the Contact header field. If not, it skips to the last step. If the Contact header field is present, the registrar checks if there is one Contact field value that contains the special value "*" and an Expires field. If the request has additional Contact fields or an expiration time other than zero, the request is invalid, and the server must return a 400 (Invalid Request) and skip the remaining steps. If not, the registrar checks whether the Call-ID agrees with the value stored for each binding. If not, it must remove the binding. If it does agree, it must remove the binding only if the CSeq in the request is higher than the value stored for that binding. Otherwise, the update must be aborted and the request fails.

7. The registrar now processes each contact address in the Contact header field in turn. For each address, it determines the expiration interval as follows:
   a. If the field value has an "expires" parameter, that value must be taken as the requested expiration.
   b. If there is no such parameter, but the request has an Expires header field, that value must be taken as the requested expiration.
   c. If there is neither, a locally-configured default value must be taken as the requested expiration.

The registrar may choose an expiration less than the requested expiration interval. If and only if the requested expiration interval is greater than zero and smaller than one hour and less than a registrar-configured minimum, the registrar may reject the registration with a response of 423 (Interval Too Brief). This response must contain a Min-Expires header field that states the minimum expiration interval the registrar is willing to honor. It then skips the remaining steps.

Allowing the registrar to set the registration interval protects it against excessively frequent registration refreshes while limiting the state that it needs to maintain and decreasing the likelihood of registrations going stale. The expiration interval of a registration is frequently used in the creation of services. An example is a follow-me service, where the user may only be available at a terminal for a brief period. Therefore, registrars should accept brief registrations; a request should only be rejected if the interval is so short that the refreshes would degrade registrar performance.

For each address, the registrar then searches the list of current bindings using the URI comparison rules. If the binding does not exist, it is tentatively added. If the binding does exist, the registrar checks the Call-ID value. If the Call-ID value in the existing binding differs from the Call-ID value in the request, the binding must be removed if the expiration time is zero and updated otherwise. If they are the same, the registrar compares the CSeq value. If the value is higher than that of the existing binding, it must update or remove the binding as above. If not, the update must be aborted and the request fails.

This algorithm ensures that out-of-order requests from the same UA are ignored.

Each binding record records the Call-ID and CSeq values from the request.

The binding updates must be committed (that is, made visible to the proxy or redirect server) if, and only if, all binding updates and additions succeed. If any one of them fails (for example, because the back-end database commit failed), the request must fail with a 500 (Server Error) response and all tentative binding updates must be removed.

8. The registrar returns a 200 (OK) response. The response must contain Contact header field values enumerating all current bindings. Each Contact value must feature an "expires" parameter indicating its expiration interval chosen by the registrar. The response should include a Date header field.

A.6 Querying for Capabilities

The SIP method OPTIONS allows a UA to query another UA or a proxy server as to its capabilities. This allows a client to discover information about the supported methods, content types, extensions, codecs, etc. without "ringing" the other party. For example, before a client inserts a Require header field into an INVITE listing an option that it is not certain the destination UAS supports, the client can query the destination UAS with an OPTIONS to see if this option is returned in a Supported header field. All UAs must support the OPTIONS method.

The target of the OPTIONS request is identified by the Request-URI, which could identify another UA or a SIP server. If the OPTIONS is addressed to a proxy server, the Request-URI is set without a user part, similar to the way a Request-URI is set for a REGISTER request.

Alternatively, a server receiving an OPTIONS request with a Max-Forwards header field value of 0 may respond to the request regardless of the Request-URI.

This behavior is common with HTTP/1.1. This behavior can be used as a "traceroute" functionality to check the capabilities of individual hop servers by sending a series of OPTIONS requests with incremented Max-Forwards values.

As is the case for general UA behavior, the transaction layer can return a timeout error if the OPTIONS yields no response. This may indicate that the target is unreachable and hence unavailable.

An OPTIONS request may be sent as part of an established dialog to query the peer on capabilities that may be utilized later in the dialog.

A.6.1 Construction of OPTIONS Request

An OPTIONS request is constructed using the standard rules for a SIP request. A Contact header field may be present in an OPTIONS. An Accept header field should be included to indicate the type of message body the UAC wishes to receive in the response. Typically, this is set to a format that is used to describe the media capabilities of a UA, such as SDP (application/sdp).

The response to an OPTIONS request is assumed to be scoped to the Request-URI in the original request. However, only when an OPTIONS is sent as part of an established dialog is it guaranteed that future requests will be received by the server that generated the OPTIONS response.

Example OPTIONS request:

```
OPTIONS sip:carol@chicago.com SIP/2.0
Via: SIP/2.0/UDP pc33.atlanta.com;branch=z9hG4bKhjhs8ass877
Max-Forwards: 70
To: <sip:carol@chicago.com>
From: Alice <sip:alice@atlanta.com>;tag=1928301774
Call-ID: a84b4c76e66710
CSeq: 63104 OPTIONS
Contact: <sip:alice@pc33.atlanta.com>
Accept: application/sdp
Content-Length: 0
```

A.6.2 Processing of OPTIONS Request

The response to an OPTIONS is constructed using the standard rules for a SIP response. The response code chosen must be the same that would have been chosen had the request been an INVITE. That is, a 200 (OK) would be returned if the UAS is ready to accept a call, a 486 (Busy Here) would be returned if the UAS is busy, etc. This allows an OPTIONS request to be used to determine the basic state of a UAS, which can be an indication of whether the UAS will accept an INVITE request.

An OPTIONS request received within a dialog generates a 200 (OK) response that is identical to one constructed outside a dialog and does not have any impact on the dialog.

This use of OPTIONS has limitations due to the differences in proxy handling of OPTIONS and INVITE requests. While a forked INVITE can result in multiple 200 (OK) responses being returned, a forked OPTIONS will only result in a single 200 (OK) response, since it is treated by proxies using the non-INVITE handling.

If the response to an OPTIONS is generated by a proxy server, the proxy returns a 200 (OK), listing the capabilities of the server. The response does not contain a message body.

Allow, Accept, Accept-Encoding, Accept-Language, and Supported header fields should be present in a 200 (OK) response to an OPTIONS request. If the response is generated by a proxy, the Allow header field should be omitted as it is ambiguous since a proxy is method agnostic. Contact header fields may be present in a 200 (OK) response and have the same semantics as in a 3xx response. That is, they may list a set of alternative names and methods of reaching the user. A Warning header field may be present.

A message body may be sent, the type of which is determined by the Accept header field in the OPTIONS request (application/sdp is the default if the Accept header field is not present). If the types include one that can describe media capabilities, the UAS should include a body in the response for that purpose.

Example OPTIONS response generated by a UAS (corresponding to the request in Section A.6.1):

```
SIP/2.0 200 OK
Via: SIP/2.0/UDP pc33.atlanta.com;branch=z9hG4bKhjhs8ass877
 ;received=192.0.2.4
To: <sip:carol@chicago.com>;tag=93810874
From: Alice <sip:alice@atlanta.com>;tag=1928301774
```

```
Call-ID: a84b4c76e66710
CSeq: 63104 OPTIONS
Contact: <sip:carol@chicago.com>
Contact: <mailto:carol@chicago.com>
Allow: INVITE, ACK, CANCEL, OPTIONS, BYE
Accept: application/sdp
Accept-Encoding: gzip
Accept-Language: en
Supported: foo
Content-Type: application/sdp
Content-Length: 274
```

(SDP not shown)

A.7 Dialogs

A key concept for a user agent is that of a dialog. A dialog represents a peer-to-peer SIP relationship between two user agents that persists for some time. The dialog facilitates sequencing of messages between the user agents and proper routing of requests between both of them. The dialog represents a context in which to interpret SIP messages. Section A.3 discussed method independent UA processing for requests and responses outside of a dialog. This section discusses how those requests and responses are used to construct a dialog, and then how subsequent requests and responses are sent within a dialog.

A dialog is identified at each UA with a dialog ID, which consists of a Call-ID value, a local tag and a remote tag. The dialog ID at each UA involved in the dialog is not the same. Specifically, the local tag at one UA is identical to the remote tag at the peer UA. The tags are opaque tokens that facilitate the generation of unique dialog IDs.

A dialog ID is also associated with all responses and with any request that contains a tag in the To field. The rules for computing the dialog ID of a message depend on whether the SIP element is a UAC or UAS. For a UAC, the Call-ID value of the dialog ID is set to the Call-ID of the message, the remote tag is set to the tag in the To field of the message, and the local tag is set to the tag in the From field of the message (these rules apply to both requests and responses). As one would expect for a UAS, the Call-ID value of the dialog ID is set to the Call-ID of the message, the remote tag is set to the tag in the From field of the message, and the local tag is set to the tag in the To field of the message.

A dialog contains certain pieces of state needed for further message transmissions within the dialog. This state consists of the dialog ID, a local sequence number (used to order requests from the UA to its peer), a remote sequence number (used to order requests from its peer to the UA), a local URI, a remote URI, remote target, a boolean flag called *secure*, and a route set, which is an ordered list of URIs. The route set is the list of servers that need to be traversed to send a request to the peer. A dialog can also be in the "early" state, which occurs when it is created with a provisional response, and then transition to the "confirmed" state when a 2xx final response arrives. For other responses, or if no response arrives at all on that dialog, the early dialog terminates.

A.7.1 Creation of a Dialog

Dialogs are created through the generation of nonfailure responses to requests with specific methods. Within this specification, only 2xx and 101–199 responses with a To tag, where the request was INVITE, will establish a dialog. A dialog established by a nonfinal response to a request is in the "early" state and it is called an *early dialog*. Extensions may define other means for creating dialogs. Section A.8 gives more details that are specific to the INVITE method. Here, we describe the process for creation of dialog state that is not dependent on the method. UAs must assign values to the dialog ID components as described below.

A.7.1.1 UAS Behavior

When a UAS responds to a request with a response that establishes a dialog (such as a 2xx to INVITE), the UAS must copy all Record-Route header field values from the request into the response (including the URIs, URI parameters, and any Record-Route header field parameters, whether they are known or unknown to the UAS) and must maintain the order of those values. The UAS must add a Contact header field to the response. The Contact header field contains an address where the UAS would like to be contacted for subsequent requests in the dialog (which includes the ACK for a 2xx response in the case of an INVITE). Generally, the host portion of this URI is the IP address or FQDN of the host. The URI provided in the Contact header field must be a SIP or SIPS URI. If the request that initiated the dialog contained a SIPS URI in the Request-URI or in the top Record-Route header field value (if there was any), or the Contact header field (if there was no Record-Route header field), the Contact header field in the response must be a SIPS URI. The URI should have global scope (that is, the same URI can be used in messages outside this dialog). The same way, the scope of the URI in the Contact header field of the INVITE is not limited to this dialog either. It can therefore be used in messages to the UAC even outside this dialog.

The UAS then constructs the state of the dialog. This state must be maintained for the duration of the dialog. If the request arrived over TLS, and the Request-URI contained a SIPS URI, the "secure" flag is set to TRUE.

The route set must be set to the list of URIs in the Record-Route header field from the request, taken in order and preserving all URI parameters. If no Record-Route header field is present in the request, the route set must be set to the empty set. This route set, even if empty, overrides any pre-existing route set for future requests in this dialog. The remote target must be set to the URI from the Contact header field of the request.

The remote sequence number must be set to the value of the sequence number in the CSeq header field of the request. The local sequence number must be empty. The call identifier component of the dialog ID must be set to the value of the Call-ID in the request. The local tag component of the dialog ID must be set to the tag in the To field in the response to the request (which always includes a tag), and the remote tag component of the dialog ID must be set to the tag from the From field in the request. A UAS must be prepared to receive a request without a tag in the From field, in which case the tag is considered to have a value of null.

This is to maintain backwards compatibility with RFC 2543, which did not mandate From tags.

The remote URI must be set to the URI in the From field, and the local URI must be set to the URI in the To field.

A.7.1.2 UAC Behavior

When a UAC sends a request that can establish a dialog (such as an INVITE), it must provide a SIP or SIPS URI with global scope (i.e., the same SIP URI can be used in messages outside this dialog) in the Contact header field of the request. If the request has a Request-URI or a topmost Route header field value with a SIPS URI, the Contact header field must contain a SIPS URI.

When a UAC receives a response that establishes a dialog, it constructs the state of the dialog. This state must be maintained for the duration of the dialog. If the request was sent over TLS, and the Request-URI contained a SIPS URI, the "secure" flag is set to TRUE.

The route set must be set to the list of URIs in the Record-Route header field from the response, taken in reverse order and preserving all URI parameters. If no Record-Route header field is present in the response, the route set must be set to the empty set. This route set, even if empty, overrides any preexisting route set for future requests in this dialog. The remote target must be set to the URI from the Contact header field of the response.

The local sequence number must be set to the value of the sequence number in the CSeq header field of the request. The remote sequence number must be empty (it is established when the remote UA sends a request within the dialog). The call identifier component of the dialog ID must be set to the value of the Call-ID in the request. The local tag component of the dialog ID must be set to the tag in the From field in the request, and the remote tag component of the dialog ID must be set to the tag in the To field of the response. A UAC must be prepared to receive a response without a tag in the To field, in which case the tag is considered to have a value of null.

This is to maintain backwards compatibility with RFC 2543, which did not mandate To tags. The remote URI must be set to the URI in the To field, and the local URI must be set to the URI in the From field.

A.7.2 Requests within a Dialog

Once a dialog has been established between two UAs, either of them may initiate new transactions as needed within the dialog. The UA sending the request will take the UAC role for the transaction. The UA receiving the request will take the UAS role. Note that these may be different roles than the UAs held during the transaction that established the dialog.

Requests within a dialog may contain Record-Route and Contact header fields. However, these requests do not cause the dialog's route set to be modified, although they may modify the remote target URI. Specifically, requests that are not target refresh requests do not modify the dialog's remote target URI, and requests that are target refresh requests do. For dialogs that have been established with an INVITE, the only target refresh request defined is re-INVITE. Other extensions may define different target refresh requests for dialogs established in other ways. Note that an ACK is not a target refresh request. Target refresh requests only update the dialog's remote target URI, and not the route set formed from the Record-Route. Updating the latter would introduce severe backwards compatibility problems with RFC 2543-compliant systems.

A.7.2.1 UAC Behavior

Generating the Request

A request within a dialog is constructed by using many of the components of the state stored as part of the dialog. The URI in the To field of the request must be set to the remote URI from the dialog state. The tag in the To header field of the request must be set to the remote tag of the dialog ID. The From URI of the request must be set to the local URI from the dialog state. The tag in the From header field of the request must be set to the local tag of the dialog ID. If the value of the remote or local tags is null, the tag parameter must be omitted from the To or From header fields, respectively.

Usage of the URI from the To and From fields in the original request within subsequent requests is done for backwards compatibility with RFC 2543, which used the URI for dialog identification. In this specification, only the tags are used for dialog identification. It is expected that mandatory reflection of the original To and From URI in mid-dialog requests will be deprecated in a subsequent revision of this specification.

The Call-ID of the request must be set to the Call-ID of the dialog. Requests within a dialog must contain strictly monotonically increasing and contiguous CSeq sequence numbers (increasing-by-one) in each direction (excepting ACK and CANCEL, of course, whose numbers equal the requests being acknowledged or cancelled). Therefore, if the local sequence number is not empty, the value of the local sequence number must be incremented by one, and this value must be placed into the CSeq header field. If the local sequence number is empty, an initial value must be chosen. The method field in the CSeq header field value must match the method of the request.

With a length of 32 bits, a client could generate, within a single call, one request a second for about 136 years before needing to wrap around. The initial value of the sequence number is chosen so that subsequent requests within the same call will not wrap around. A nonzero initial value allows clients to use a time-based initial sequence number. A client could, for example, choose the 31 most significant bits of a 32-bit second clock as an initial sequence number.

The UAC uses the remote target and route set to build the Request-URI and Route header field of the request. If the route set is empty, the UAC must place the remote target URI into the Request-URI. The UAC must not add a Route header field to the request. If the route set is not empty, and the first URI in the route set contains the lr parameter, the UAC must place the remote target URI into the Request-URI and must include a Route header field containing the route set values in order, including all parameters.

If the route set is not empty, and its first URI does not contain the lr parameter, the UAC must place the first URI from the route set into the Request-URI, stripping any parameters that are not allowed in a Request-URI. The UAC must add a Route header field containing the remainder of the route set values in order, including all parameters. The UAC must then place the remote target URI into the Route header field as the last value.

For example, if the remote target is sip:user@remoteua and the route set contains:

```
<sip:proxy1>,<sip:proxy2>,<sip:proxy3;lr>,<sip:proxy4>
```

The request will be formed with the following Request-URI and Route header field:

```
METHOD sip:proxy1
Route: <sip:proxy2>,<sip:proxy3;lr>,<sip:proxy4>,<sip:user@remoteua>
```

If the first URI of the route set does not contain the lr parameter, the proxy indicated does not understand the routing mechanisms described in this RFC and will act as specified in RFC 2543, replacing the Request-URI with the first Route header field value it receives while forwarding the message. Placing the Request-URI at the end of the Route header field preserves the information in that Request-URI across the strict router (it will be returned to the Request-URI when the request reaches a loose-router).

A UAC should include a Contact header field in any target refresh requests within a dialog, and unless there is a need to change it, the URI should be the same as used in previous requests within the dialog. If the "secure" flag is true, that URI must be a SIPS URI. A Contact header field in a target refresh request updates the remote target URI. This allows a UA to provide a new contact address, should its address change during the duration of the dialog.

However, requests that are not target refresh requests do not affect the remote target URI for the dialog. Once the request has been constructed, the address of the server is computed and the request is sent, using the same procedures for requests outside of a dialog. The procedures in Section A.3.1.2 will normally result in the request being sent to the address indicated by the topmost Route header field value or the Request-URI if no Route header field is present. Subject to certain restrictions, they allow the request to be sent to an alternate address (such as a default outbound proxy not represented in the route set).

Processing the Responses

The UAC will receive responses to the request from the transaction layer. If the client transaction returns a timeout, this is treated as a 408 (Request Timeout) response. The behavior of a UAC that receives a 3xx response for a request sent within a dialog is the same as if the request had been sent outside a dialog. Note, however, that when the UAC tries alternative locations, it still uses the route set for the dialog to build the Route header of the request.

When a UAC receives a 2xx response to a target refresh request, it must replace the dialog's remote target URI with the URI from the Contact header field in that response, if present.

If the response for a request within a dialog is a 481 (Call/Transaction Does Not Exist) or a 408 (Request Timeout), the UAC should terminate the dialog. A UAC should also terminate a dialog if no response at all is received for the request (the client transaction would inform the TU about the timeout.)

For INVITE initiated dialogs, terminating the dialog consists of sending a BYE.

A.7.2.2   UAS Behavior

Requests sent within a dialog, as any other requests, are atomic. If a particular request is accepted by the UAS, all the state changes associated with it are performed. If the request is rejected, none of the state changes are performed. Note that some requests, such as INVITEs, affect several pieces of state.

The UAS will receive the request from the transaction layer. If the request has a tag in the To header field, the UAS core computes the dialog identifier corresponding to the request and compares it with existing dialogs. If there is a match, this is a mid-dialog request. In that case, the UAS first applies the same processing rules for requests outside of a dialog.

If the request has a tag in the To header field, but the dialog identifier does not match any existing dialogs, the UAS may have crashed and restarted, or it may have received a request for a different (possibly failed) UAS (the UASs can construct the To tags so that a UAS can identify that the tag was for a UAS for which it is providing recovery). Another possibility is that the incoming request has been simply misrouted. Based on the To tag, the UAS may either accept or reject the request. Accepting the request for acceptable To tags provides robustness, so that dialogs can persist even through crashes. UAs wishing to support this capability must take into consideration some issues such as choosing monotonically increasing CSeq sequence numbers even across reboots, reconstructing the route set, and accepting out-of-range RTP timestamps and sequence numbers.

If the UAS wishes to reject the request because it does not wish to recreate the dialog, it must respond to the request with a 481 (Call/Transaction Does Not Exist) status code and pass that to the server transaction.

Requests that do not change in any way the state of a dialog may be received within a dialog (for example, an OPTIONS request). They are processed as if they had been received outside the dialog.

If the remote sequence number is empty, it must be set to the value of the sequence number in the CSeq header field value in the request. If the remote sequence number was not empty, but the sequence number of the request is lower than the remote sequence number, the request is out of order and must be rejected with a 500 (Server Internal Error) response. If the remote sequence number was not empty, and the sequence number of the request is greater than the remote sequence number, the request is in order. It is possible for the CSeq sequence number to be higher than the remote sequence number by more than one. This is not an error condition, and a UAS should be prepared to receive and process requests with CSeq values more than one higher than the previous received request. The UAS must then set the remote sequence number to the value of the sequence number in the CSeq header field value in the request.

If a proxy challenges a request generated by the UAC, the UAC has to resubmit the request with credentials. The resubmitted request will have a new CSeq number. The UAS will never see the first request, and thus, it will notice a gap in the CSeq number space. Such a gap does not represent any error condition.

When a UAS receives a target refresh request, it must replace the dialog's remote target URI with the URI from the Contact header field in that request, if present.

A.7.3 Termination of a Dialog

Independent of the method, if a request outside of a dialog generates a non-2xx final response, any early dialogs created through provisional responses to that request are terminated. The mechanism for terminating confirmed dialogs is method-specific. In this specification, the BYE method terminates a session and the dialog associated with it.

A.8 Initiating a Session

A.8.1 Overview

When a user agent client desires to initiate a session (for example, audio, video, or a game), it formulates an INVITE request. The INVITE request asks a server to establish a session. This request may be forwarded by proxies, eventually arriving at one or more UAS that can potentially accept the invitation. These UASs will frequently need to query the user about whether to accept the invitation. After some time, those UASs can accept the invitation (meaning the session is to be established) by sending a 2xx response. If the invitation is not accepted, a 3xx, 4xx, 5xx or 6xx response is sent, depending on the reason for the rejection. Before sending a final response, the UAS can also send provisional responses (1xx) to advise the UAC of progress in contacting the called user.

After possibly receiving one or more provisional responses, the UAC will get one or more 2xx responses or one non-2xx final response. Because of the protracted amount of time it can take to receive final responses to INVITE, the reliability mechanisms for INVITE transactions differ from those of other requests (like OPTIONS). Once it receives a final response, the UAC needs to send an ACK for every final response it receives. The procedure for sending this ACK depends on the type of response. For final responses between 300 and 699, the ACK processing is done in the transaction layer and follows one set of rules. For 2xx responses, the ACK is generated by the UAC core.

A 2xx response to an INVITE establishes a session, and it also creates a dialog between the UA that issued the INVITE and the UA that generated the 2xx response. Therefore, when multiple 2xx responses are received from different remote UAs (because the INVITE forked), each 2xx establishes a different dialog. All these dialogs are part of the same call.

This section provides details on the establishment of a session using INVITE. A UA that supports INVITE must also support ACK, CANCEL and BYE.

A.8.2 UAC Processing

A.8.2.1 Creating the Initial INVITE

Since the initial INVITE represents a request outside of a dialog, its construction follows the procedures of Section A.3.1.1. Additional processing is required for the specific case of INVITE. An Allow header field should be present in the INVITE. It indicates what methods can be invoked within a dialog, on the UA sending the INVITE, for the duration of the dialog. For example, a UA capable of receiving INFO requests within a dialog should include an Allow header field listing the INFO method. A Supported header field should be present in the INVITE. It enumerates all the extensions understood by the UAC.

An Accept header field may be present in the INVITE. It indicates which Content-Types are acceptable to the UA, in both the response received by it, and in any subsequent requests sent to it within dialogs established by the INVITE. The Accept header field is especially useful for indicating support of various session description formats.

The UAC may add an Expires header field to limit the validity of the invitation. If the time indicated in the Expires header field is reached and no final answer for the INVITE has been received, the UAC core should generate a CANCEL request for the INVITE. A UAC may also find it useful to add, among others, Subject, Organization and User-Agent header fields. They all contain information related to the INVITE. The UAC may choose to add a message body to the INVITE.

There are special rules for message bodies that contain a session description—their corresponding Content-Disposition is "session." SIP uses an offer/answer model where one UA sends a session description, called the *offer*, which contains a proposed description of the session. The offer indicates the desired communications means (audio, video, games), parameters of those means (such as codec types) and addresses for receiving media from the answerer. The other UA responds with another session description, called the *answer*, which indicates which communications means are accepted, the parameters that apply to those means, and addresses for receiving media from the offerer. An offer/answer exchange is within the context of a dialog, so that if a SIP INVITE results in multiple dialogs, each is a separate offer/answer exchange. The offer/answer model defines restrictions on when offers and answers can be made (for example, you cannot make a new offer while one is in progress). This results in restrictions on where the offers and answers can appear in SIP messages. In this specification, offers and answers can only appear in INVITE requests and responses, and ACK. The usage of offers and answers is further restricted. For the initial INVITE transaction, the rules are:

- The initial offer must be in either an INVITE or, if not there, in the first reliable nonfailure message from the UAS back to the UAC. In this specification, that is the final 2xx response.
- If the initial offer is in an INVITE, the answer must be in a reliable nonfailure message from UAS back to UAC which is correlated to that INVITE. For this specification, that is only the final 2xx response to that INVITE. That same exact answer may also be placed in any provisional responses sent prior to the answer. The UAC must treat the first session description it receives as the answer, and must ignore any session descriptions in subsequent responses to the initial INVITE.
- If the initial offer is in the first reliable nonfailure message from the UAS back to UAC, the answer must be in the acknowledgment for that message (in this specification, ACK for a 2xx response).
- After having sent or received an answer to the first offer, the UAC may generate subsequent offers in requests based on rules specified for that method, but only if it has received answers to any previous offers, and has not sent any offers to which it has not gotten an answer.
- Once the UAS has sent or received an answer to the initial offer, it must not generate subsequent offers in any responses to the initial INVITE. This means that a UAS based on this specification alone can never generate subsequent offers until completion of the initial transaction.

Concretely, the above rules specify two exchanges for UAs compliant to this specification alone—the offer is in the INVITE, and the answer in the 2xx (and possibly in a 1xx as well, with the same value), or the offer is in the 2xx, and the answer is in the ACK. All user agents that support INVITE must support these two exchanges.

The Session Description Protocol (SDP) (RFC 2327) must be supported by all user agents as a means to describe sessions. The restrictions of the offer-answer model just described only apply to bodies whose Content-Disposition header field value is "session." Therefore, it is possible that both the INVITE and the ACK contain a body message (for example, the INVITE carries a photo (Content-Disposition: render) and the ACK a session description (Content-Disposition: session)). If the Content-Disposition header field is missing, bodies of Content-Type application/sdp imply the disposition "session," while other content types imply "render." Once the INVITE has been created, the UAC follows the procedures defined for sending requests outside of a dialog. This results in the construction of a client transaction that will ultimately send the request and deliver responses to the UAC.

A.8.2.2 Processing INVITE Responses

Once the INVITE has been passed to the INVITE client transaction, the UAC waits for responses for the INVITE. If the INVITE client transaction returns a timeout rather than a response, the TU acts as if a 408 (Request Timeout) response had been received, as described in Section A.3.1.3.

1xx Responses

Zero, one, or multiple provisional responses may arrive before one or more final responses are received. Provisional responses for an INVITE request can create "early dialogs." If a provisional response has a tag in the To field, and if the dialog ID of the response does not match an existing dialog, one is constructed using the procedures defined in Section A.7.1.2.

The early dialog will only be needed if the UAC needs to send a request to its peer within the dialog before the initial INVITE transaction completes. Header fields present in a provisional response are applicable as long as the dialog is in the early state (for example, an Allow header field in a provisional response contains the methods that can be used in the dialog while this is in the early state).

3xx Responses

A 3xx response may contain one or more Contact header field values providing new addresses where the callee might be reachable. Depending on the status code of the 3xx response, the UAC may choose to try those new addresses.

4xx, 5xx and 6xx Responses

A single non-2xx final response may be received for the INVITE. 4xx, 5xx and 6xx responses may contain a Contact header field value indicating the location where additional information about the error can be found. Subsequent final responses (which would only arrive under error conditions) must be ignored. All early dialogs are considered terminated upon reception of the non-2xx final response. After having received the non-2xx final response the UAC core considers the INVITE transaction completed. The INVITE client transaction handles the generation of ACKs for the response.

2xx Responses

Multiple 2xx responses may arrive at the UAC for a single INVITE request due to a forking proxy. Each response is distinguished by the tag parameter in the To header field, and each represents a distinct dialog, with a distinct dialog identifier.

If the dialog identifier in the 2xx response matches the dialog identifier of an existing dialog, the dialog must be transitioned to the "confirmed" state, and the route set for the dialog must be recomputed based on the 2xx response using the procedures of Section A.7.2.1. Otherwise, a new dialog in the "confirmed" state must be constructed using the procedures of Section A.7.1.2.

Note that the only piece of state that is recomputed is the route set. Other pieces of state such as the highest sequence numbers (remote and local) sent within the dialog are not recomputed. The route set only is recomputed for backwards compatibility. RFC 2543 did not mandate mirroring of the Record-Route header field in a 1xx, only 2xx. However, we cannot update the entire state of the dialog, since mid-dialog requests may have been sent within the early dialog, modifying the sequence numbers, for example.

The UAC core must generate an ACK request for each 2xx received from the transaction layer. The header fields of the ACK are constructed in the same way as for any request sent within a dialog with the exception of the CSeq and the header fields related to authentication. The sequence number of the CSeq header field must be the same as the INVITE being acknowledged, but the CSeq method must be ACK. The ACK must contain the same credentials as the INVITE. If the 2xx contains an offer (based on the rules above), the ACK must carry an answer in its body. If the offer in the 2xx response is not acceptable, the UAC core must generate a valid answer in the ACK and then send a BYE immediately.

Once the ACK has been constructed, there is a need to determine the destination address, port and transport. However, the request is passed to the transport layer directly for transmission, rather than a client transaction. This is because the UAC core handles retransmissions of the ACK, not the transaction layer. The ACK must be passed to the client transport every time a retransmission of the 2xx final response that triggered the ACK arrives.

The UAC core considers the INVITE transaction completed 64*T1 seconds after the reception of the first 2xx response. At this point all the early dialogs that have not transitioned to established dialogs are terminated. Once the INVITE transaction is considered completed by the UAC core, no more new 2xx responses are expected to arrive.

If, after acknowledging any 2xx response to an INVITE, the UAC does not want to continue with that dialog, then the UAC must terminate the dialog by sending a BYE request.

A.8.3 UAS Processing

A.8.3.1   Processing of the INVITE

The UAS core will receive INVITE requests from the transaction layer. It first performs the request processing procedures of Section A.3.2 which are applied for both requests inside and outside of a dialog.

Assuming these processing states are completed without generating a response, the UAS core performs the additional processing steps:

1. If the request is an INVITE that contains an Expires header field, the UAS core sets a timer for the number of seconds indicated in the header field value. When the timer fires, the invitation is considered to be expired. If the invitation expires before the UAS has generated a final response, a 487 (Request Terminated) response should be generated.
2. If the request is a mid-dialog request, the method-independent processing described in Section A.7.2.2 is first applied. It might also modify the session.
3. If the request has a tag in the To header field but the dialog identifier does not match any of the existing dialogs, the UAS may have crashed and restarted, or may have received a request for a different (possibly failed) UAS.

Processing from here forward assumes that the INVITE is outside of a dialog, and is thus for the purposes of establishing a new session.

The INVITE may contain a session description, in which case the UAS is being presented with an offer for that session. It is possible that the user is already a participant in that session, even though the INVITE is outside of a dialog. This can happen when a user is invited to the same multicast conference by multiple other participants. If desired, the UAS may use identifiers within the session description to detect this duplication. For example, SDP contains a session ID and version number in the origin (o) field. If the user is already a member of the session, and the session parameters contained in the session description have not changed, the UAS may silently accept the INVITE (that is, send a 2xx response without prompting the user).

If the INVITE does not contain a session description, the UAS is being asked to participate in a session, and the UAC has asked that the UAS provide the offer of the session. It must provide the offer in its first nonfailure reliable message back to the UAC. In this specification, that is a 2xx response to the INVITE. The UAS can indicate progress, accept, redirect, or reject the invitation. In all of these cases, it formulates a response using the procedures described in Section A.3.2.6.

Progress

If the UAS is not able to answer the invitation immediately, it can choose to indicate some kind of progress to the UAC (for example, an indication that a phone is ringing). This is accomplished with a provisional response between 101 and 199. These provisional responses establish early dialogs and therefore follow the procedures of Section A.7.1.1 in addition to those of Section A.3.2.6. A UAS may send as many provisional responses as it likes. Each of these must indicate the same dialog ID. However, these will not be delivered reliably.

If the UAS desires an extended period of time to answer the INVITE, it will need to ask for an "extension" in order to prevent proxies from canceling the transaction. A proxy has the option of canceling a transaction when there is a gap of 3 minutes between responses in a transaction. To prevent cancellation, the UAS must send a non-100 provisional response at every minute, to handle the possibility of lost provisional responses.

An INVITE transaction can go on for extended durations when the user is placed on hold, or when inter-working with PSTN systems which allow communications to take place without answering the call. The latter is common in Interactive Voice Response (IVR) systems.

The INVITE is Redirected

If the UAS decides to redirect the call, a 3xx response is sent. A 300 (Multiple Choices), 301 (Moved Permanently) or 302 (Moved Temporarily) response should contain a Contact header field containing one or more URIs of new addresses to be tried. The response is passed to the INVITE server transaction, which will deal with its retransmissions.

The INVITE is Rejected

A common scenario occurs when the callee is currently not willing or able to take additional calls at this end system. A 486 (Busy Here) should be returned in such a scenario. If the UAS knows that no other end system will be able to accept this call, a 600 (Busy Everywhere) response should be sent instead. However, it is unlikely that a UAS will be able to know this in general, and thus this response will not usually be used. The response is passed to the INVITE server transaction, which will deal with its retransmissions.

A UAS rejecting an offer contained in an INVITE should return a 488 (Not Acceptable Here) response. Such a response should include a Warning header field value explaining why the offer was rejected.

The INVITE is Accepted

The UAS core generates a 2xx response. This response establishes a dialog, and therefore follows the procedures of Section A.7.1.1 in addition to those of Section A.3.2.6.

A 2xx response to an INVITE should contain the Allow header field and the Supported header field, and may contain the Accept header field. Including these header fields allows the UAC to determine the features and extensions supported by the UAS for the duration of the call, without probing.

If the INVITE request contained an offer, and the UAS had not yet sent an answer, the 2xx must contain an answer. If the INVITE did not contain an offer, the 2xx must contain an offer if the UAS had not yet sent an offer.

Once the response has been constructed, it is passed to the INVITE server transaction. Note, however, that the INVITE server transaction will be destroyed as soon as it receives this final response and passes it to the transport. Therefore, it is necessary to periodically pass the response directly to the transport until the ACK arrives. The 2xx response is passed to the transport with an interval that starts at T1 seconds and doubles for each retransmission until it reaches T2 seconds (T1 and T2 are defined in the RFC). Response retransmissions cease when an ACK request for the response is received. This is independent of whatever transport protocols are used to send the response.

Since 2xx is retransmitted end-to-end, there may be hops between UAS and UAC that are UDP. To ensure reliable delivery across these hops, the response is retransmitted periodically even if the transport at the UAS is reliable.

If the server retransmits the 2xx response for 64*T1 seconds without receiving an ACK, the dialog is confirmed, but the session should be terminated. This is accomplished with a BYE.

A.9 Modifying an Existing Session

A successful INVITE request establishes both a dialog between two user agents and a session using the offer-answer model. Section A.7 explains how to modify an existing dialog using a target refresh request (for example, changing the remote target URI of the dialog). This section describes how to modify the actual session. This modification can involve changing addresses or ports, adding a media stream, deleting a media stream, and so on. This is accomplished by sending a new INVITE request within the same dialog that established the session. An INVITE request sent within an existing dialog is known as a re-INVITE.

Note that a single re-INVITE can modify the dialog and the parameters of the session at the same time. Either the caller or callee can modify an existing session. The behavior of a UA on detection of media failure is a matter of local policy. However, automated generation of re-INVITE or BYE is not recommended to avoid flooding the network with traffic when there is congestion. In any case, if these messages are sent automatically, they should be sent after some randomized interval. Note that the paragraph above refers to automatically generated BYEs and re-INVITEs. If the user hangs up upon media failure, the UA would send a BYE request as usual.

A.9.1 UAC Behavior

The same offer-answer model that applies to session descriptions in INVITEs applies to re-INVITEs. As a result, a UAC that wants to add a media stream, for example, will create a new offer that contains this media stream, and send that in an INVITE request to its peer. It is important to note that the full description of the session, not just the change, is sent. This supports stateless session processing in various elements, and supports failover and recovery capabilities. Of course, a UAC may send a re-INVITE with no session description, in which case the first reliable nonfailure response to the re-INVITE will contain the offer (in this specification, that is a 2xx response).

If the session description format has the capability for version numbers, the offerer should indicate that the version of the session description has changed.

The To, From, Call-ID, CSeq, and Request-URI of a re-INVITE are set following the same rules as for regular requests within an existing dialog, described in Section A.7.

A UAC may choose not to add an Alert-Info header field or a body with Content-Disposition "alert" to re-INVITEs because UASs do not typically alert the user upon reception of a re-INVITE. Unlike an INVITE, which can fork, a re-INVITE will never fork, and therefore, only ever generate a single final response. The reason a re-INVITE will never fork is that the Request-URI identifies the target as the UA instance it established the dialog with, rather than identifying an address-of-record for the user. Note that a UAC must not initiate a new INVITE transaction within a dialog while another INVITE transaction is in progress in either direction.

1. If there is an ongoing INVITE client transaction, the TU must wait until the transaction reaches the completed or terminated state before initiating the new INVITE.
2. If there is an ongoing INVITE server transaction, the TU must wait until the transaction reaches the confirmed or terminated state before initiating the new INVITE.

However, a UA may initiate a regular transaction while an INVITE transaction is in progress. A UA may also initiate an INVITE transaction while a regular transaction is in progress.

If a UA receives a non-2xx final response to a re-INVITE, the session parameters must remain unchanged, as if no re-INVITE had been issued. Note that, as stated in Section A.7.2.1, if the non-2xx final response is a 481 (Call/Transaction Does Not Exist), or a 408 (Request Timeout), or no response at all is received for the re-INVITE (that is, a timeout is returned by the INVITE client transaction), the UAC will terminate the dialog.

If a UAC receives a 491 response to a re-INVITE, it should start a timer with a value T chosen as follows:

1. If the UAC is the owner of the Call-ID of the dialog ID (meaning it generated the value), T has a randomly chosen value between 2.1 and 4 seconds in units of 10 ms.
2. If the UAC is not the owner of the Call-ID of the dialog ID, T has a randomly chosen value of between 0 and 2 seconds in units of 10 ms.

When the timer fires, the UAC should attempt the re-INVITE once more, if it still desires for that session modification to take place. For example, if the call was already hung up with a BYE, the re-INVITE would not take place. The rules for transmitting a re-INVITE and for generating an ACK for a 2xx response to re-INVITE are the same as for the initial INVITE.

A.9.2 UAS Behavior

Section A.8.3.1 describes the procedure for distinguishing incoming re-INVITEs from incoming initial INVITEs and handling a re-INVITE for an existing dialog.

A UAS that receives a second INVITE before it sends the final response to a first INVITE with a lower CSeq sequence number on the same dialog must return a 500 (Server Internal Error) response to the second INVITE and must include a Retry-After header field with a randomly chosen value of between 0 and 10 seconds.

A UAS that receives an INVITE on a dialog while an INVITE it had sent on that dialog is in progress must return a 491 (Request Pending) response to the received INVITE. If a UA receives a re-INVITE for an existing dialog, it must check any version identifiers in the session description or, if there are no version identifiers, the content of the session description to see if it has changed. If the session description has changed, the UAS must adjust the session parameters accordingly, possibly after asking the user for confirmation. Versioning of the session description can be used to accommodate the capabilities of new arrivals to a conference, add or delete media, or change from a unicast to a multicast conference. If the new session description is not acceptable, the UAS can reject it by returning a 488 (Not Acceptable Here) response for the re-INVITE. This response should include a Warning header field.

If a UAS generates a 2xx response and never receives an ACK, it should generate a BYE to terminate the dialog.

A UAS may choose not to generate 180 (Ringing) responses for a re-INVITE because UACs do not typically render this information to the user. For the same reason, UASs may choose not to use an Alert-Info header field or a body with Content-Disposition "alert" in responses to a re-INVITE.

A UAS providing an offer in a 2xx (because the INVITE did not contain an offer) should construct the offer as if the UAS were making a brand new call, subject to the constraints of sending an offer that updates an existing session. Specifically, this means that it should include as many media formats and media types that the UA is willing to support. The UAS must ensure that the session description overlaps with its previous session description in media formats, transports, or other parameters that require support from the peer. This is to avoid the need for the peer to reject the session description. If, however, it is unacceptable to the UAC, the UAC should generate an answer with a valid session description, and then send a BYE to terminate the session.

A.10 Terminating a Session

This section describes the procedures for terminating a session established by SIP. The state of the session and the state of the dialog are very closely related. When a session is initiated with an INVITE, each

1xx or 2xx response from a distinct UAS creates a dialog, and if that response completes the offer/answer exchange, it also creates a session. As a result, each session is "associated" with a single dialog—the one which resulted in its creation. If an initial INVITE generates a non-2xx final response, that terminates all sessions (if any) and all dialogs (if any) that were created through responses to the request. By virtue of completing the transaction, a non-2xx final response also prevents further sessions from being created as a result of the INVITE. The BYE request is used to terminate a specific session or attempted session. In this case, the specific session is the one with the peer UA on the other side of the dialog. When a BYE is received on a dialog, any session associated with that dialog should terminate. A UA must not send a BYE outside of a dialog. The caller's UA may send a BYE for either confirmed or early dialogs, and the callee's UA may send a BYE on confirmed dialogs, but must not send a BYE on early dialogs.

However, the callee's UA must not send a BYE on a confirmed dialog until it has received an ACK for its 2xx response or until the server transaction times out. If no SIP extensions have defined other application layer states associated with the dialog, the BYE also terminates the dialog.

The impact of a non-2xx final response to INVITE on dialogs and sessions makes the use of CANCEL attractive. The CANCEL attempts to force a non-2xx response to the INVITE (in particular, a 487). Therefore, if a UAC wishes to give up on its call attempt entirely, it can send a CANCEL. If the INVITE results in 2xx final response(s) to the INVITE, this means that a UAS accepted the invitation while the CANCEL was in progress. The UAC may continue with the sessions established by any 2xx responses or may terminate them with BYE.

The notion of "hanging up" is not well-defined within SIP. It is specific to a particular, albeit common, user interface. Typically, when the user hangs up, it indicates a desire to terminate the attempt to establish a session, and to terminate any sessions already created. For the caller's UA, this would imply a CANCEL request if the initial INVITE has not generated a final response, and a BYE to all confirmed dialogs after a final response. For the callee's UA, it would typically imply a BYE; presumably, when the user picked up the phone, a 2xx was generated, and so hanging up would result in a BYE after the ACK is received. This does not mean a user cannot hang up before receipt of the ACK, it just means that the software in his phone needs to maintain state for a short while in order to clean up properly. If the particular UI allows for the user to reject a call before it's answered, a 403 (Forbidden) is a good way to express that. As per the rules above, a BYE cannot be sent.

A.10.1 Terminating a Session with a BYE Request

A.10.1.1 UAC Behavior

A BYE request is constructed as would any other request within a dialog, as described in Section A.7.

Once the BYE is constructed, the UAC core creates a new non-INVITE client transaction and passes it the BYE request. The UAC must consider the session terminated (and therefore stop sending or listening for media) as soon as the BYE request is passed to the client transaction. If the response for the BYE is a 481 (Call/Transaction Does Not Exist) or a 408 (Request Timeout) or no response at all is received for the BYE (that is, a timeout is returned by the client transaction), the UAC must consider the session and the dialog terminated.

A.10.1.2 UAS Behavior

A UAS first processes the BYE request according to the general UAS processing described in Section A.3.2. A UAS core receiving a BYE request checks if it matches an existing dialog. If the BYE does not match an existing dialog, the UAS core should generate a 481 (Call/Transaction Does Not Exist) response and pass that to the server transaction.

This rule means that a BYE sent without tags by a UAC will be rejected. This is a change from RFC 2543, which allowed BYE without tags. A UAS core receiving a BYE request for an existing dialog must follow the procedures of Section A.7.2.2 to process the request. Once done, the UAS should terminate the session (and therefore stop sending and listening for media). The only case where it can elect not to are multicast sessions, where participation is possible even if the other participant in the dialog has terminated its involvement in the session. Whether or not it ends its participation on the session, the UAS core must generate a 2xx response to the BYE, and must pass that to the server transaction for transmission.

The UAS must still respond to any pending requests received for that dialog. It is recommended that a 487 (Request Terminated) response be generated to those pending requests.

A.11 Proxy Behavior

A.11.1 Overview

SIP proxies are elements that route SIP requests to user agent servers and SIP responses to user agent clients. A request may traverse several proxies on its way to a UAS. Each will make routing decisions, modifying the request before forwarding it to the next element. Responses will route through the same set of proxies traversed by the request in the reverse order.

Being a proxy is a logical role for a SIP element. When a request arrives, an element that can play the role of a proxy first decides if it needs to respond to the request on its own. For instance, the request may be malformed or the element may need credentials from the client before acting as a proxy. The element may respond with any appropriate error code. When responding directly to a request, the element is playing the role of a UAS and must behave as described in Section A.3.2.

A proxy can operate in either a stateful or stateless mode for each new request. When stateless, a proxy acts as a simple forwarding element. It forwards each request downstream to a single element determined by making a targeting and routing decision based on the request. It simply forwards every response it receives upstream. A stateless proxy discards information about a message once the message has been forwarded. A stateful proxy remembers information (specifically, transaction state) about each incoming request and any requests it sends as a result of processing the incoming request. It uses this information to affect the processing of future messages associated with that request. A stateful proxy may choose to "fork" a request, routing it to multiple destinations. Any request that is forwarded to more than one location must be handled statefully.

In some circumstances, a proxy may forward requests using stateful transports (such as TCP) without being transaction-stateful. For instance, a proxy may forward a request from one TCP connection to another transaction statelessly as long as it places enough information in the message to be able to forward the response down the same connection the request arrived on. Requests forwarded between different types of transports where the proxy's TU must take an active role in ensuring reliable delivery on one of the transports must be forwarded transaction statefully.

A stateful proxy may transition to stateless operation at any time during the processing of a request, so long as it did not do anything that would otherwise prevent it from being stateless initially (forking, for example, or generation of a 100 response). When performing such a transition, all state is simply discarded. The proxy should not initiate a CANCEL request.

Much of the processing involved when acting statelessly or statefully for a request is identical. The next several subsections are written from the point of view of a stateful proxy. The last section calls out those places where a stateless proxy behaves differently.

A.11.2 Stateful Proxy

When stateful, a proxy is purely a SIP transaction processing engine. A stateful proxy has a server transaction associated with one or more client transactions by a higher layer proxy processing component (see Figure 3.3), known as a proxy core. An incoming request is processed by a server transaction. Requests from the server transaction are passed to a proxy core. The proxy core determines where to route the request, choosing one or more next-hop locations. An outgoing request for each next-hop location is processed by its own associated client transaction. The proxy core collects the responses from the client transactions and uses them to send responses to the server transaction.

A stateful proxy creates a new server transaction for each new request received. Any retransmissions of the request will then be handled by that server transaction per Section A.12. The proxy core must behave as a UAS with respect to sending an immediate provisional on that server transaction (such as 100 Trying) as described in Section A.3.2.6. Thus, a stateful proxy should not generate 100 (Trying) responses to non-INVITE requests.

This is a model of proxy behavior, not of software. An implementation is free to take any approach that replicates the external behavior this model defines.

For all new requests, including any with unknown methods, an element intending to proxy the request must:

1. Validate the request;
2. Preprocess routing information;
3. Determine target(s) for the request;
4. Forward the request to each target;
5. Process all responses.

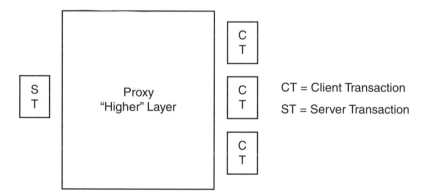

Figure 3.3: Stateful proxy model.

See RFC 3261 for details.

A.11.3 Summary of Proxy Route Processing

In the absence of local policy to the contrary, the processing a proxy performs on a request containing a Route header field can be summarized in the following steps.

1. The proxy will inspect the Request-URI. If it indicates a resource owned by this proxy, the proxy will replace it with the results of running a location service. Otherwise, the proxy will not change the Request-URI.

2. The proxy will inspect the URI in the topmost Route header field value. If it indicates this proxy, the proxy removes it from the Route header field (this route node has been reached).

3. The proxy will forward the request to the resource indicated by the URI in the topmost Route header field value or in the Request-URI if no Route header field is present. The proxy determines the address, port and transport to use when forwarding the request.

If no strict-routing elements are encountered on the path of the request, the Request-URI will always indicate the target of the request.

A.11.3.1 Examples

Basic SIP Trapezoid

This scenario is the basic SIP trapezoid, U1 → P1 → P2 → U2, with both proxies record-routing. Here is the flow.

U1 sends:

```
INVITE sip:callee@domain.com SIP/2.0
Contact: sip:caller@u1.example.com
```

to P1. P1 is an outbound proxy. P1 is not responsible for domain.com, so it looks it up in DNS and sends it there. It also adds a Record-Route header field value:

```
INVITE sip:callee@domain.com SIP/2.0
Contact: sip:caller@u1.example.com
Record-Route: <sip:p1.example.com;lr>
```

P2 gets this. It is responsible for domain.com so it runs a location service and rewrites the Request-URI. It also adds a Record-Route header field value. There is no Route header field, so it resolves the new Request-URI to determine where to send the request:

```
INVITE sip:callee@u2.domain.com SIP/2.0
Contact: sip:caller@u1.example.com
Record-Route: <sip:p2.domain.com;lr>
Record-Route: <sip:p1.example.com;lr>
```

The callee at u2.domain.com gets this and responds with a 200 OK:

```
SIP/2.0 200 OK
Contact: sip:callee@u2.domain.com
Record-Route: <sip:p2.domain.com;lr>
Record-Route: <sip:p1.example.com;lr>
```

The callee at u2 also sets its dialog state's remote target URI to sip:caller@u1.example.com and its route set to:

```
(<sip:p2.domain.com;lr>,<sip:p1.example.com;lr>)
```

This is forwarded by P2 to P1 to U1 as normal. Now, U1 sets its dialog state's remote target URI to sip:callee@u2.domain.com and its route set to:

```
(<sip:p1.example.com;lr>,<sip:p2.domain.com;lr>)
```

Since all the route set elements contain the lr parameter, U1 constructs the following BYE request:

```
BYE sip:callee@u2.domain.com SIP/2.0
Route: <sip:p1.example.com;lr>,<sip:p2.domain.com;lr>
```

As any other element (including proxies) would do, it resolves the URI in the topmost Route header field value using DNS to determine where to send the request. This goes to P1. P1 notices that it is not responsible for the resource indicated in the Request-URI so it does not change it. It does see that it is the first value in the Route header field, so it removes that value, and forwards the request to P2:

```
BYE sip:callee@u2.domain.com SIP/2.0
Route: <sip:p2.domain.com;lr>
```

P2 also notices it is not responsible for the resource indicated by the Request-URI (it is responsible for domain. com, not u2.domain.com), so it does not change it. It does see itself in the first Route header field value, so it removes it and forwards the following to u2.domain.com based on a DNS lookup against the Request-URI:

```
BYE sip:callee@u2.domain.com SIP/2.0
```

Traversing a Strict-Routing Proxy

In this scenario, a dialog is established across four proxies, each of which adds Record-Route header field values. The third proxy implements the strict-routing procedures specified in RFC 2543 and many works in progress.

U1→P1→P2→P3→P4→U2

The INVITE arriving at U2 contains:

```
INVITE sip:callee@u2.domain.com SIP/2.0
Contact: sip:caller@u1.example.com
Record-Route: <sip:p4.domain.com;lr>
Record-Route: <sip:p3.middle.com>
Record-Route: <sip:p2.example.com;lr>
Record-Route: <sip:p1.example.com;lr>
```

to which U2 responds to with a 200 OK. Later, U2 sends the following BYE request to P4 based on the first Route header field value.

```
BYE sip:caller@u1.example.com SIP/2.0
Route: <sip:p4.domain.com;lr>
Route: <sip:p3.middle.com>
Route: <sip:p2.example.com;lr>
Route: <sip:p1.example.com;lr>
```

P4 is not responsible for the resource indicated in the Request-URI so it will leave it alone. It notices that it is the element in the first Route header field value so it removes it. It then prepares to send the request based on the now first Route header field value of sip:p3.middle.com, but it notices that this URI does not contain the lr parameter, so, before sending, it reformats the request to be:

```
BYE sip:p3.middle.com SIP/2.0
Route: <sip:p2.example.com;lr>
```

```
Route: <sip:p1.example.com;lr>
Route: <sip:caller@u1.example.com>
```

P3 is a strict router, so it forwards the following to P2:

```
BYE sip:p2.example.com;lr SIP/2.0
Route: <sip:p1.example.com;lr>
Route: <sip:caller@u1.example.com>
```

P2 sees the request-URI is a value it placed into a Record-Route header field, so, before further processing, it rewrites the request to be:

```
BYE sip:caller@u1.example.com SIP/2.0
Route: <sip:p1.example.com;lr>
```

P2 is not responsible for u1.example.com, so it sends the request to P1 based on the resolution of the Route header field value.

P1 notices itself in the topmost Route header field value, so it removes it, resulting in:

```
BYE sip:caller@u1.example.com SIP/2.0
```

Since P1 is not responsible for u1.example.com and there is no Route header field, P1 will forward the request to u1.example.com based on the Request-URI.

Rewriting Record-Route Header Field Values

In this scenario, U1 and U2 are in different private namespaces and they enter a dialog through a proxy P1, which acts as a gateway between the namespaces.

U1→P1→U2

U1 sends:

```
INVITE sip:callee@gateway.leftprivatespace.com SIP/2.0
Contact: <sip:caller@u1.leftprivatespace.com>
```

P1 uses its location service and sends the following to U2:

```
INVITE sip:callee@rightprivatespace.com SIP/2.0
Contact: <sip:caller@u1.leftprivatespace.com>
Record-Route: <sip:gateway.rightprivatespace.com;lr>
```

U2 sends this 200 (OK) back to P1:

```
SIP/2.0 200 OK
Contact: <sip:callee@u2.rightprivatespace.com>
Record-Route: <sip:gateway.rightprivatespace.com;lr>
```

P1 rewrites its Record-Route header parameter to provide a value that U1 will find useful, and sends the following to U1:

```
SIP/2.0 200 OK
```

```
Contact: <sip:callee@u2.rightprivatespace.com>
Record-Route: <sip:gateway.leftprivatespace.com;lr>
```

Later, U1 sends the following BYE request to P1:

```
BYE sip:callee@u2.rightprivatespace.com SIP/2.0
Route: <sip:gateway.leftprivatespace.com;lr>
```

which P1 forwards to U2 as:

```
BYE sip:callee@u2.rightprivatespace.com SIP/2.0
```

A.12 Transactions

SIP is a transactional protocol: interactions between components take place in a series of independent message exchanges. Specifically, a SIP transaction consists of a single request and any responses to that request, which include zero or more provisional responses and one or more final responses. In the case of a transaction where the request was an INVITE (known as an INVITE transaction), the transaction also includes the ACK only if the final response was not a 2xx response. If the response was a 2xx, the ACK is not considered part of the transaction.

The reason for this separation is rooted in the importance of delivering all 200 (OK) responses to an INVITE to the UAC. To deliver them all to the UAC, the UAS alone takes responsibility for retransmitting them and the UAC alone takes responsibility for acknowledging them with ACK. Since this ACK is retransmitted only by the UAC, it is effectively considered its own transaction.

Transactions have a client side and a server side. The client side is known as a client transaction and the server side as a server transaction. The client transaction sends the request, and the server transaction sends the response. The client and server transactions are logical functions that are embedded in any number of elements. Specifically, they exist within user agents and stateful proxy servers. Consider the example presented at the beginning of this chapter. In this example, the UAC executes the client transaction and its outbound proxy executes the server transaction. The outbound proxy also executes a client transaction, which sends the request to a server transaction in the inbound proxy. That proxy also executes a client transaction, which in turn sends the request to a server transaction in the UAS. This is shown in Figure 3.4.

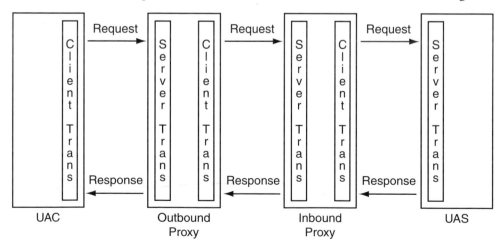

Figure 3.4: Transaction relationships.

A stateless proxy does not contain a client or server transaction. The transaction exists between the UA or stateful proxy on one side, and the UA or stateful proxy on the other side. As far as SIP transactions are concerned, stateless proxies are effectively transparent. The purpose of the client transaction is to receive a request from the element in which the client is embedded (call this element the "Transaction User" or TU; it can be a UA or a stateful proxy), and reliably deliver the request to a server transaction.

The client transaction is also responsible for receiving responses and delivering them to the TU, filtering out any response retransmissions or disallowed responses (such as a response to ACK). Additionally, in the case of an INVITE request, the client transaction is responsible for generating the ACK request for any final response accepting a 2xx response.

Similarly, the purpose of the server transaction is to receive requests from the transport layer and deliver them to the TU. The server transaction filters any request retransmissions from the network. The server transaction accepts responses from the TU and delivers them to the transport layer for transmission over the network. In the case of an INVITE transaction, it absorbs the ACK request for any final response excepting a 2xx response.

The 2xx response and its ACK receive special treatment. This response is retransmitted only by a UAS, and its ACK generated only by the UAC. This end-to-end treatment is needed so that a caller knows the entire set of users that have accepted the call. Because of this special handling, retransmissions of the 2xx response are handled by the UA core, not the transaction layer. Similarly, generation of the ACK for the 2xx is handled by the UA core. Each proxy along the path merely forwards each 2xx response to INVITE and its corresponding ACK.

A.12.1 Client Transaction

The client transaction provides its functionality through the maintenance of a state machine.

The TU communicates with the client transaction through a simple interface. When the TU wishes to initiate a new transaction, it creates a client transaction and passes it the SIP request to send and an IP address, port, and transport to which to send it. The client transaction begins execution of its state machine. Valid responses are passed up to the TU from the client transaction.

There are two types of client transaction state machines, depending on the method of the request passed by the TU. One handles client transactions for INVITE requests. This type of machine is referred to as an IN-VITE client transaction. Another type handles client transactions for all requests except INVITE and ACK. This is referred to as a non-INVITE client transaction. There is no client transaction for ACK. If the TU wishes to send an ACK, it passes one directly to the transport layer for transmission.

The INVITE transaction is different from those of other methods because of its extended duration. Normally, human input is required in order to respond to an INVITE. The long delays expected for sending a response argue for a three-way handshake. On the other hand, requests of other methods are expected to complete rapidly. Because of the non-INVITE transaction's reliance on a two-way handshake, TUs should respond immediately to non-INVITE requests.

See RFC 3261 for details.

A.12.2 Server Transaction

The server transaction is responsible for the delivery of requests to the TU and the reliable transmission of responses. It accomplishes this through a state machine. Server transactions are created by the core when a request is received, and transaction handling is desired for that request (this is not always the case). As with the client transactions, the state machine depends on whether the received request is an INVITE request.

See RFC 3261 for details.

A.13 Transport

The transport layer is responsible for the actual transmission of requests and responses over network transports. This includes determination of the connection to use for a request or response in the case of connection-oriented transports.

The transport layer is responsible for managing persistent connections for transport protocols like TCP and SCTP, or TLS over those, including ones opened to the transport layer. This includes connections opened by the client or server transports, so that connections are shared between client and server transport functions. These connections are indexed by the tuple formed from the address, port, and transport protocol at the far end of the connection. When a connection is opened by the transport layer, this index is set to the destination IP, port, and transport. When the connection is accepted by the transport layer, this index is set to the source IP address, port number, and transport. Note that, because the source port is often ephemeral, but it cannot be known whether it is ephemeral or selected through procedures described in RFC 3261, connections accepted by the transport layer will frequently not be reused. The result is that two proxies in a "peering" relationship using a connection-oriented transport frequently will have two connections in use—one for transactions initiated in each direction.

It is recommended that connections be kept open for some implementation-defined duration after the last message was sent or received over that connection. This duration should at least equal the longest amount of time the element would need in order to bring a transaction from instantiation to the terminated state. This is to make it likely that transactions are completed over the same connection on which they are initiated (for example, request, response, and, in the case of INVITE, ACK for non-2xx responses). This usually means at least 64*T1. However, it could be larger, for example, in an element that has a TU using a large value for timer C.

All SIP elements must implement UDP and TCP. SIP elements may implement other protocols. Making TCP mandatory for the UA is a substantial change from RFC 2543. It has arisen out of the need to handle larger messages, which must use TCP, as discussed below. Thus, even if an element never sends large messages, it may receive one and needs to be able to handle them.

A.13.1 Clients

A.13.1.1 Sending Requests

The client side of the transport layer is responsible for sending the request and receiving responses. The user of the transport layer passes the client transport the request, an IP address, port, transport, and possibly TTL for multicast destinations.

If a request is within 200 bytes of the path MTU, or if it is larger than 1300 bytes and the path MTU is unknown, the request must be sent using an RFC 2914 congestion-controlled transport protocol, such as TCP. If this causes a change in the transport protocol from the one indicated in the top Via, the value in the top Via must be changed. This prevents fragmentation of messages over UDP and provides congestion control for larger messages. However, implementations must be able to handle messages up to the maximum datagram packet size. For UDP, this size is 65,535 bytes, including IP and UDP headers.

The 200 byte "buffer" between the message size and the MTU accommodates the fact that the response in SIP can be larger than the request. This happens due to the addition of Record-Route header field values to the responses to INVITE, for example. With the extra buffer, the response can be about 170 bytes larger than the request, and still not be fragmented on IPv4 (about 30 bytes is consumed by IP/UDP, assuming no IPSec). 1300 is chosen when path MTU is not known, based on the assumption of a 1500 byte Ethernet MTU.

If an element sends a request over TCP because of these message size constraints, and that request would have otherwise been sent over UDP, if the attempt to establish the connection generates either an ICMP Protocol Not Supported, or results in a TCP reset, the element should retry the request, using UDP. This is only to provide backwards compatibility with RFC 2543 compliant implementations that do not support TCP. It is anticipated that this behavior will be deprecated in a future revision of this specification.

A client that sends a request to a multicast address must add the "maddr" parameter to its Via header field value containing the destination multicast address, and for IPv4, should add the "ttl" parameter with a value of 1. Usage of IPv6 multicast is not defined in this specification, and will be a subject of future standardization when the need arises.

These rules result in a purposeful limitation of multicast in SIP. Its primary function is to provide a "single-hop-discovery-like" service, delivering a request to a group of homogeneous servers, where it is only required to process the response from any one of them. This functionality is most useful for registrations. In fact, based on the transaction processing rules in RFC 3261, the client transaction will accept the first response, and view any others as retransmissions because they all contain the same Via branch identifier.

Before a request is sent, the client transport must insert a value of the "sent-by" field into the Via header field. This field contains an IP address or host name, and port. The usage of an FQDN is recommended. This field is used for sending responses under certain conditions, described below. If the port is absent, the default value depends on the transport. It is 5060 for UDP, TCP and SCTP, 5061 for TLS.

For reliable transports, the response is normally sent on the connection on which the request was received. Therefore, the client transport must be prepared to receive the response on the same connection used to send the request. Under error conditions, the server may attempt to open a new connection to send the response. To handle this case, the transport layer must also be prepared to receive an incoming connection on the source IP address from which the request was sent and port number in the "sent-by" field. It also must be prepared to receive incoming connections on any address and port that would be selected by a server based on the procedures described in RFC 3261.

For unreliable unicast transports, the client transport must be prepared to receive responses on the source IP address from which the request is sent (as responses are sent back to the source address) and the port number in the "sent-by" field. Furthermore, as with reliable transports, in certain cases the response will be sent elsewhere. The client must be prepared to receive responses on any address and port that would be selected by a server based on the procedures described in RFC 3261.

For multicast, the client transport must be prepared to receive responses on the same multicast group and port to which the request is sent (that is, it needs to be a member of the multicast group it sent the request to.)

If a request is destined for an IP address, port, and transport to which an existing connection is open, it is recommended that this connection be used to send the request, but another connection may be opened and used. If a request is sent using multicast, it is sent to the group address, port, and TTL provided by the transport user. If a request is sent using unicast unreliable transports, it is sent to the IP address and port provided by the transport user.

A.13.1.2 Receiving Responses

When a response is received, the client transport examines the top Via header field value. If the value of the "sent-by" parameter in that header field value does not correspond to a value that the client transport is configured to insert into requests, the response must be silently discarded.

If there are any client transactions in existence, the client transport uses the matching procedures of Section A.12.1.3 to attempt to match the response to an existing transaction. If there is a match, the response must be passed to that transaction. Otherwise, the response must be passed to the core (whether it be stateless proxy, stateful proxy, or UA) for further processing. Handling of these "stray" responses is dependent on the core (a proxy will forward them, while a UA will discard, for example).

A.13.2 Servers

A.13.2.1 Receiving Requests

A server should be prepared to receive requests on any IP address, port, and transport combination that can be the result of a DNS lookup on a SIP or SIPS URI that is handed out for the purposes of communicating with that server. In this context, "handing out" includes placing a URI in a Contact header field in a REGISTER request or a redirect response, or in a Record-Route header field in a request or response. A URI can also be "handed out" by placing it on a web page or business card. It is also recommended that a server listen for requests on the default SIP ports (5060 for TCP and UDP, 5061 for TLS over TCP) on all public interfaces. The typical exception would be private networks, or when multiple server instances are running on the same host. For any port and interface that a server listens on for UDP, it must listen on that same port and interface for TCP. This is because a message may need to be sent using TCP, rather than UDP, if it is too large. As a result, the converse is not true. A server need not listen for UDP on a particular address and port just because it is listening on that same address and port for TCP. There may, of course, be other reasons why a server needs to listen for UDP on a particular address and port.

When the server transport receives a request over any transport, it must examine the value of the "sent-by" parameter in the top Via header field value. If the host portion of the "sent-by" parameter contains a domain name, or if it contains an IP address that differs from the packet source address, the server must add a "received" parameter to that Via header field value. This parameter must contain the source address from which the packet was received. This is to assist the server transport layer in sending the response, since it must be sent to the source IP address from which the request came.

Consider a request received by the server transport which looks like, in part:

```
INVITE sip:bob@Biloxi.com SIP/2.0  Via: SIP/2.0/UDP bobspc.biloxi.com:5060
```

The request is received with a source IP address of 192.0.2.4. Before passing the request up, the transport adds a "received" parameter, so that the request would look like, in part:

```
INVITE sip:bob@Biloxi.com SIP/2.0  Via: SIP/2.0/UDP bobspc.biloxi.com:5060;
received=192.0.2.4
```

Next, the server transport attempts to match the request to a server transaction. It does so using the matching rules described in Section A.12.2.3. If a matching server transaction is found, the request is passed to that transaction for processing. If no match is found, the request is passed to the core, which may decide to construct a new server transaction for that request. Note that when a UAS core sends a 2xx response to INVITE, the server transaction is destroyed. This means that when the ACK arrives, there will be no matching server transaction, and based on this rule, the ACK is passed to the UAS core, where it is processed.

A.13.2.2 Sending Responses

The server transport uses the value of the top Via header field in order to determine where to send a response. It must follow the following process:

- If the "sent-protocol" is a reliable transport protocol such as TCP or SCTP, or TLS over those, the response must be sent using the existing connection to the source of the original request that created the transaction, provided that connection is still open. This requires the server transport to maintain an association between server transactions and transport connections. If that connection is no longer open, the server should open a connection to the IP address in the "received" parameter, if present, using the port in the "sent-by" value, or the default port for that transport, if no port is specified. If that connection attempt fails, the server should use the procedures in [4] for servers in order to determine the IP address and port to open the connection and send the response to.

- Otherwise, if the Via header field value contains a "maddr" parameter, the response must be forwarded to the address listed there, using the port indicated in "sent-by," or port 5060 if none is present. If the address is a multicast address, the response should be sent using the TTL indicated in the "ttl" parameter, or with a TTL of 1 if that parameter is not present.

- Otherwise (for unreliable unicast transports), if the top Via has a "received" parameter, the response must be sent to the address in the "received" parameter, using the port indicated in the "sent-by" value, or using port 5060 if none is specified explicitly. If this fails—for example, elicits an ICMP "port unreachable" response—the procedures described in RFC 3261 should be used to determine where to send the response.

- Otherwise, if it is not receiver-tagged, the response must be sent to the address indicated by the "sent-by" value using the procedures described in RFC 3261.

A.13.3 Framing

In the case of message-oriented transports (such as UDP), if the message has a Content-Length header field, the message body is assumed to contain that many bytes. If there are additional bytes in the transport packet beyond the end of the body, they must be discarded. When the transport packet ends before the end of the message body, this is considered an error. If the message is a response, it must be discarded. If the message is a request, the element should generate a 400 (Bad Request) response. When the message has no Content-Length header field, the message body is assumed to end at the end of the transport packet. In the case of stream-oriented transports such as TCP, the Content-Length header field indicates the size of the body. The Content-Length header field must be used with stream oriented transports.

A.13.4 Error Handling

Error handling is independent of whether the message was a request or response. If the transport user asks for a message to be sent over an unreliable transport, and the result is an ICMP error, the behavior depends on the type of ICMP error. Host, network, port or protocol unreachable errors, or parameter problem errors should cause the transport layer to inform the transport user of a failure in sending. Source quench and TTL exceeded ICMP errors should be ignored. If the transport user asks for a request to be sent over a reliable transport, and the result is a connection failure, the transport layer should inform the transport user of a failure in sending.

A.14 Additional Details

Refer to RFC 3261 for a multitude of additional details.

Basic "Presence" Concepts

4.1 Introduction

This chapter describes, in a general manner, presence concepts and services. Presence and Instant Messaging have recently emerged as a new medium of communications over enterprise networks, extranets, and the Internet. Presence is a means for finding, retrieving, and subscribing to changes in the presence information (e.g., "online" or "offline") of other users, typically in the context of messaging (e-mail), Unified Messaging, and/or VoIP. It can be defined as the willingness and ability of a user to communicate with other users on the network [ROS200401].

Historically, presence has been limited to "on-line" and "off-line" indicators. Instant messaging is a means for sending small, simple messages that are delivered immediately to online users [AGG200001]. The notion of presence is now taking on a broader definition/role. VoIP can be used in support of, or in conjunction with, presence: it is expected that the sophisticated voice environments made possible by VoIP, including Unified Messaging, will make have use of presence capabilities for one-to-one conversations, roaming applications, and voice/video-conferencing. Specifically, the anticipation is that presence concepts and services will play a key role in 3G VoIP networks based on IPv6 as described in this text; hence, the coverage we allocate to this topic.

While the protocol linkage between presence and IPv6 is a pragmatic one at this juncture, what makes the two have a symbiotic affinity is that converged VoIP, multimedia (music, video, radio, TV streaming), Ubiquitous Computing, and presence are expected to be key services of the near-term future (next 3–6 years) that will be delivered over IPv6 networks (services based on IPv4 are already available but the scalability and end-to-end robustness and connection reliability are impacted by the current infrastructure). Presence, particularly as supported by SIP, is described in a series of IETF RFCs, as follows in Table 4.1:

Table 4.1: Series of IETF RFCs on "Presence."

RFC 3953	Telephone Number Mapping (ENUM) Service Registration for Presence Services, *J. Peterson (January 2005)*
RFC 3923	End-to-End Signing and Object Encryption for the Extensible Messaging and Presence Protocol (XMPP), *P. Saint-Andre (October 2004)*
RFC 3922	Mapping the Extensible Messaging and Presence Protocol (XMPP) to Common Presence and Instant Messaging (CPIM), *P. Saint-Andre (October 2004)*
RFC 3921	Extensible Messaging and Presence Protocol (XMPP): Instant Messaging and Presence, *P. Saint-Andre, Ed. (October 2004)*
RFC 3920	Extensible Messaging and Presence Protocol (XMPP): Core, *P. Saint-Andre, Ed. (October 2004)*
RFC 3863	Presence Information Data Format (PIDF), *H. Sugano, S. Fujimoto, G. Klyne, A. Bateman, W. Carr, J. Peterson (August 2004)*
RFC 3862	Common Presence and Instant Messaging (CPIM): Message Format, *G. Klyne, D. Atkins (August 2004)*

RFC 3861	Address Resolution for Instant Messaging and Presence. *J. Peterson (August 2004)*
RFC 3860	Common Profile for Instant Messaging (CPIM), *J. Peterson (August 2004)*
RFC 3859	Common Profile for Presence (CPP), *J. Peterson (August 2004)*
RFC 3856	A Presence Event Package for the Session Initiation Protocol (SIP), *J. Rosenberg (August 2004)*
RFC 3343	The Application Exchange (APEX) Presence Service, *M. Rose, G. Klyne, D. Crocker (April 2003)*
RFC 2779	Instant Messaging / Presence Protocol Requirements, *M. Day, S. Aggarwal, G. Mohr, J. Vincent (February 2000)*
RFC 2778	A Model for Presence and Instant Messaging, *M. Day, J. Rosenberg, H. Sugano (February 2000)*

The treatment of this topic in this chapter is based on IETF RFC 2778, RFC 2779, and RFC 3856 [ROS200001], [ROS200401], [AGG200001]. This discussion is strictly for pedagogical purposes. All normative and/ or development work should make direct and explicit reference to the latest IETF/RFC documentation.

4.2 Abstract Model for a Presence and Instant Messaging

To begin with there is a need for an abstract model for a presence and instant messaging system. This section defines such a model. The section defines the various entities involved, defines terminology, and outlines the services provided by the system. The goal is to arrive at a common vocabulary for work on requirements for protocols and markup for presence and instant messaging; the purpose of the model is to provide a common baseline for defining and implementing interoperable presence and instant messaging protocols. This treatment is based on IETF RFC 2778 [ROS200001].

4.2.1 Introduction

A presence and instant messaging system allows users to subscribe to each other and be notified of changes in state, and for users to send each other short instant messages. To facilitate the development of a suite of protocols providing this service, it is valuable to first develop a model for the system. The model consists of the various partaking entities, descriptions of the basic functions they provide, and definition of nomenclature that can be used to facilitate discussion.

The purpose of this model is to be descriptive and universal: one wants the model to map reasonably onto all of the systems that are informally described as presence or instant messaging systems. The mode, however, is not intended to be prescriptive or achieve interoperability: an element that appears in the model will not necessarily be an element of an interoperable protocol.

In RFC 2778, each element of the model appears in upper case (e.g., PRESENCE SERVICE). (no term in lower case or mixed case is intended to be a term of the model.) The first part of RFC 2278 is intended as an overview of the model; terms are presented in an order intended to help the reader understand the relationship between elements. The second part of the RFC is the actual definition of the model, with terms defined for ease of reference. The overview is intended to be helpful but is not definitive.

4.2.2 Overview

The model is intended to provide a means for understanding, comparing, and describing systems that support the services typically referred to as presence and instant messaging. It consists of a number of named entities that appear, in some form, in existing systems. No actual implementation is likely to have every entity of the model as a distinct part. Instead, there will almost always be parts of the implementation that embody two or more entities of the model. However, different implementations may combine entities in different ways.

The model defines two services: a PRESENCE SERVICE and an INSTANT MESSAGE SERVICE. The PRESENCE SERVICE serves to accept information, store it, and distribute it. The information stored is (unsurprisingly) PRESENCE INFORMATION. The INSTANT MESSAGE SERVICE serves to accept and deliver INSTANT MESSAGES to INSTANT INBOXES.

4.2.2.1 PRESENCE SERVICE

The PRESENCE SERVICE (see Figure 4.1) has two distinct sets of "clients" (remember, these may be combined in an implementation, but treated separately in the model). One set of clients, called *PRESENTITIES*, provides PRESENCE INFORMATION to be stored and distributed. The other set of clients, called *WATCHERS*, receives PRESENCE INFORMATION from the service.

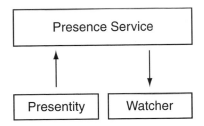

Figure 4.1: Overview of PRESENCE SERVICE.

There are two kinds of WATCHERS, called *FETCHERS* and *SUBSCRIBERS* (see Figure 4.2). A FETCHER simply requests the current value of some PRESENTITY's PRESENCE INFORMATION from the PRESENCE SERVICE. In contrast, a SUBSCRIBER requests notification from the PRESENCE SERVICE of (future) changes in some PRESENTITY's PRESENCE INFORMATION. A special kind of FETCHER is one that fetches information on a regular basis; this is called a *POLLER*.

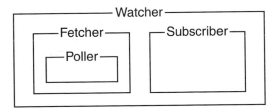

Figure 4.2: Varieties of WATCHER.

The PRESENCE SERVICE also has WATCHER INFORMATION about WATCHERS and their activities in terms of fetching or subscribing to PRESENCE INFORMATION. The PRESENCE SERVICE may also distribute WATCHER INFORMATION to some WATCHERS using the same mechanisms that are available for distributing PRESENCE INFORMATION.

Changes to PRESENCE INFORMATION are distributed to SUBSCRIBERS via NOTIFICATIONS. Figures 4.3a through 4.3c show the flow of information as a piece of PRESENCE INFORMATION is changed from P1 to P2.

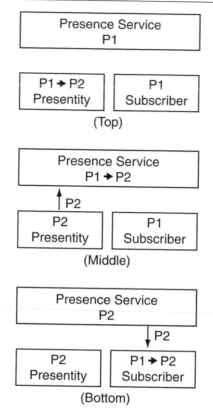

Figure 4.3: NOTIFICATION.
(Step 1: Top) (Step 2: Middle) (Step 3: Bottom)

4.2.2.2 INSTANT MESSAGE SERVICE

The INSTANT MESSAGE SERVICE (see Figure 4.4) also has two distinct sets of "clients": SENDERS and INSTANT INBOXES. A SENDER provides INSTANT MESSAGES to the INSTANT MESSAGE SERVICE for delivery. Each INSTANT MESSAGE is addressed to a particular INSTANT INBOX ADDRESS, and the INSTANT MESSAGE SERVICE attempts to deliver the message to a corresponding INSTANT INBOX.

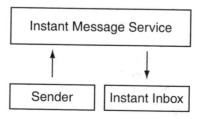

Figure 4.4: Overview of INSTANT MESSAGE SERVICE.

4.2.2.3 Protocols

A PRESENCE PROTOCOL defines the interaction between PRESENCE SERVICE, PRESENTITIES, and WATCHERS. PRESENCE INFORMATION is carried by the PRESENCE PROTOCOL.

An INSTANT MESSAGE PROTOCOL defines the interaction between INSTANT MESSAGE SERVICE, SENDERS, and INSTANT INBOXES. INSTANT MESSAGES are carried by the INSTANT MESSAGE PROTOCOL.

In terms of this model, there is a desire to develop detailed requirements and specifications for the structure and formats of the PRESENCE PROTOCOL, PRESENCE INFORMATION, INSTANT MESSAGE PROTOCOL, and INSTANT MESSAGES.

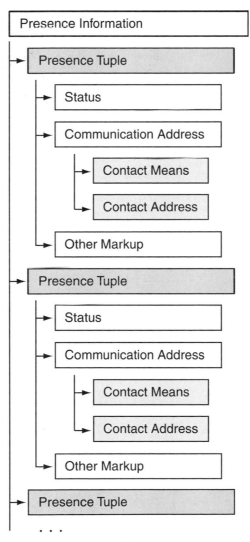

Figure 4.5: The structure of PRESENCE INFORMATION.

4.2.2.4 Formats

The model defines the PRESENCE INFORMATION (also see Figure 4.5) to consist of an arbitrary number of elements, called *PRESENCE TUPLES*. Each such element consists of a STATUS marker (which might convey information such as online/offline/busy/away/do not disturb), an optional COMMUNICATION ADDRESS, and optional OTHER PRESENCE MARKUP. A COMMUNICATION ADDRESS includes a COMMUNICATION MEANS and a CONTACT ADDRESS. One type of COMMUNICATION MEANS, and the only one defined by this model, is INSTANT MESSAGE SERVICE. One type of CONTACT ADDRESS, and the only one defined by this model, is INSTANT INBOX ADDRESS. However, other possibilities exist: a COMMUNICATION MEANS might indicate some form of telephony, for example, with the corresponding CONTACT ADDRESS containing a telephone number.

STATUS is further defined by the model to have at least two states that interact with INSTANT MESSAGE delivery—OPEN, in which INSTANT MESSAGES will be accepted, and CLOSED, in which INSTANT MESSAGES will not be accepted. OPEN and CLOSED may also be applicable to other COMMUNICA-TION MEANS—OPEN mapping to some state meaning "available" or "open for business" while CLOSED means "unavailable" or "closed to business." The model allows STATUS to include other values, which may be interpretable by programs or only by persons. The model also allows STATUS to consist of single or multiple values.

4.2.2.5 Presence and Its Effect on INSTANT MESSAGES

An INSTANT INBOX is a receptacle for INSTANT MESSAGES. Its INSTANT INBOX ADDRESS is the information that can be included in PRESENCE INFORMATION to define how an INSTANT MES-SAGE should be delivered to that INSTANT INBOX. As noted above, certain values of the STATUS marker indicate whether INSTANT MESSAGES will be accepted at the INSTANT INBOX. The model does not otherwise constrain the delivery mechanism or format for instant messages. Reasonable people can disagree about whether this omission is a strength or a weakness of this model.

4.2.2.6 PRINCIPALS and Their Agents

This model includes other elements that are useful in characterizing how the protocol and markup work. PRINCIPALS are the people, groups, and/or software in the "real world" outside the system that use the system as a means of coordination and communication. It is entirely outside the model how the real world maps onto PRINCIPALS—the system of model entities knows only that two distinct PRINCIPALS are distinct, and two identical PRINCIPALS are identical.

A PRINCIPAL interacts with the system via one of several user agents (INBOX USER AGENT; SENDER USER AGENT; PRESENCE USER AGENT; WATCHER USER AGENT). As usual, the different kinds of user agents are split apart in this model even though most implementations will combine at least some of them. A user agent is purely coupling between a PRINCIPAL and some core entity of the system (respectively, INSTANT INBOX; SENDER; PRESENTITY; WATCHER); (see Figures 4.6 and 4.7.)

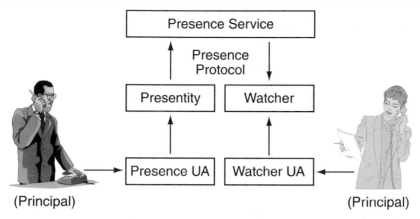

Figure 4.6: A PRESENCE SYSTEM.

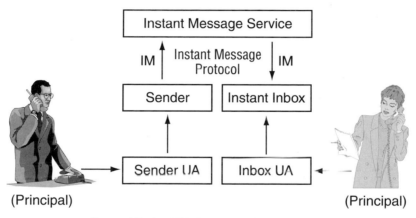

Figure 4.7: An INSTANT MESSAGING SYSTEM.

4.2.2.7 Examples

A simple example of the model is a generic "buddy list" application. These applications typically expose the user's presence to others, and make it possible to see the presence of others. So we could describe a buddy list as the combination of a PRESENCE USER AGENT and WATCHER USER AGENT for a single PRINCIPAL, using a single PRESENTITY and a single SUBSCRIBER.

One could then extend the example to instant messaging and describe a generic "instant messenger" as essentially a buddy list with additional capabilities for sending and receiving instant messages. Hence, an instant messenger would be the combination of a PRESENCE USER AGENT, WATCHER USER AGENT, INBOX USER AGENT, and SENDER USER AGENT for a single PRINCIPAL, using a single PRESENTITY, single SUBSCRIBER, and single INSTANT INBOX, with the PRESENTITY's PRESENCE INFORMATION including an INSTANT INBOX ADDRESS that leads to the INSTANT INBOX.

4.2.3 Model

As noted, the model entails establishing basic nomenclature. Such nomenclature follows:

ACCESS RULES: Constraints on how a PRESENCE SERVICE makes PRESENCE INFORMATION available to WATCHERS. For each PRESENTITY's PRESENCE INFORMATION, the applicable ACCESS RULES are manipulated by the PRESENCE USER AGENT of a PRINCIPAL that controls the PRESENTITY. Motivation: One needs some way of talking about hiding presence information from people.

CLOSED: A distinguished value of the STATUS marker. In the context of INSTANT MESSAGES, this value means that the associated INSTANT INBOX ADDRESS, if any, corresponds to an INSTANT INBOX that is unable to accept an INSTANT MESSAGE. This value may have an analogous meaning for other COMMUNICATION MEANS, but any such meaning is not defined by this model. Contrast with OPEN.

COMMUNICATION ADDRESS: Consists of COMMUNICATION MEANS and CONTACT ADDRESS.

COMMUNICATION MEANS: Indicates a method whereby communication can take place. INSTANT MESSAGE SERVICE is one example of a COMMUNICATION MEANS.

CONTACT ADDRESS: A specific point of contact via some COMMUNICATION MEANS. When using an INSTANT MESSAGE SERVICE, the CONTACT ADDRESS is an INSTANT INBOX ADDRESS.

DELIVERY RULES: Constraints on how an INSTANT MESSAGE SERVICE delivers received INSTANT MESSAGES to INSTANT INBOXES. For each INSTANT INBOX, the applicable DELIVERY RULES are manipulated by the INBOX USER AGENT of a PRINCIPAL that controls the INSTANT INBOX. Motivation: One needs a way of talking about filtering instant messages.

FETCHER: A form of WATCHER that has asked the PRESENCE SERVICE to for the PRESENCE INFORMATION of one or more PRESENTITIES, but has not asked for a SUBSCRIPTION to be created.

INBOX USER AGENT: Means for a PRINCIPAL to manipulate zero or more INSTANT INBOXES controlled by that PRINCIPAL. Motivation: This is intended to isolate the core functionality of an INSTANT INBOX from how it might appear to be manipulated by a product. This manipulation includes fetching messages, deleting messages, and setting DELIVERY RULES. The INBOX USER AGENT, INSTANT INBOX, and INSTANT MESSAGE SERVICE can be colocated can be or distributed across machines.

INSTANT INBOX: Receptacle for INSTANT MESSAGES intended to be read by the INSTANT INBOX's PRINCIPAL.

INSTANT INBOX ADDRESS: Indicates whether and how the PRESENTITY's PRINCIPAL can receive an INSTANT MESSAGE in an INSTANT INBOX. The STATUS and INSTANT INBOX ADDRESS information are sufficient to determine whether the PRINCIPAL appears ready to accept the INSTANT MESSAGE. Motivation: The definition is pretty loose about exactly how any of this works, even leaving open the possibility of reusing parts of the email infrastructure for instant messaging.

INSTANT MESSAGE: An identifiable unit of data, of small size, to be sent to an INSTANT INBOX (this definition seeks to avoid the possibility of transporting an arbitrary-length stream labelled as an "instant message.")

INSTANT MESSAGE PROTOCOL: The messages that can be exchanged between a SENDER USER AGENT and an INSTANT MESSAGE SERVICE, or between an INSTANT MESSAGE SERVICE and an INSTANT INBOX.

INSTANT MESSAGE SERVICE: Accepts and delivers INSTANT MESSAGES.
- May require authentication of SENDER USER AGENTS and/or INSTANT INBOXES.
- May have different authentication requirements for different INSTANT INBOXES, and may also have different authentication requirements for different INSTANT INBOXES controlled by a single PRINCIPAL.

- May have an internal structure involving multiple SERVERS and/or PROXIES. There may be complex patterns of redirection and/or proxying while retaining logical connectivity to a single INSTANT MESSAGE SERVICE. Note that an INSTANT MESSAGE SERVICE does not require having a distinct SERVER—the service may be implemented as direct communication between SENDER and INSTANT INBOX.
- May have an internal structure involving other INSTANT MESSAGE SERVICES, which may be independently accessible in their own right as well as being reachable through the initial INSTANT MESSAGE SERVICE.

NOTIFICATION: A message sent from the PRESENCE SERVICE to a SUBSCRIBER when there is a change in the PRESENCE INFORMATION of some PRESENTITY of interest, as recorded in one or more SUBSCRIPTIONS.

OPEN: A distinguished value of the STATUS marker. In the context of INSTANT MESSAGES, this value means that the associated INSTANT INBOX ADDRESS, if any, corresponds to an INSTANT INBOX that is ready to accept an INSTANT MESSAGE. This value may have an analogous meaning for other COMMUNICATION MEANS, but any such meaning is not defined by this model. Contrast with CLOSED.

OTHER PRESENCE MARKUP: Any additional information included in the PRESENCE INFORMATION of a PRESENTITY. The model does not define this further.

POLLER: A FETCHER that requests PRESENCE INFORMATION on a regular basis.

PRESENCE INFORMATION: Consists of one or more PRESENCE TUPLES.

PRESENCE PROTOCOL: The messages that can be exchanged between a PRESENTITY and a PRESENCE SERVICE, or a WATCHER and a PRESENCE SERVICE.

PRESENCE SERVICE: Accepts, stores, and distributes PRESENCE INFORMATION.
- May require authentication of PRESENTITIES, and/or WATCHERS.
- May have different authentication requirements for different PRESENTITIES.
- May have different authentication requirements for different WATCHERS, and may also have different authentication requirements for different PRESENTITIES being watched by a single WATCHER.
- May have an internal structure involving multiple SERVERS and/or PROXIES. There may be complex patterns of redirection and/or proxying while retaining logical connectivity to a single PRESENCE SERVICE. Note that a PRESENCE SERVICE does not require having a distinct SERVER—the service may be implemented as direct communication among PRESENTITY and WATCHERS.
- May have an internal structure involving other PRESENCE SERVICES, which may be independently accessible in their own right as well as being reachable through the initial PRESENCE SERVICE.

PRESENCE TUPLE: Consists of a STATUS, an optional COMMUNICATION ADDRESS, and optional OTHER PRESENCE MARKUP.

PRESENCE USER AGENT: Means for a PRINCIPAL to manipulate zero or more PRESENTITIES. Motivation: This is essentially a "model/view" distinction. The PRESENTITY is the model of the presence being exposed and is independent of its manifestation in any user interface. The PRESENCE USER AGENT, PRESENTITY, and PRESENCE SERVICE can be colocated or can be distributed across machines.

PRESENTITY (presence entity): Provides PRESENCE INFORMATION to a PRESENCE SERVICE. Note that the presentity is not (usually) located in the presence service: The presence service only has a recent version of the presentity's presence information. The presentity initiates changes in the presence information to be distributed by the presence service.

PRINCIPAL: Human, program, or collection of humans and/or programs that chooses to appear to the PRESENCE SERVICE as a single actor, distinct from all other PRINCIPALS. Motivation: One needs a clear notion of the actors outside the system. "Principal" seems as good a term as any.

PROXY: A SERVER that communicates PRESENCE INFORMATION, INSTANT MESSAGES, SUBSCRIPTIONS and/or NOTIFICATIONS to another SERVER. Sometimes a PROXY acts on behalf of a PRESENTITY, WATCHER, or INSTANT INBOX.

SENDER: Source of INSTANT MESSAGES to be delivered by the INSTANT MESSAGE SERVICE.

SENDER USER AGENT: Means for a PRINCIPAL to manipulate zero or more SENDERS.

SERVER: An indivisible unit of a PRESENCE SERVICE or INSTANT MESSAGE SERVICE.

SPAM: Unwanted INSTANT MESSAGES.

SPOOFING: A PRINCIPAL improperly imitating another PRINCIPAL.

STALKING: Using PRESENCE INFORMATION to infer the whereabouts of a PRINCIPAL, especially for malicious or illegal purposes.

STATUS: A distinguished part of the PRESENCE INFORMATION of a PRESENTITY. STATUS has at least the mutually-exclusive values OPEN and CLOSED, which have meaning for the acceptance of INSTANT MESSAGES, and may have meaning for other COMMUNICATION MEANS. There may be other values of STATUS that do not imply anything about INSTANT MESSAGE acceptance. These other values of STATUS may be combined with OPEN and CLOSED or they may be mutually exclusive with those values.

Some implementations may combine STATUS with other entities. For example, an implementation might make an INSTANT INBOX ADDRESS visible only when the INSTANT INBOX can accept an INSTANT MESSAGE. Then, the existence of an INSTANT INBOX ADDRESS implies OPEN, while its absence implies CLOSED.

SUBSCRIBER: A form of WATCHER that has asked the PRESENCE SERVICE to notify it immediately of changes in the PRESENCE INFORMATION of one or more PRESENTITIES.

SUBSCRIPTION: The information kept by the PRESENCE SERVICE about a SUBSCRIBER's request to be notified of changes in the PRESENCE INFORMATION of one or more PRESENTITIES.

VISIBILITY RULES: Constraints on how a PRESENCE SERVICE makes WATCHER INFORMATION available to WATCHERS. For each WATCHER's WATCHER INFORMATION, the applicable VISIBILITY RULES are manipulated by the WATCHER USER AGENT of a PRINCIPAL that controls the WATCHER.

WATCHER: Requests PRESENCE INFORMATION about a PRESENTITY, or WATCHER INFORMATION about a WATCHER, from the PRESENCE SERVICE. Special types of WATCHER are FETCHER, POLLER, and SUBSCRIBER.

WATCHER INFORMATION: Information about WATCHERS that have received PRESENCE INFORMATION about a particular PRESENTITY within a particular recent span of time. WATCHER INFORMATION is maintained by the PRESENCE SERVICE, which may choose to present it in the same form as PRESENCE INFORMATION; that is, the service may choose to make WATCHERS look like a special form of PRESENTITY. Motivation: If a PRESENTITY wants to know who knows about it, it is not enough to examine only information about SUBSCRIPTIONS. A WATCHER might repeatedly fetch information without ever subscribing. Alternately, a WATCHER might repeatedly subscribe, then cancel the SUBSCRIPTION. Such WATCHERS should be visible to the PRESENTITY if the PRESENCE SERVICE offers WATCHER INFORMATION, but will not be appropriately visible if the WATCHER INFORMATION includes only SUBSCRIPTIONS.

WATCHER USER AGENT: Means for a PRINCIPAL to manipulate zero or more WATCHERS controlled by that PRINCIPAL. Motivation: As with PRESENCE USER AGENT and PRESENTITY, the distinction here is intended to isolate the core functionality of a WATCHER from how it might appear to be manipulated by a product. WATCHER USER AGENT, WATCHER, and PRESENCE SERVICE can be colocatcd or can be distributed across machines.

4.3 Instant Messaging/Presence Protocol Requirements

As noted at the beginning of this chapter, presence is a means for finding, retrieving, and subscribing to changes in the presence information (e.g., "online" or "offline") of other users. Instant messaging is a means for sending small, simple messages that are delivered immediately to online users. VoIP applications are envisioned. Unfortunately, up-to-now, applications of presence and instant messaging currently use independent, nonstandard and noninteroperable protocols developed by various vendors [AGG200001]. The goal of the Instant Messaging and Presence Protocol (IMPP) is to define a standard protocol so that independently developed applications of instant messaging and/or presencc can interoperate across the Internet. RFC 2779 defines a minimal set of requirements that IMPP must meet. This treatment is based on RFC 2779 [AGG200001].

4.3.1 Machinery

The terms are defined above (from RFC 2778) are used in this discussion. Additionally, the following terms are used:

ADMINISTRATOR: A PRINCIPAL with authority over local computer and network resources, who manages local DOMAINS or FIREWALLS. For security and other purposes, an ADMINISTRA-TOR often needs or wants to impose restrictions on network usage based on traffic type, content, volume, or endpoints. A PRINCIPAL's ADMINISTRATOR has authority over some or all of that PRINCIPAL's computer and network resources.

DOMAIN: A portion of a NAMESPACE.

ENTITY: Any of PRESENTITY, SUBSCRIBER, FETCHER, POLLER, or WATCHER (all defined above as per RFC 2778).

FIREWALL: A point of administrative control over connectivity. Depending on the policies being enforced, parties may need to take unusual measures to establish communications through the FIREWALL.

IDENTIFIER: A means of indicating a point of contact, intended for public use such as on a business card. Telephone numbers, email addresses, and typical home page URLs are all examples of IDENTIFIERS in other systems. Numeric IP addresses like 10.0.0.26 are not, and ncither are URLs containing numerous CGI parameters or long arbitrary identifiers.

INTENDED RECIPIENT: The PRINCIPAL to whom the sender of an INSTANT MESSAGE is sending it.

NAMESPACE: The system that maps from a name of an ENTITY to the concrete implementation of that ENTITY. A NAMESPACE may be composed of a number of distinct DOMAINS.

OUT OF CONTACT: A situation in which some ENTITY and the PRESENCE SERVICE cannot communicate.

SUCCESSFUL DELIVERY: A situation in which an INSTANT MESSAGE was transmitted to an INSTANT INBOX for the INTENDED RECIPIENT, and the INSTANT INBOX acknowledged its receipt. SUCCESSFUL DELIVERY usually also implies that an INBOX USER AGENT has handled the message in a way chosen by the PRINCIPAL. However, SUCCESSFUL DELIVERY does not imply that the message was actually seen by that PRINCIPAL.

4.3.2 Shared Requirements

This section describes requirements that are common to both an PRESENCE SERVICE and an INSTANT MESSAGE SERVICE. Section 4.3.3 describes requirements specific to a PRESENCE SERVICE, while Section 4.3.4 describes requirements specific to an INSTANT MESSAGE SERVICE (sccurity requirements are not covered herewith—refer to the RFC for details.)

It is expected that Presence and Instant Messaging services will be particularly valuable to users over mobile IP wireless access devices. Indeed the number of devices connected to the Internet via wireless means is expected to grow substantially in the coming years. It is not reasonable to assume that separate protocols will be available for the wireless portions of the Internet. In addition, we note that wireless infrastructure is maturing rapidly; the work undertaken by this group should take into account the expected state of the maturity of the technology in the time frame in which the Presence and Instant Messaging protocols are expected to be deployed.

To this end, the protocols designed by this Working Group must be suitable for operation in a context typically associated with mobile wireless access devices, viz. high latency, low bandwidth and possibly intermittent connectivity (which lead to a desire to minimize round-trip delays), modest computing power, battery constraints, small displays, etc. In particular, the protocols must be designed to be reasonably efficient for small payloads.

4.3.2.1 Namespace and Administration

Requirements are as follows:

1. The protocols must allow a PRESENCE SERVICE to be available independent of whether an INSTANT MESSAGE SERVICE is available, and vice-versa;
2. The protocols must not assume that an INSTANT INBOX is necessarily reached by the same IDENTIFIER as that of a PRESENTITY. Specifically, the protocols must assume that some INSTANT INBOXes may have no associated PRESENTITIES, and vice versa;
3. The protocols must also allow an INSTANT INBOX to be reached via the same IDENTIFIER as the IDENTIFIER of some PRESENTITY;
4. The administration and naming of ENTITIES within a given DOMAIN must be able to operate independently of actions in any other DOMAIN;
5. The protocol must allow for an arbitrary number of DOMAINS within the NAMESPACE.

4.3.2.2 Scalability

Requirements are as follows:

1. It must be possible for ENTITIES in one DOMAIN to interoperate with ENTITIES in another DOMAIN, without the DOMAINS having previously been aware of each other;
2. The protocol must be capable of meeting its other functional and performance requirements even when;
   - There are millions of ENTITIES within a single DOMAIN;
   - There are millions of DOMAINS within the singleNAMESPACE;
   - Every single SUBSCRIBER has SUBSCRIPTIONS to hundreds of PRESENTITIES;
   - Hundreds of distinct SUBSCRIBERS have SUBSCRIPTIONS to a single PRESENTITY;
   - Every single SUBSCRIBER has SUBSCRIPTIONS to PRESENTITIES in hundreds of distinct DOMAINS.

These are protocol design goals; implementations may choose to place lower limits.

4.3.2.3 Access Control

Requirements are as follows:

1. The PRINCIPAL controlling a PRESENTITY MUST be able to control:
   - Which WATCHERS can observe that PRESENTITY's PRESENCE INFORMATION;
   - Which WATCHERS can have SUBSCRIPTIONS to that PRESENTITY's PRESENCE INFORMATION;
   - What PRESENCE INFORMATION a particular WATCHER will see for that PRESENTITY,

regardless of whether the WATCHER gets it by fetching or NOTIFICATION;

- Which other PRINCIPALS, if any, can update the PRESENCE INFORMATION of that PRESENTITY.

2. The PRINCIPAL controlling an INSTANT INBOX must be able to control:
   - Which other PRINCIPALS, if any, can send INSTANT MESSAGES to that INSTANT INBOX;
   - Which other PRINCIPALS, if any, can read INSTANTMESSAGES from that INSTANT INBOX.

Access control MUST be independent of presence: the PRESENCE SERVICE MUST be able to make access control decisions even when the PRESENTITY is OUT OF CONTACT.

4.3.2.4 Network Topology

Note that intermediaries such as PROXIES may be necessitated between IP and non-IP networks, and by an end-user's desire to provide anonymity and hide their IP address.

Requirements are as follows:

1. The protocol must allow the creation of a SUBSCRIPTION both directly and via intermediaries, such as PROXIES;
2. The protocol must allow the sending of a NOTIFICATION both directly and via intermediaries, such as PROXIES;
3. The protocol must allow the sending of an INSTANT MESSAGE both directly and via intermediaries, such as PROXIES.

The protocol proxying facilities and transport practices must allow ADMINISTRATORS ways to enable and disable protocol activity through existing and commonly-deployed FIREWALLS. The protocol must specify how it can be effectively filtered by such FIREWALLS.

Message Encryption and Authentication

- The protocol must provide means to ensure confidence that a received message (NOTIFICATION or INSTANT MESSAGE) has not been corrupted or tampered with.
- The protocol must provide means to ensure confidence that a received message (NOTIFICATION or INSTANT MESSAGE) has not been recorded and played back by an adversary.
- The protocol must provide means to ensure that a sent message (NOTIFICATION or INSTANT MESSAGE) is only readable by ENTITIES that the sender allows.
- The protocol must allow any client to use the means to ensure noncorruption, nonplayback, and privacy, but the protocol must not require that all clients use these means at all times.

4.3.3 Additional Requirements for PRESENCE INFORMATION

The requirements in this section are applicable only to PRESENCE INFORMATION and not to INSTANT MESSAGES.

Requirements are as follows:

Common Presence Format

- All ENTITIES must produce and consume at least a common base format for PRESENCE INFORMATION.
- The common presence format must include a means to uniquely identify the PRESENTITY whose PRESENCE INFORMATION is reported.
- The common presence format must include a means to encapsulate contact information for the PRESENTITY's PRINCIPAL (if applicable), such as email address, telephone number, postal address, or the like.

- There must be a means of extending the common presence format to represent additional information not included in the common format without undermining or rendering invalid the fields of the common format.
- The working group must define the extension and registration mechanisms for presence information schema, including new STATUS conditions and new forms for OTHER PRESENCE MARKUP.
- The presence format should be based on IETF standards such as vCard (RFC 2426) if possible.

Presence Lookup and Notification

- A FETCHER MUST be able to fetch a PRESENTITY's PRESENCE INFORMATION even when the PRESENTITY is OUT OF CONTACT.
- A SUBSCRIBER MUST be able to request a SUBSCRIPTION to a PRESENTITY's PRESENCE INFORMATION, even when the PRESENTITY is OUT OF CONTACT.
- If the PRESENCE SERVICE has SUBSCRIPTIONS for a PRESENTITY's PRESENCE INFORMATION and that PRESENCE INFORMATION changes, the PRESENCE SERVICE MUST deliver a NOTIFICATION to each SUBSCRIBER, unless prevented by the PRESENTITY's ACCESS RULES.
- The protocol must provide a mechanism for detecting when a PRESENTITY or SUBSCRIBER has gone OUT OF CONTACT.
- The protocol must not depend on a PRESENTITY or SUBSCRIBER gracefully telling the service that it will no longer be in communication since a PRESENTITY or SUBSCRIBER may go OUT OF CONTACT due to unanticipated failures.

Presence Caching and Replication

- The protocol must include mechanisms to allow PRESENCE INFORMATION to be cached.
- The protocol must include mechanisms to allow cached PRESENCE INFORMATION to be updated when the master copy changes.
- The protocol caching facilities must not circumvent established ACCESS RULES or restrict choice of authentication/encryption mechanisms.

Performance

- When a PRESENTITY changes its PRESENCE INFORMATION, any SUBSCRIBER to that information MUST be notified of the changed information rapidly, except when such notification is entirely prevented by ACCESS RULES. This requirement is met if each SUBSCRIBER's NOTIFICATION is transported as rapidly as an INSTANT MESSAGE would be transported to an INSTANT INBOX.

4.3.4 Additional Requirements for INSTANT MESSAGES

The requirements in section 4.3.4 are applicable only to INSTANT MESSAGES and not to PRESENCE INFORMATION.

Common Message Format

- All ENTITIES sending and receiving INSTANT MESSAGES must implement at least a common base format for INSTANT MESSAGES.
- The common base format for an INSTANT MESSAGE must identify the sender and intended recipient.
- The common message format must include a return address for the receiver to reply to the sender with another INSTANT MESSAGE.
- The common message format should include standard forms of addresses or contact means for media other than INSTANT MESSAGES, such as telephone numbers or email addresses.

- The common message format must permit the encoding and identification of the message payload to allow for non-ASCII or encrypted content.
- The protocol must reflect best current practices related to internationalization.
- The protocol must reflect best current practices related to accessibility.
- The working group must define the extension and registration mechanisms for the message format including new fields and new schemes for INSTANT INBOX ADDRESSES.
- The working group must determine whether the common message format includes fields for numbering or identifying messages. If there are such fields, the working group must define the scope within which such identifiers are unique and the acceptable means of generating such identifiers.
- The common message format should be based on IETF-standard MIME (RFC 2045).

Reliability

- The protocol must include mechanisms so that a sender can be informed of the SUCCESSFUL DELIVERY of an INSTANT MESSAGE or reasons for failure. The working group must determine what mechanisms apply when final delivery status is unknown, such as when a message is relayed to non-IMPP systems.

Performance

- The transport of INSTANT MESSAGES must be sufficiently rapid to allow for comfortable conversational exchanges of short messages.

Presence Format

- The common presence format must define a minimum standard presence schema suitable for INSTANT MESSAGE SERVICES.
- When used in a system supporting INSTANT MESSAGES, the common presence format must include a means to represent the STATUS conditions OPEN and CLOSED.
- The STATUS conditions OPEN and CLOSED may also be applied to messaging or communication modes other than INSTANT MESSAGE SERVICES.

4.4 SIP Applications

This section briefly describes the usage of the Session Initiation Protocol (SIP) for subscriptions and notifications of presence. As noted, presence is defined as the willingness and ability of a user to communicate with other users on the network. Historically, presence has been limited to "on-line" and "off-line" indicators; the notion of presence here is broader. Subscriptions and notifications of presence are supported by defining an event package within the general SIP event notification framework. This protocol is also compliant with the Common Presence Profile (CPP) framework. This discussion is based on IETF RFC 3856 [ROS200401].

4.4.1 Introduction

RFC 2778 discussed earlier defines a model and terminology for describing systems that provide presence information. In that model, a presence service is a system that accepts, stores, and distributes presence information to interested parties, called *watchers*. A presence protocol is a protocol for providing a presence service over the Internet or any IP network.

RFC 3856 proposed the usage of SIP as a presence protocol. This is accomplished through a concrete instantiation of the general event notification framework defined for SIP, and as such, makes use of the SUBSCRIBE and NOTIFY methods. SIP is particularly well suited as a presence protocol. SIP location services already contain presence information in the form of registrations. Furthermore, SIP networks are capable of routing requests from any user on the network to the server that holds the registration state for a user. As this state is a key component of user presence, those SIP networks can allow SUBSCRIBE requests to be routed to the same server. This means that SIP networks can be reused to establish global connectivity for presence subscriptions and notifications.

4.4.2 Terminology

This event package is based on the concept of a presence agent which is a new logical entity that is capable of accepting subscriptions, storing subscription state, and generating notifications when there are changes in presence. The entity is defined as a logical one since it is generally co-resident with another entity. This event package is also compliant with the CPP framework that has been defined in RFC 3859. This allows SIP for presence to easily interwork with other presence systems compliant to CPP.

4.4.3 Definitions

This section uses the terms as defined in RFC 2778. Additionally, the following terms are defined and/or additionally clarified:

Presence User Agent (PUA): A Presence User Agent manipulates presence information for a presentity. This manipulation can be the side effect of some other action (such as sending a SIP REGISTER request to add a new Contact) or can be done explicitly through the publication of presence documents. We explicitly allow multiple PUAs per presentity. This means that a user can have many devices, such as a cell phone and Personal Digital Assistant (PDA), each of which is independently generating a component of the overall presence information for a presentity. PUAs push data into the presence system but are outside of it in that they do not receive SUBSCRIBE messages or send NOTIFY messages.

Presence Agent (PA): A presence agent is a SIP user agent which is capable of receiving SUBSCRIBE requests, responding to them, and generating notifications of changes in presence state. A presence agent must have knowledge of the presence state of a presentity. This means that it must have access to presence data manipulated by PUAs for the presentity. One way to do this is by co-locating the PA with the proxy/registrar. Another way is to co-locate it with the presence user agent of the presentity. However, these are not the only ways, and this specification makes no recommendations about where the PA function should be located. A PA is always addressable with a SIP URI that uniquely identifies the presentity (i.e., sip:joe@example.com). There can be multiple PAs for a particular presentity, each of which handles some subset of the total subscriptions currently active for the presentity. A PA is also a notifier (defined in RFC 3265) that supports the presence event package.

Presence Server: A presence server is a physical entity that can act as either a presence agent or as a proxy server for SUBSCRIBE requests. When acting as a PA, it is aware of the presence information of the presentity through some protocol means. When acting as a proxy, the SUBSCRIBE requests are proxied to another entity that may act as a PA.

Edge Presence Server: An edge presence server is a presence agent that is co-located with a PUA. It is aware of the presence information of the presentity because it is co-located with the entity that manipulates this presence information.

4.4.4 Overview of Operation

In this section, we present an overview of the operation of this event package. The overview describes behavior that is documented in part here, in part within the SIP event framework (RFC 3265), and in part in the SIP specification (RFC 3261), in order to provide clarity on this package for readers only casually familiar with those specifications. However, the detailed semantics of this package require the reader to be familiar with SIP events and the SIP specification itself.

When an entity—the subscriber—wishes to learn about presence information from some user, it creates a SUBSCRIBE request. This request identifies the desired presentity in the Request-URI using a SIP URI, SIPS URI (RFC 3261), or a presence (pres) URI (RFC 3859). The SUBSCRIBE request is carried along SIP proxies as any other SIP request would be. In most cases, it eventually arrives at a presence server which can either generate a response to the request (in which case it acts as the presence agent for the presentity) or proxy it on to an edge presence server. If the edge presence server handles the subscription, it is acting as the presence agent for the presentity. The decision at a presence server about whether to proxy or terminate the SUBSCRIBE is a local matter; however, we describe one way to effect such a configuration using REGISTER.

The presence agent (whether in the presence server or edge presence server) first authenticates the subscription, then authorizes it. The means for authorization are outside the scope of this protocol, and we expect that many mechanisms will be used. If authorized, a 200 OK response is returned. If authorization could not be obtained at this time, the subscription is considered "pending," and a 202 response is returned. In both cases, the PA sends an immediate NOTIFY message containing the state of the presentity and of the subscription. The presentity state may be bogus in the case of a pending subscription; for example, indicating offline no matter what the actual state of the presentity. This is to protect the privacy of the presentity who may not want to reveal that they have not provided authorization for the subscriber. As the state of the presentity changes, the PA generates NOTIFYs containing those state changes to all subscribers with authorized subscriptions. Changes in the state of the subscription itself can also trigger NOTIFY requests; that state is carried in the Subscription-State header field of the NOTIFY, and would typically indicate whether the subscription is active or pending.

The SUBSCRIBE message establishes a "dialog" with the presence agent. A dialog is defined in RFC 3261, and it represents the SIP state between a pair of entities to facilitate peer-to-peer message exchanges. This state includes the sequence numbers for messages in both directions (SUBSCRIBE from the subscriber, NOTIFY from the presence agent) in addition to a route set and remote target URI. The route set is a list of SIP (or SIPS) URIs which identify SIP proxy servers that are to be visited along the path of SUBSCRIBE refreshes or NOTIFY requests. The remote target URI is the SIP or SIPS URI that identifies the target of the message—the subscriber in the case of NOTIFY, or the presence agent in the case of a SUBSCRIBE refresh.

SIP provides a procedure called *record-routing* that allows for proxy servers to request to be on the path of NOTIFY messages and SUBSCRIBE refreshes. This is accomplished by inserting a URI into the Record-Route header field in the initial SUBSCRIBE request.

The subscription persists for a duration that is negotiated as part of the initial SUBSCRIBE. The subscriber will need to refresh the subscription before its expiration if they wish to retain the subscription. This is accomplished by sending a SUBSCRIBE refresh within the same dialog established by the initial SUB-SCRIBE. This SUBSCRIBE is nearly identical to the initial one but contains a tag in the To header field, a higher CSeq header field value, and possibly a set of Route header field values that identify the path of proxies the request is to take.

The subscriber can terminate the subscription by sending a SUBSCRIBE within the dialog with an Expires header field (which indicates duration of the subscription) value of zero. This causes an immediate termination of the subscription. A NOTIFY request is then generated by the presence agent with the most recent state. In fact, behavior of the presence agent for handling a SUBSCRIBE request with Expires of zero is no different than for any other expiration value; pending or authorized SUBSCRIBE requests result in a triggered NOTIFY with the current presentity and subscription state.

The presence agent can terminate the subscription at any time. To do so, it sends a NOTIFY request with a Subscription-State header field indicating that the subscription has been terminated. A reason parameter can be supplied which provides the reason. It is also possible to fetch the current presence state resulting in a one-time notification containing the current state. This is accomplished by sending a SUBSCRIBE request with an immediate expiration.

4.4.5 Usage of Presence URIs

A presentity is identified in the most general way through a presence URI, which is of the form pres: user@domain. These URIs are resolved to protocol specific URIs, such as the SIP or SIPS URI, through domain-specific mapping policies maintained on a server.

It is very possible that a user will have both a SIP (and/or SIPS) URI and a pres URI to identify both themself and other users. This leads to questions about how these URI relate and which are to be used.

In some instances, a user starts with one URI format, such as the pres URI, and learns a URI in a different format through some protocol means. As an example, a SUBSCRIBE request sent to a pres URI will result in learning a SIP or SIPS URI for the presentity from the Contact header field of the 200 OK to the SUBSCRIBE request. As another example, a DNS mechanism might be defined that would allow lookup of a pres URI to obtain a SIP or SIPS URI. In cases where one URI is learned from another through protocol means, those means will often provide some kind of scoping that limit the lifetime of the learned URI. DNS, for example, provides a TTL which would limit the scope of the URI. These scopes are very useful to avoid stale or conflicting URIs for identifying the same resource. To ensure that a user can always determine whether a learned URI is still valid, it is recommended that systems which provide lookup services for presence URIs have some kind of scoping mechanism.

If a subscriber is only aware of the protocol-independent pres URI for a presentity, it follows the procedures defined in RFC 3861. These procedures will result in the placement of the pres URI in the Request-URI of the SIP request, followed by the usage of the DNS procedures defined in RFC 3861 to determine the host to send the SIP request to. Of course, a local outbound proxy may alternatively be used as specified in RFC 3261. If the subscriber is aware of both the protocol-independent pres URI and the SIP or SIPS URI for the same presentity and both are valid (as discussed above), it should use the pres URI format. Of course, if the subscriber only knows the SIP URI for the presentity, that URI is used and standard RFC 3261 processing will occur. When the pres URI is used, any proxies along the path of the SUBSCRIBE request which do not understand the URI scheme will reject the request. As such, it is expected that many systems will be initially deployed that only provide users with a SIP URI.

SUBSCRIBE messages also contain logical identifiers that define the originator and recipient of the subscription (the To and From header fields). These headers can take either a pres or SIP URI. When the subscriber is aware of both a pres and SIP URI for its own identity, it should use the pres URI in the From header field. Similarly, when the subscriber is aware of both a pres and a SIP URI for the desired presentity, it should use the pres URI in the To header field.

The usage of the pres URI instead of the SIP URI within the SIP message supports interoperability through gateways to other CPP-compliant systems. It provides a protocol-independent form of identification which can be passed between systems. Without such an identity, gateways would be forced to map SIP URIs into the addressing format of other protocols. Generally, this is done by converting the SIP URI to the form <foreign-protocol-scheme>:<encoded SIP URI>@<gateway>. This is commonly done in email systems, and has many known problems. The usage of the pres URI is a *should*, and not a *must*, to allow for cases where it is known that there are no gateways present or where the usage of the pres URI will cause interoperability problems with SIP components that do not support the pres URI.

The Contact, Record-Route, and Route fields do not identify logical entities, but rather concrete ones used for SIP messaging. SIP specifies rules for their construction.

4.4.6 Presence Event Package

The SIP event framework defines a SIP extension for subscribing to, and receiving notifications of, events. It leaves the definition of many aspects of these events to concrete extensions known as event packages. This RFC qualifies as an event package. This section fills in the information required for all event packages by RFC 3265.

4.4.6.1 Package Name

The name of this package is "presence." As specified in RFC 3265, this value appears in the Event header field present in SUBSCRIBE and NOTIFY requests.

Example: Event: presence

4.4.6.2 Event Package Parameters

The SIP event framework allows event packages to define additional parameters carried in the Event header field. This package, "presence," does not define any additional parameters.

4.4.6.3 SUBSCRIBE Bodies

A SUBSCRIBE request MAY contain a body. The purpose of the body depends on its type. Subscriptions will normally not contain bodies. The Request-URI, which identifies the presentity, combined with the event package name, is sufficient for presence.

One type of body that can be included in a SUBSCRIBE request is a filter document. These filters request that only certain presence events generate notifications, or request for a restriction on the set of data returned in NOTIFY requests. For example, a presence filter might specify that the notifications should only be generated when the status of the user's instant inbox changes. It might also say that the content of these notifications should only contain the status of the instant inbox. Filter documents are not specified in this RFC, and at the time of writing, are expected to be the subject of future standardization activity.

Honoring of these filters is at the policy discretion of the PA.

If the SUBSCRIBE request does not contain a filter, this tells the PA that no filter is to be applied. The PA should send NOTIFY requests at the discretion of its own policy.

4.4.6.4 Subscription Duration

User presence changes as a result of many events. Some examples are:

- Turning on and off of a cell phone;
- Modifying the registration from a softphone;
- Changing the status on an instant messaging tool.

These events are usually triggered by human intervention, and occur with a frequency on the order of seconds to hours. As such, subscriptions should have an expiration in the middle of this range which is roughly one hour. Therefore, the default expiration time for subscriptions within this package is 3600 seconds. As per RFC 3265, the subscriber MAY specify an alternate expiration in the Expires header field.

4.4.6.5 NOTIFY Bodies

As described in RFC 3265, the NOTIFY message will contain bodies that describe the state of the subscribed resource. This body is in a format listed in the Accept header field of the SUBSCRIBE or a package-specific default if the Accept header field was omitted from the SUBSCRIBE.

In this event package, the body of the notification contains a presence document. This RFC describes the presence of the presentity that was subscribed to. All subscribers and notifiers MUST support the "application/pidf+xml" presence data format described in RFC 3863. The subscribe request may contain an Accept header field. If no such header field is present, it has a default value of "application/pidf+xml." If the header field is present, it must include "application/pidf+xml," and may include any other types capable of representing user presence.

4.4.6.6 Notifier Processing of SUBSCRIBE Requests

Based on the proxy routing procedures defined in the SIP specification, the SUBSCRIBE request will arrive at a Presence Agent (PA). This subsection defines package-specific processing at the PA of a SUBSCRIBE request. General processing rules for requests are covered in Section 8.2 of RFC 3261, in addition to general SUBSCRIBE processing in RFC 3265.

User presence is highly sensitive information. Because the implications of divulging presence information can be severe, strong requirements are imposed on the PA regarding subscription processing, especially related to authentication and authorization.

4.4.6.6.1 Authentication

A presence agent must authenticate all subscription requests. This authentication can be done using any of the mechanisms defined in RFC 3261. Note that digest is mandatory to implement as specified in RFC 3261. In single-domain systems where the subscribers all have shared secrets with the PA, the combination of digest authentication over Transport Layer Security (TLS) provides a secure and workable solution for authentication. This use case is described in Section 26.3.2.1 of RFC 3261. In interdomain scenarios, establishing an authenticated identity of the subscriber is harder. It is anticipated that authentication will often be established through transitive trust. SIP mechanisms for network asserted identity can be applied to establish the identity of the subscriber.

A presentity may choose to represent itself with a SIPS URI. By "represent itself," it means that the user represented by the presentity hands out, on business cards, web pages, and so on, a SIPS URI for their presentity. The semantics associated with this URI, as described in RFC 3261, require TLS usage on each hop between the subscriber and the server in the domain of the URI. This provides additional assurances (but no absolute guarantees) that identity has been verified at each hop.

Another mechanism for authentication is S/MIME. Its usage with SIP is described fully in RFC 3261. It provides an end-to-end authentication mechanism that can be used for a PA to establish the identity of the subscriber.

4.4.6.6.2 Authorization

Once authenticated, the PA makes an authorization decision. A PA must not accept a subscription unless authorization has been provided by the presentity. The means by which authorization are provided are outside the scope of this RFC. Authorization may have been provided ahead of time through access lists, perhaps specified in a web page. Authorization may have been provided by means of uploading of some kind of standardized access control list document. Back-end authorization servers, such as a DIAMETER (RFC 3588) server, can also be used. It is also useful to be able to query the user for authorization following the receipt of a subscription request for which no authorization information has been provided. The "watcherinfo" event template package for SIP defines a means by which a presentity can become aware that a user has attempted to subscribe to it so that it can then provide an authorization decision.

Authorization decisions can be very complex. Ultimately, all authorization decisions can be mapped into one of three states: rejected, successful, and pending. Any subscription for which the client is authorized to receive information about some subset of presence state at some points in time is a successful subscription. Any subscription for which the client will never receive any information about any subset of the presence state is a rejected subscription. Any subscription for which it is not yet known whether it is successful or rejected is pending. Generally, a pending subscription occurs when the server cannot obtain authorization at the time of the subscription, but may be able to do so at a later time, perhaps when the presentity becomes available.

The appropriate response codes for conveying a successful, rejected, or pending subscription (200, 403 or 603, and 202, respectively) are described in RFC 3265.

If the resource is not in a meaningful state, RFC 3265 allows the body of the initial NOTIFY to be empty. In the case of presence, that NOTIFY may contain a presence document. This RFC would indicate whatever presence state the subscriber has been authorized to see; it is interpreted by the subscriber as the current presence state of the presentity. For pending subscriptions, the state of the presentity should include some kind of textual note that indicates a pending status.

Polite blocking is possible by generating a 200 OK to the subscription even though it has been rejected (or marked pending). Of course, an immediate NOTIFY will still be sent. The contents of the presence document in such a NOTIFY are at the discretion of the implementor, but should be constructed in such a way as to not reveal to the subscriber that their request has actually been blocked. Typically, this is done by indicating "offline" or equivalent status for a single contact address.

4.4.6.7 Notifier Generation of NOTIFY Requests

RFC 3265 details the formatting and structure of NOTIFY messages. However, packages are mandated to provide detailed information on when to send a NOTIFY, how to compute the state of the resource, how to generate neutral or fake state information, and whether state information is complete or partial. This section describes those details for the presence event package.

A PA may send a NOTIFY at any time. Typically, it will send one when the state of the presentity changes. The NOTIFY request may contain a body indicating the state of the presentity. The times at which the NOTIFY is sent for a particular subscriber, and the contents of the body within that notification, are subject to any rules specified by the authorization policy that governs the subscription. This protocol in no way limits the scope of such policies. As a baseline, a reasonable policy is to generate notifications when the state of any of the presence tuples changes. These notifications would contain the complete and current presence state of the presentity as known to the presence agent. Future extensions can be defined that allow a subscriber to request that the notifications contain changes in presence information only, rather than complete state.

In the case of a pending subscription, when final authorization is determined, a NOTIFY can be sent. If the result of the authorization decision was success, a NOTIFY should be sent and should contain a presence document with the current state of the presentity. If the subscription is rejected, a NOTIFY may be sent. As described in RFC 3265, the Subscription-State header field indicates the state of the subscription.

The body of the NOTIFY must be sent using one of the types listed in the Accept header field in the most recent SUBSCRIBE request, or using the type "application/pidf+xml" if no Accept header field was present.

The means by which the PA learns the state of the presentity are also outside the scope of this recommendation. Registrations can provide a component of the presentity state. However, the means by which a PA uses registrations to construct a presence document are an implementation choice. If a PUA wishes to explicitly inform the presence agent of its presence state, it should explicitly publish the presence document (or its piece of it) rather than attempting to manipulate their registrations to achieve the desired result.

For reasons of privacy, it will frequently be necessary to encrypt the contents of the notifications. This can be accomplished using S/MIME. The encryption can be performed using the key of the subscriber as identified in the From field of the SUBSCRIBE request. Similarly, integrity of the notifications is important to subscribers. As such, the contents of the notifications MAY provide authentication and message integrity using S/MIME. Since the NOTIFY is generated by the presence server, which may not have access to the key of the user represented by the presentity, it will frequently be the case that the NOTIFY is signed by a third party. It is recommended that the signature be by an authority over the domain of the presentity. In other words, for a user pres:user@example.com, the signator of the NOTIFY should be the authority for example.com.

4.4.6.8 Subscriber Processing of NOTIFY Requests

RFC 3265 leaves it to event packages to describe the process followed by the subscriber upon receipt of a NOTIFY request including any logic required to form a coherent resource state.

In RFC 3856, each NOTIFY contains either no presence document or a document representing the complete and coherent state of the presentity. Within a dialog, the presence document in the NOTIFY request with the highest CSeq header field value is the current one. When no document is present in that NOTIFY, the presence document present in the NOTIFY with the next highest CSeq value is used. Extensions which specify the use of partial state for presentities will need to dictate how coherent state is achieved.

4.4.6.9 Handling of Forked Requests

RFC 3265requires each package to describe handling of forked SUBSCRIBE requests.

This specification only allows a single dialog to be constructed as a result of emitting an initial SUBSCRIBE request. This guarantees that only a single PA is generating notifications for a particular subscription to a particular presentity. The result of this is that a presentity can have multiple PAs active, but these should be homogeneous, so that each can generate the same set of notifications for the presentity. Supporting heterogeneous PAs, each of which generates notifications for a subset of the presence data, is complex and difficult to manage. Doing so would require the subscriber to act as the aggregator for presence data. This aggregation function can only reasonably be performed by agents representing the presentity. Therefore, if aggregation is needed, it must be done in a PA representing the presentity.

Section 4.4.9 of RFC 3265 describes the processing that is required to guarantee the creation of a single dialog in response to a SUBSCRIBE request.

4.4.6.10 Rate of Notifications

RFC 3265 requires each package to specify the maximum rate at which notifications can be sent.

A PA should not generate notifications for a single presentity at a rate of more than once every five seconds.

4.4.6.11 State Agents

RFC 3265 requires each package to consider the role of state agents in the package, and if they are used, to specify how authentication and authorization are done.

State agents are core to this package. Whenever the PA is not co-located with the PUA for the presentity, the PA is acting as a state agent. It collects presence state from the PUA, and aggregates it into a presence document. Because there can be multiple PUAs, a centralized state agent is needed to perform this aggregation. That is why state agents are fundamental to presence. Indeed, they have a specific term that describes them—a presence server.

4.4.6.11.1 Aggregation, Authentication, and Authorization

The means by which aggregation is done in the state agent is purely a matter of policy. The policy will typically combine the desires of the presentity along with the desires of the provider. This RFC in no way restricts the set of policies which may be applied.

However, there is clearly a need for the state agent to have access to the state of the presentity. This state is manipulated by the PUA. One way in which the state agent can obtain this state is to subscribe to it. As a result, if there were five PUA manipulating presence state for a single presentity, the state agent would generate five subscriptions, one to each PUA. For this mechanism to be effective, all PUA should be capable of acting as a PA for the state that they manipulate, and that they authorize subscriptions that can be authenticated as coming from the domain of the presentity.

The usage of state agents does not significantly alter the way in which authentication is done by the PA. Any of the SIP authentication mechanisms can be used by a state agent. However, digest authentication will require the state agent to be aware of the shared secret between the presentity and the subscriber. This will require some means to securely transfer the shared secrets from the presentity to the state agent.

The usage of state agents does, however, have a significant impact on authorization. As stated in Section 4.4.6.6, a PA is required to authorize all subscriptions. If no explicit authorization policy has been defined, the PA will need to query the user for authorization. In a presence edge server (where the PA is co-located with the PUA), this is trivially accomplished. However, when state agents are used (i.e., a presence server), a means is needed to alert the user that an authorization decision is required. This is the reason for the watcherinfo event template-package. All state agents should support the watcherinfo template-package.

4.4.6.11.2 Migration

On occasion, it makes sense for the PA function to migrate from one server to another. For example—for reasons of scale—the PA function may reside in the presence server when the PUA is not running, but when the PUA connects to the network, the PA migrates subscriptions to it in order to reduce state in the network. The mechanism for accomplishing the migration is described in Section 3.3.5 of RFC 3265. However, packages need to define under what conditions such a migration would take place.

A PA may choose to migrate subscriptions at any time, through configuration, or through dynamic means. The REGISTER request provides one dynamic means for a presence server to discover that the function can migrate to a PUA. Specifically, if a PUA wishes to indicate support for the PA function, it should use the callee capabilities specification [9] to indicate that it supports the SUBSCRIBE request method and the presence event package. The combination of these two define a PA. Of course, a presence server can always attempt a migration without these explicit hints. If it fails with either a 405 or 489 response code, the server knows that the PUA does not support the PA function. In this case, the server itself will need to act as a PA for that subscription request. Once such a failure has occurred, the server should not attempt further migrations to that PUA for the duration of its registration. However, to avoid the extra traffic generated by these failed requests, a presence server should support the callee capabilities extension.

Furthermore, indication of support for the SUBSCRIBE request and the presence event package is not sufficient for migration of subscriptions. A PA should not migrate the subscription if it is composing aggregated presence documents received from multiple PUA.

4.4.7 Learning Presence State

Presence information can be obtained by the PA in many ways. No specific mechanism is mandated by this specification. This section overviews some of the options, for informational purposes only.

4.4.7.1 Co-location

When the PA function is co-located with the PUA, presence is known directly by the PA.

4.4.7.2 REGISTER

A UA uses the SIP REGISTER method to inform the SIP network of its current communications addresses (i.e., Contact addresses). Multiple UA can independently register Contact addresses for the same address-of-record. This registration state represents an important piece of the overall presence information for a presentity. It is an indication of basic reachability for communications.

Usage of REGISTER information to construct presence is only possible if the PA has access to the registration database, and can be informed of changes to that database. One way to accomplish that is to co-locate the PA with the registrar.

The means by which registration state is converted into presence state is a matter of local policy and beyond the scope of this specification. However, some general guidelines can be provided. The address-of-record in the registration (the To header field) identifies the presentity. Each registered Contact header field identifies a point of communications for that presentity, which can be modeled using a tuple. Note that the contact address in the tuple need not be the same as the registered contact address. Using an address-of-record instead allows subsequent communications from a watcher to pass through proxies. This is useful for policy processing on behalf of the presentity and the provider.

A PUA that uses registrations to manipulate presence state should make use of the SIP callee capabilities extension. This allows the PUA to provide the PA with richer information about itself. For example, the presence of the methods parameter listing the method "MESSAGE" indicates support for instant messaging.

The q values from the Contact header field can be used to establish relative priorities amongst the various communications addresses in the Contact header fields.

The usage of registrations to obtain presence information increases the requirements for authenticity and integrity of registrations. Therefore, REGISTER requests used by presence user agents must be authenticated.

4.4.7.3 Uploading Presence Documents

If a means exists to upload presence documents from PUA to the PA, the PA can act as an aggregator and redistributor of those documents. The PA, in this case, would take the presence documents received from each PUA for the same presentity, and merge the tuples across all of those PUA into a single presence document. Typically, this aggregation would be accomplished through administrator or user-defined policies about how the aggregation should be done.

The specific means by which a presence document is uploaded to a presence agent are outside the scope of this specification. When a PUA wishes to have direct manipulation of the presence that is distributed to subscribers, direct uploading of presence documents is recommended.

4.4.8 Example Message Flow

The message flow of Figure 4.8 illustrates how the presence server can be responsible for sending notifications for a presentity. This flow assumes that the watcher has previously been authorized to subscribe to this resource at the server. In this flow, the PUA informs the server about the updated presence information through some non-SIP means. When the value of the Content-Length header field is "..." this means that the value should be whatever the computed length of the body is.

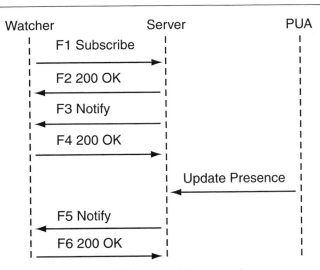

Figure 4.8: Message flow example.

Message Details

```
F1 SUBSCRIBE   watcher->example.com server

   SUBSCRIBE sip:resource@example.com SIP/2.0
   Via: SIP/2.0/TCP watcherhost.example.com;branch=z9hG4bKnashds7
   To: <sip:resource@example.com>
   From: <sip:user@example.com>;tag=xfg9
   Call-ID: 2010@watcherhost.example.com
   CSeq: 17766 SUBSCRIBE
   Max-Forwards: 70
   Event: presence
   Accept: application/pidf+xml
   Contact: <sip:user@watcherhost.example.com>
   Expires: 600
   Content-Length: 0
```

4.5 Conclusion

This chapter widened the discussion of pure VoIP to include an assessment of presence capabilities. Currently there are an estimated 3×10^6 IP phones in the world compared with an estimated 3×10^9 landline and 2G/2.5G cellular phones; it follows that so far the penetration of VoIP is 0.1%. In order to make rapid penetration and achieve market share (beyond the hype stage), new services are needed. Presence-related services represent one possible such service. Ubiquitous computing ("wearable computers"), may be another application (such application also would make use of presence concepts); hence, our coverage of this topic.

Issues with Current VoIP Technologies

This chapter examines some of the issues that are faced by VoIP systems, particularly systems that would be used by carriers for true end-to-end anytime-anyplace connectivity, comparable to what one enjoys today with traditional PSTN voice telephony. We only focus on issues and opportunities that can be addressed by IPv6, namely scalability and end-to-end robustness. The flow mechanism of IPv6 can be employed to manage QoS-specific paths which is critical to VoIP support—the value of flows is already highlighted in MPLS and the deployment/applications it is already experiencing at this time. (To be fair, it should be noted that not all issues faced by VoIP are addressed by IPv6—it is not a complete panacea. Examples here include: general security concerns; equipment interworking; carrier interworking; VoWi-Fi to cellular and/or 3G interworking; and, straightforward and reliable Unified Messaging deployments).

In the sections that follow we first briefly introduce the issue of security (Section 5.1); then we look at the NAT issue (Section 5.2). As a potential solution to some of the NAT problems, we then look in Section 5.3 at Simple Traversal of User Datagram Protocol Through Network Address Translators (STUN). STUN is a lightweight protocol that allows applications to discover the presence and types of NATs and firewalls between them and the public Internet. It also provides the ability for applications to determine the public IPv4 addresses allocated to them by the NAT. STUN works with many existing NATs and does not require any special behavior from them. As a result, it allows a variety of applications to work through existing NAT infrastructure [ROS200301] (however, up to now STUN has not experienced major acceptance/deployment). In Section 5.4 we look at Middlebox Communication (MIDCOM) as a possible other approach to dealing with the issues at hand. Finally, we look at some pragmatic short-term approaches, as embodied in the Session Border Controller (SBC) technology (Section 5.5). None of these solutions are optimal in all factors, hence, the utility of IPv6-based solutions.

5.1 General Enterprise Security Issues

Network and host security continue to be major concerns for enterprise-, institutional-, and service-provider environments. Well-documented recent studies show that cyber attacks continue to remain a substantial threat to organizations of all types. On average, companies experience several dozen attacks per week on their Information Technology resources. About 20% of large companies suffer at least two severe events a year. The challenge to corporate planners just continues to get more onerous. It has been conservatively forecasted that in 2010, around 100,000 new vulnerabilities will be discovered in software applications in that year alone; this will force companies to assess and mitigate one new risk every few minutes of every hour each day.

Considering that each vulnerability instance has the potential to disrupt or bring a company's business to a complete halt, organizations must take risk assessment seriously and determine how each risk will be handled. The increased number of vulnerabilities being discovered also drives up the number of security incidents worldwide and it will increase to a point where 8,000 incidents a week will affect organizations that have not properly addressed and mitigated their risks. It is estimated that the worldwide financial impact of malicious code is around $100 B year. Beyond the original venue of proving technical bravado, in the recent past attacks have been aimed at stealing customer data, obtaining proprietary information, and deliberately hampering a corporation's ability to do business [POL200401].

If a company loses its information technology (computer and/or voice/data networking) resources for more than a day or two, the company may well find itself in financial trouble. Obviously brokerage firms, banks, airports, medical establishments, and homeland security concerns would be impacted faster than, say, a manufacturing firm or a book publishing firm. However, the general concern is universal. If a company is unable to conduct business for more than a week, the company may well be permanently incapacitated. Therefore, there is a clear need to protect the enterprises from random, negligent, malicious, or planned attacks on its Information Technology resources. As more and more companies send their IT business abroad under the rubric of "outsourcing," the potential IT (and, hence, corporate) risks are arguably growing at a geometric pace; these risks can have ultimate negative implications, particularly in view of cumulative exposures to risks which, in the aggregate, take on nontrivial probability.

Many companies are (now) shifting to a highly mobile work force. To support this mobility firms are upgrading their network architectures to support remote workforces. Mobile users need access to centrally located applications and data over the Internet; voice is also an issue. This, once again, raises the issue of security.

5.1.1 Typical Enterprise Network Approaches

Firewalls are a basic mechanism to support perimeter security, even if by themselves they tend to be inadequate. See Figure 5.1 for a typical environment. Firewalls provide a method of guarding a private network by analyzing the data leaving and entering the intranet. Typically they are implemented as a network appliance (dedicated/standalone hardware), although it can also be a just a software program (for example, for a PC client). [CSO200501]. The majority of packet-inspection firewalls are designed to secure and apply policy to the transport level. Firewalls range in functionality from basic protocol/port inspection, to stateful session-oriented packet inspection, to sophisticated application-layer proxy firewalls. A typical firewall may support the following functions: packet filtering, object grouping, proxy services, URL filtering, stateful inspection, and inline authentication (with or without access to a RADIUS (remote access dial-in user service) server. Firewalls can also provide network address translation, so the actual IP addresses of devices inside the firewall stay hidden from public view; but this is precisely one of the issues of concern for end-to-end connectivity.

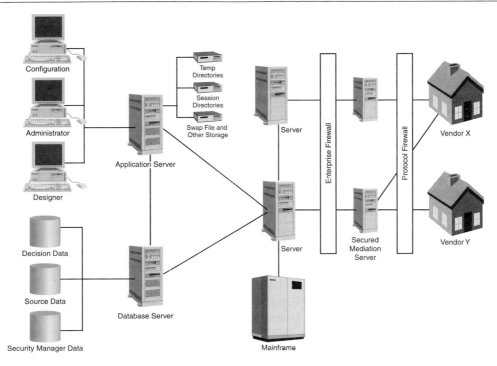

Figure 5.1: Typical firewall environment.

Most companies implement security in layers. The layering can be in terms of domains or in terms of assets categories. It is not effective to rely on a single point-of-protection when addressing the panoply of threats that can impact an IT environment; robust information security requires a multilayered approach.

Companies typically see the environment as being comprised of the following zones (also known as domains). (See Figure 5.2, which depicts both a logical view and an example of a physical view):

Externally-Controlled Zone (ECZ) (such as a particular extranet or 3rd-party environment with an established business relationship): Here the physical access, the IT administration, and the security authority are controlled by a third party.

Uncontrolled Zone (UZ) (such as the Internet and also carrier networks): No established business relationship exists where the firm can assess the security of the environment. Here the physical access, the IT administration, and the security authority are basically unknown.

Controlled Zone (CZ): Network point (zone) where all inbound and outbound communications are mediated (such as the firewall complex). Here the physical access, the IT administration, and the security authority are controlled by the firm in question. This domain separates the ECZ and UZ from the Restricted Zone (typically the intranet) of the firm.

Restricted Zone (RZ): Here the physical access, the IT administration, and the security authority are controlled by the firm in question. Access is granted only to authorized/authenticated users or systems.

Secure(d) Zone (SZ): Network location (zone) that provides isolation from the RZ. This zone may contain more critical assets such as the firm's data warehouse, the Directory, or specialized applications (such as financials, payroll, etc.). Here the physical access, the IT administration, and the security authority are controlled by the firm in question.

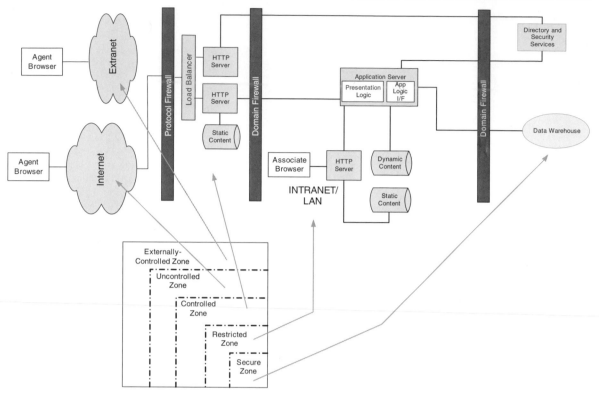

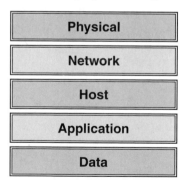

Figure 5.2: Layered security apparatus for typical enterprise environment.

It is also useful to look at layers from an asset category perspective. One example of this is Microsoft's Defense-in-Depth Model, as shown for illustrative purposes in Figure 5.3 [MIC200501].

Physical
Network
Host
Application
Data

Figure 5.3: Asset category layering per Microsoft's Defense-in-Depth Model.

A very basic firewall glossary as included in Table 5.1 (for a more extensive glossary the reader may refer to [MIN200601]).

Most corporations today address security with a number of technical solutions ranging from login/password, hardware tokens, and RADIUS servers for authentication to Virtual Private Networks (VPNs) for data encryption; hardware (appliance) firewalls at corporate locations for data packet filtering; antivirus software on remote PCs; and encrypted storage (e.g., per the new IEEE standard P1619) [POL200401]. Hardware firewalls (routers and/or appliances), generally protect the corporate network from external attacks but cannot provide protection against attacks originating from within the corporate network (as noted above, however, the Secure Zone (Domain) is delimited by firewalls that are inside of the corporate intranet itself). As noted, increasingly enterprises make use of a "layered" security approach. While authentication mechanisms ensure user/machine authorization and VPNs ensure data privacy in transit, the conventional security tools (e.g., hardware firewalls and antivirus software) cannot fully protect the environment. Malicious code, such as "spyware" can use peer-to-peer file sharing, instant messaging, and file downloading as a vehicle and enter the corporate network to create damage or hog network bandwidth. These are the reasons why XML firewalls (which inspect deep into the transmitted text) can be useful [POL200401].

TCP/IP-based networking uses the TCP-Port apparatus to identify the protocol and/or applications with which a given TCP session should be associated. Firewall technology is very much dependent on this arrangement for proper functioning (other/supplementary techniques such as specifying an IP address or IP address range are also utilized). Two general observations are useful:

- Applications using TCP are easier to manage through a firewall than applications using UDP;
- Protocols/applications that have a smaller range of allowed ports are easier to manage through a firewall than applications using a larger range—those using a single port are the easiest of all.

As it can be seen in Table 5.1, RTP and H.323 have some wire ranges making VoIP based on these protocols something of a challenge (the RTP issue is the same whether one uses SIP or H.323).

In the context of "layered" security, it should be mentioned that many organizations end up using the mechanism of NAT as part of the "toolkit" of available techniques by providing what some call *security through obscurity*. This entails keeping outside entities "unaware" of what the address of internal devices (servers, etc.) is, so that these entities cannot then launch a direct attack against said devices (for example, via a TELNET or a specifically-targeted flow of PDUs and-the-like). Clearly, NAT is a means-to-an-end; hence, if every device has a globally-unique address as in IPv6, then other methods will have to be put in place to provide a layer of security comparable to that provided by the previous state of "obscurity."

Figure 5.4 [ISL200501] depicts today's security environment as compared to what is possible/desirable in an IPv6 future state. The new (NAT-free) security mechanisms facilitate end-to-end connectivity, mobility, and collaboration, under a VoIP and/or 3G wireless environment in the coming years. Today's environment is very different, as discussed in the section that follows.

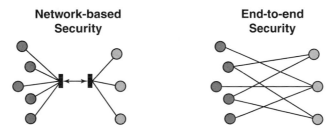

Figure 5.4: End-to-end security environment in IPv6.

Table 5.1: Basic security glossary.

Demilitarized Zone (DMZ)	*(We prefer the expansion "demarcation zone.") An area of an intranet that is a barrier, or a buffer, between a company's internal network and resources connected to the network, and the outside public network. That portion of the intranet-to-extranet or intranet-to-Internet interface apparatus that supports a highly constrained access environment. An area between the hostile Internet and protected services; may be implemented as a Layer 2 switch that support a number of Ethernet-attached devices "sandwiched" between a front-end and a back-end firewall. The purpose of the DMZ is to prevent external users from getting direct access to a server or other corporate IT resources. A DMZ is usually comprised of routers, packet filters, firewalls, proxies, and/or mediation devices.* *A neutral zone, or buffer, that separates the internal and external networks. The DMZ usually exists between two firewalls. External users can access servers in the DMZ, but not the computers on the internal network. The servers in the DMZ act as an intermediary for both incoming and outgoing traffic [BRA200501].* *The DMZ designates the area of protection that lies between the corporate computing environment and the Internet or publicly-accessible network. The DMZ is typically where the firewalls, gateways, application proxies, and other protective computing devices are connected, and employs protective software such as filtering and intrusion detection applications.*
Filter	*Packet matching information that identifies a set of packets to be treated a certain way by a middlebox (security mediation device). A set of terms and/or criteria used for the purpose of separating or categorizing. This is accomplished via single- or multifield matching of traffic header and/or payload data. 5-Tuple specification of packets in the case of a firewall and 5-tuple specification of a session in the case of a NAT middlebox function are examples of a filter [SRI200201].*
Firewall	*A method of guarding a private network by analyzing the data leaving and entering. Typically implemented as a network appliance (dedicated/standalone hardware), although it can also be a just a software program (for example for a PC client.) [CSO200501]. The majority of packet-inspection firewalls are designed to secure and apply policy to the transport level. Firewalls range in functionality from basic protocol/port filtering devices to stateful session-level packet-inspection systems and sophisticated application-layer proxy firewalls. Firewalls can also provide network address translation, so the actual IP addresses of devices inside the firewall stay hidden from public view.* *A policy-based packet filtering middlebox function, typically used for restricting access to/from specific devices and applications. The policies are often termed Access Control Lists (ACLs) [SRI200201].* *Includes four basic types: (1) Application-layer gateway; (2) Stateful-inspection firewall at the Session Layer; (3) Circuit-level gateway at the Network Layer; and (4) Packet-filtering firewall. Firewalls form the fundamental gateway that controls (at different layers of the OSI protocol stack) traffic entering and leaving the network, and all security issues of this type (such as Denial of Service attacks) come under this heading [LIG200501].* *Packet-filtering firewalls use rules based on basic information, such as a packet's source, destination, or port, to determine whether or not to allow it into the network. More advanced stateful packet-filtering firewalls have access to more information from which to make their decisions. Stateful firewalls examine related inbound-outbound traffic for expected/predicted patterns.)* *Proxy firewalls that look at content and can involve authentication and encryption can be more flexible and secure but also tend to be slower. Although firewalls require configuration expertise they are a critical component of network security [INF200501], [CSO200501].*
Layer 2	*The protocol layer below Layer 3 (that therefore offers the services used by Layer 3). Forwarding, when done by the swapping of short fixed length labels, occurs at layer 2 regardless of whether the label being examined is an ATM VPI/VCI, a frame relay DLCI, or a Multiprotocol Label Switching (MPLS) label.*
Layer 2 VPN (L2VPN)	*(aka L2 VPN) Three types of L2VPNs are currently defined [AND200501]: Virtual Private Wire Service (VPWS); Virtual Private LAN Service (VPLS); and IP-only LAN-like Service (IPLS).*
Layer 3	*The protocol layer at which IP and its associated routing protocols operate.*
Layer 3 Security Mechanisms	*Encryption mechanisms such as IPsec or Multilayer IPSec (ML-IPsec).*

Layer 3 VPN (L3VPN)	(a.k.a L3 VPN) An L3VPN interconnects sets of hosts and routers based on Layer 3 addresses; see [CAL200301].
Middlebox	A middlebox is a network intermediate device (in IETF parlance) that implements one or more of the middlebox services. A NAT middlebox is a middlebox implementing NAT service. A firewall middlebox is a middlebox implementing firewall service. Traditional middleboxes embed application intelligence within the device to support specific application traversal. Proposed middleboxes supporting the Middlebox Communications (MIDCOM) protocol, as defined in RFC 3303, will be able to externalize application intelligence into MIDCOM agents. In reality, some of the middleboxes may continue to embed application intelligence for certain applications and depend on MIDCOM protocol and MIDCOM agents for the support of remaining applications [SRI200201].
Proxy	An intermediary program (system) that acts both as a server and as a client for the purpose of making requests on behalf of other clients. Requests are serviced internally or by passing them on, with possible translation, to other servers. A software agent that acts on behalf of a user, typical proxies accept a connection from a user, make a decision as to whether or not the user or client IP address is permitted to use the proxy, perhaps does additional authentication, and then completes a connection on behalf of the user to a remote destination [INF200501].
	An intermediate relay agent between clients and servers of an application, relaying application messages between the two. Proxies use special protocol mechanisms to communicate with proxy clients and relay client data to servers and vice versa. A Proxy terminates sessions with both the client and the server, acting as server to the end-host client and as client to the end-host server. Applications such as FTP, SIP, and RTSP use a control session to establish data sessions. These control and data sessions can take divergent paths. While a proxy can intercept both the control and data sessions, it might intercept only the control session. This is often the case with real-time streaming applications such as SIP and RTSP [SRI200201].
	May include a function that replaces the IP address of a host on the internal (protected) network with its own IP address for all traffic passing through it.
Proxy Firewall	Unlike packet-filtering, this type of firewall does more than simply block port access. Instead, it acts as a proxy server, processing access requests on behalf of the network on which it is located. This protects individual computers on the network because they never interact directly with incoming client requests [CSO200501].
	Firewalls that look at content and can involve authentication and encryption can be more flexible and secure but may require more processing power [INF200501], [CSO200501].
Proxy Servers	Specialized application or server programs that run on a firewall host or on a dedicated appliance: either a dual-homed host with an interface on the internal network and one on the external network, or some other bastion host that has access to the Internet and is accessible from the internal devices. These programs take users' requests for Internet services (such as FTP and Telnet) and forward them, as appropriate according to the site's security policy, to the actual services. The proxies provide replacement connections and act as gateways to the services. For this reason, proxies are sometimes known as application-level gateways. Proxy services intervene, often transparently, between a user on the inside (on the internal network) and a service on the outside (on the Internet). Instead of talking to each other directly, each talks to a proxy. Proxies handle all the communication between users and Internet services behind the scenes. To the user, a proxy server gives the appearance that the user is dealing directly with the real server. To the real server, the proxy server presents the illusion that the real server is dealing directly with a user on the proxy host (as opposed to the user's real host). Proxy servers have two main purposes:
	Improve Performance: Proxy servers can improve performance for groups of users. This is because it saves the results of all requests for a certain amount of time. Consider the case where both user X and user Y access the World Wide Web through a proxy server. First user X requests a certain Web page, say Page 1. Sometime later, user Y requests the same page. Instead of forwarding the request to the Web server where Page 1 resides, which can be a time-consuming operation, the proxy server simply returns the Page 1 that it already fetched for user X. Since the proxy server is often on the same network as the user, this is a much faster operation.
	Filter Requests: Proxy servers can also be used to filter requests. For example, a company might use a proxy server to prevent its employees from accessing a specific set of websites.

Proxy Services	Proxy services intervene, often transparently, between a user on the inside (on the internal network) and a service on the outside (on the Internet). Proxy services are effective only when they are used in conjunction with a mechanism that restricts direct communications between the internal and external hosts. Dual-homed hosts and packet filtering are two such mechanisms. If internal hosts are able to communicate directly with external hosts, there is no need for users to use proxy services, and so (in general) they will not; such bypass, however, is typically not in accordance with an organization's security policy.
	A proxy service requires two components: a proxy server and a proxy client. In this situation, the proxy server runs on the dual-homed host. A proxy client is a special version of a normal client program (i.e., a Telnet or FTP client) that talks to the proxy server rather than to the "real" server out on the Internet; in addition, if users are taught special procedures to follow, normal client programs can often be used as proxy clients. The proxy server evaluates requests from the proxy client and decides which to approve and which to deny. If a request is approved, the proxy server contacts the real server on behalf of the client (thus the term "proxy"), and proceeds to relay requests from the proxy client to the real server, and responses from the real server to the proxy client. In some proxy systems instead of installing custom client proxy software, one employs standard software, but set up custom user procedures for using it. A proxy service is not a firewall architecture; proxy services are used in conjunction with a firewall architecture.
Proxying	Approach that involves mediating a connection at an intermediate point. In this case the TCP connection is not between the client and the (application) host, but from the client to the intermediate proxy-server/gateway. In turn, the proxy will decide (based on some criteria) if/where a companion session to the ultimate (application) host needs to be established. Proxy servers can also be used to filter requests.
	Companies use proxy servers to improve performance (through caching Web pages and graphics), to filter requests to certain sites, to make sure that only certain users can get to the Internet, or as a way of accounting for Web use (logging sites that users visit). Most proxy servers can perform all of these tasks.
TCP Ports	Transport layer end-to-end protocol identifiers of traffic being carrier in a network. Ports of interest to VoIP include (but are not limited to):

H.323 RAS	**TCP 1719**	
H.323 (H.225)	**TCP 1720**	**GW → CM, Call Setup**
MGCP	**TCP 2427/2428**	**GW → CM, Call Setup**
RTP	**UDP 16384–32767**	**Bearer Channel**
SIP	**TCP 5060**	
Skinny Client	**TCP 2000**	**Call Setup and Control**
Skinny GW (Digital)	**TCP 2002**	**Call Setup and Control**

GW = Gateway; CM = Call Manager (server)

(long lists of well-known ports are published by the IETF)

XML Firewall	A (relatively) new type of firewall intended to secure XML messages and Web Services (WS). Traditional firewalls are not designed to understand/interpret the XML message-level security and they cannot defend against new XML message-based attacks. The majority of packet-inspection firewalls are designed to secure and apply policy to the transport level, therefore they generally do not scan for content in Simple Object Access Protocol (SOAP), Universal Description, Discovery and Integration (UDDI), SAML or other Web services protocols. The difference between an XML firewall and other firewalls is that much of the features in an XML firewall exist at the application layer and within the data payload or content, as opposed to the transport and session layer. Many modern XML firewalls act like high-performance proxies: they can approach wire speed performance by offloading crytpo and XML validation functions to dedicated hardware (features such as message routing, encryption and forwarding are somewhat of a commodity). In this role, the XML firewall performs security services such as Authentication, Authorization, Auditing (AAA) and XML validation at a message level. The features are a separation of message-level security from transport-level security (these XML features do not act as transport-level connection security such as done in SSL) [WRE200401].

5.1.2 Typical Enterprise Network VoIP Security/Integration Approaches

This section briefly describes typical Best-in-Class designs for enterprise VoIP/converged networks. It highlights some of the architecture/design issues. This discussion is loosely based on reference [KUI200501].

Security issues affecting VoIP networks include, but are not limited to, the following: Toll fraud, packet/call eavesdropping, viruses, worms, Denial-of-Services, TCP vulnerabilities, Layer 3 exploits, rogue device in the network, man-in-the-middle issues, DHCP spoofing, DHCP starvation, and DNS spoofing.

A rogue device has access to the voice stream (packets and/or session/call) between the two communicating endpoints. Products/programs have appeared (e.g., Voice Over Misconfigured Internet Telephones (VOMIT)) that facilitate such eavesdropping by assembling tcpdumps of conversations into .WAV files. A rogue device (an unauthorized device that has been able to inject itself into the network) can undertake theft of telephone service; rogue voice gateways can cause even more harm.

Countermeasures include all IP-based security mechanisms such as (the relatively weak) VLAN switch/port management methods, Layer 3 firewalls, proxies, intrusion prevention systems, encryption/tunneling, H.245 security, certificates, authentication/RADIUS/IEEE 802.1 services, physical hardening.

Figure 5.5 depicts a typical corporate VoIP arrangement, somewhat similar to the figures included in Chapter 1 (keep in mind that a carrier arrangement would be quite different). As can be seen from this figure, there are several places (at least at the firewall locations) where the NAT/firewall issues can be problematic.

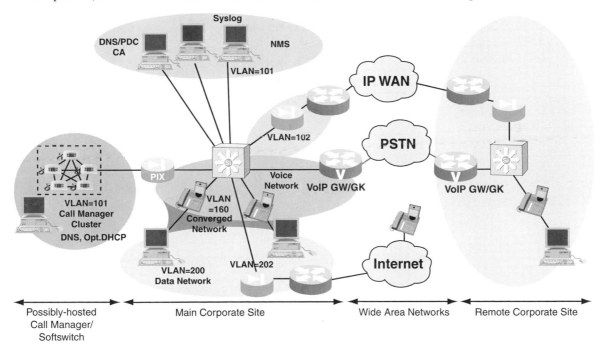

Figure 5.5: A typical corporate VoIP arrangement.

VoIP security is built in layers, as was the case for the more general intranet discussion earlier in the chapter. Note the firewall arrangement facing the IP WAN as well as facing the call manager/softswitch cluster (which can be co-located or can be at a hosted network/ASP-resident site). Again, these are impacted by NAT.

Security related to the call manager/softswitch cluster centers on hardened Operating System (OS) (such as Linux and/or a high-quality maintenance process on other less reliable OSs), IPSec tunneling (for remote/hosted arrangements), and Host-based Intrusion Prevention system. IPSec in also affected by NAT.

Security related to the firewall connecting to the call control manager deals with Access Control Lists, control of source addresses, and proper filtering (e.g., to allow only call control, directory/LDAP functions, and network management). NAT impacts the overall setup.

Connection to the outside world (Internet) is handled over a Layer 3 VPN mechanism. Network Intrusion Detection Systems/Network Protection Systems are typically used.

Endpoints (clients) use separate voice and data VLANs (in support of the already mentioned relatively weak VLAN switch/port management mechanism), authentication, and encryption (especially if over a wireless LAN or VPN, here for a softphone.) Endpoint encryption, particularly for VoWi-Fi, is still evolving in terms of broad vendor support.

The campus network typically makes use of the normal Layer 3/Application Layer firewalls, and IP filters between voice and data. NAT use should be minimized.

Most deployments today make use a distinct VLAN for voice and a VLAN for data traffic, as already mentioned and further depicted in Figure 5.6. This is done for administrative, QoS, and pseudo-security[23] considerations (the voice VLAN is called an *auxiliary* VLAN). However, firms want to use the same access, core, and distribution layers for the two segments in order to be able to make the claim and gain the operational and financial advantage of a converged network (see Figure 5.7.) This is supported by mechanisms such as Layer 3 access control and stateful firewalls (firewalls that examine related inbound-outbound traffic for expected/predicted patterns.)

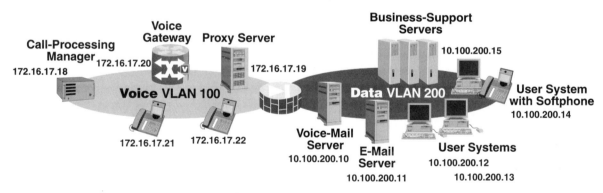

Figure 5.6: Use of two VLANs—typical 2G VoIP arrangement.

---

[23] Pseudo-security is a term we use to describe an environment or technology with a weak (and/or false sense of) protection.

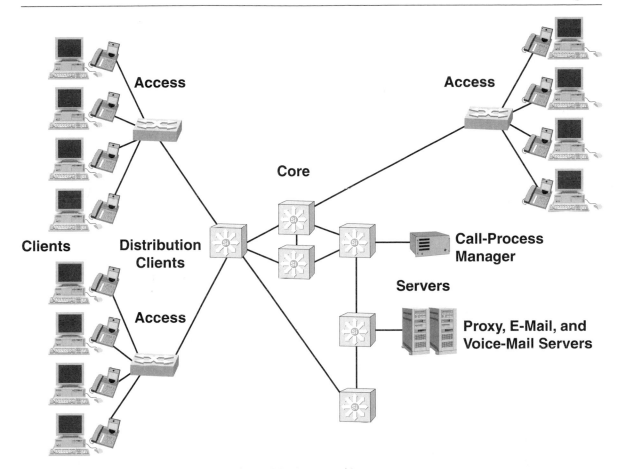

Figure 5.7: Converged intranet.

IP phones typically support access to both segments (IP phones have a "data port"/Ethernet for the local PC to connect—this uses a single Ethernet cable to the desk, often with in-line power.) Planners need to make sure that the phone supports separation of the two segments. However, firms should not rely solely on VLANs for separation: to support more robust security one needs to make provision for Layer 3 filtering between the data and voice VLANs.)

A stateful firewall between the two VLAN segments is typically used to manage the data/voice VLAN interaction. The stateful firewall provides dynamic access and mitigation against TCP connection starvation, UDP flooding, and spoofing attacks.

As seen in Figure 5.6 in 2G VoIP one typically makes use of a private address space (RFC 1918) for the data and for the voice VLAN segments. The partitioned addressing facilitates filtering and recognition. The approach in RFC 1918 does not support routability, but this can be utilized to reduce the likelihood of recon- naissance scans even if NAT happened to be misconfigured. Spoof-mitigation filtering addresses the identity issue (that nodes are who they claim to be) in local segments.

Related to the end-points, blocking PC access to the voice VLAN at the VLAN switch (even if the PC has physical access to the network or to the Layer 2 switch) greatly reduces the possibility of eavesdropping attacks (such as those that may be unleashed with VOMIT-like products); techniques also exist to prevent man-in-the-middle attacks or traffic interception. Access Control Lists (ACLs) can be used to prevent directed-TCP attacks. DHCP snooping stops DHCP spoofing and starvation attacks. Digitally-signed firmware and configuration files on clients mitigate security liabilities. Certificates can be used to prevent rogue call managers, gateways, and phone set insertion (particularly in a VoWi-Fi environment). Finally, encryption prevents interception. Similar techniques can be used to protect the servers that support VoIP, e.g., Call Managers, Gateways, Gatekeepers, etc.

5.1.3 Firewall Issues for VoIP

As noted earlier, firewalls are a basic mechanism to support perimeter security; packet-inspection firewalls are designed to apply policy to the transport level. As discussed, firewalls range in functionality from a basic stateful packet-inspection engine to sophisticated application-layer proxy firewalls; firewalls can also provide network address translation. TCP/IP-based networking uses the TCP/UDP-Port apparatus to identify the protocol and/or applications with which a given TCP session should be associated. As we already observed, two general observations are useful in a networking context that are also useful in a VoIP context:

- Applications using TCP are easier to manage through a firewall than applications using UDP;
- Protocols/applications that have a smaller range of allowed ports are easier to manage through a firewall than applications using a larger range—those using a single port are the easiest of all.

Figure 5.8 depicts the protocol stacks of interest to VoIP. As it can be seen in Table 5.1, RTP and H.323 have some wide ranges making VoIP based on these protocols something of a challenge (the RTP issue is the same whether one uses SIP or H.323.)

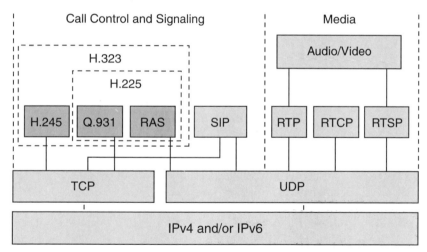

H.323 Version 3 and 4 supports H.245 over UDP/TCP, Q.931 over UDP/TCP, and RAS over UDP. SIP supports TCP and UDP.

Figure 5.8: VoIP protocol stack.

We limit the rest of this discussion to SIP. Some of the NAT-related issues are highlighted next. As we discussed in Chapter 3, the Via field in SIP indicates the path taken by the SIP request under discussion up to the present point. This prevents request looping and ensures replies take the same path as the requests, which, in principle, assists in firewall traversal and other unusual routing situations [HAN199901].

According to [HAN199901] (on which the discussion that follow is based) if a SIP proxy server[24] forwards a SIP request, it must add itself to the beginning of the list of forwarders noted in the Via headers. The Via trace ensures that replies can take the same path back, ensuring correct operation through compliant firewalls and avoiding request loops. On the response path, each host must remove its Via, so that routing internal information is hidden from the callee and outside networks. A proxy server must check that it does not generate a request to a host listed in the Via sent-by, via-received, or via-maddr parameters (the maddr parameter provides the server address to be contacted for this user, overriding the address supplied in the host field; this address is typically a multicast address but could also be the address of a backup server.)

Hence, the client originating the request inserts into the request messages a Via field containing its host name or network address and, if not the default port number, the port number at which it wishes to receive responses. (Note that this port number can differ from the UDP source port number of the request.) A fully-qualified domain name is typically used. Each subsequent proxy server that sends the request onwards must add its own additional Via field before any existing Via fields. A proxy that receives a redirection (3xx) response and then searches recursively, must use the same Via headers as on the original proxied request. A SIP proxy should check the top-most Via header field to ensure that it contains the sender's correct network address, as seen from that proxy. If the sender's address is incorrect, the proxy must add an additional "received" attribute.

A host behind a network address translator or firewall may not be able to insert a network address into the Via header that can be reached by the next hop beyond the NAT. Use of the received attribute allows SIP requests to traverse NATs that only modify the source IP address. NATs that modify port numbers, called *Network Address Port Translators (NAPTs),* will not properly pass SIP when transported on UDP, in which case an application layer gateway is required[25]. When run over TCP, SIP stands a better chance of traversing NATs, since its behavior, in this case, is similar to HTTP (but of course on different ports).

A proxy sending a request to a multicast address must add the "maddr" parameter to its Via header field, and should add the "ttl" parameter. If a server receives a request that contained an "maddr" parameter in the top-most Via field, it should send the response to the multicast address listed in the "maddr" parameter. If a SIP proxy server receives a request which contains its own address in the Via header value, it must respond with a 482 (Loop Detected) status code. A proxy server must not forward a request to a multicast group which already appears in any of the Via headers. This prevents a malfunctioning proxy server from causing loops. Also, it cannot be guaranteed that a proxy server can always detect that the address returned by a location service refers to a host listed in the Via list, as a single host may have aliases or several network interfaces.

Normally, every host that sends or forwards a SIP message adds a Via field indicating the path traversed. However, it is possible that NATs changes the source address and port of the request (e.g., from net-10 to a globally routable address), in which case the Via header field cannot be relied on to route replies. To prevent this, a proxy should check the top-most Via header field to ensure that it contains the sender's correct network address, as seen from that proxy. If the sender's address is incorrect, the proxy must add a "received" parameter to the Via header field inserted by the previous hop. Such a modified Via header field is known as a receiver-tagged Via header field. An example is:

Via: SIP/2.0/UDP erlang.bell-telephone.com:5060
Via: SIP/2.0/UDP 10.0.0.1:5060;received=199.172.136.3

In this example, the message originated from 10.0.0.1 and traversed a NAT with the external address border. ieee.org (199.172.136.3) to reach erlang.bell-telephone.com. The latter noticed the mismatch, and added a

---

[24] Refer to Chapter for definition of functionality.

[25] An example is a Border Session controller.

parameter to the previous hop's Via header field, containing the address that the packet actually came from. (Note that the NAT border.ieee.org is not a SIP server.)

Via header fields in responses are processed by a proxy or UAC according to the following rules:

1. The first Via header field should indicate the proxy or client processing this response. If it does not, discard the message. Otherwise, remove this Via field.
2. If there is no second Via header field, this response is destined for this client. Otherwise, the processing depends on whether the Via field contains a "maddr" parameter or is a receiver-tagged field:
    a. If the second Via header field contains a "maddr" parameter, send the response to the multicast address listed there, using the port indicated in "sent-by," or port 5060 if none is present. The response should be sent using the TTL indicated in the "ttl" parameter, or with a TTL of 1 if that parameter is not present. For robustness, responses must be sent to the address indicated in the "maddr" parameter even if it is not a multicast address.
    b. If the second Via header field does not contain a "maddr" parameter and is a receiver-tagged field, send the message to the address in the "received" parameter using the port indicated in the "sent-by" value, or using port 5060 if none is present.
    c. If neither of the previous cases apply, send the message to the address indicated by the "sent-by" value in the second Via header field.

This discussion implicitly highlights the private address/NAT issues faced in 2G VoIP systems. Some of these issues can be mitigated in certain IPv6 implementations.

5.2 What is NAT?

We mentioned NAT a number of times. In this section we provide some detailed information on it. Basic Network Address Translation or Basic NAT is a method by which IP addresses are mapped from one group to another, transparent to end users. Network Address Port Translation, or NAPT is a method by which many network addresses and their TCP/UDP (Transmission Control Protocol/User Datagram Protocol) ports are translated into a single network address and its TCP/UDP ports. Together, these two operations, referred to as traditional NAT, provide a mechanism to connect a realm with private addresses to an external realm with globally unique registered addresses. As discussed, NAT has impact on 2G VoIP systems; hence, the reason for our coverage. The NAT operation described in this section is based on IETF RFC 3022 [SRI200101]. Developers should refer directly to the RFC for any normative guidance.

Note: IPv4 NAT is described in RFC 2663 and RFC 3022, but has also been is extended beyond IPv4 networks to include the IPv4-v6 NAT-PT described in RFC 2766. While the IPv4 NAT translates one IPv4 address into another IPv4 address to provide routing between private v4 and external V4 address realms, IPv4-v6 NAT-PT (RFC 2766) translates an IPv4 address into an IPv6 address, and vice versa, to provide routing between a v6 address realm and an external v4 address realm. Unless specified otherwise, NAT is a proxy (middlebox) function referring to both IPv4 NAT, as well as IPv4-v6 NAT-PT [SRI200101], [TSI200001], [SRI200201].

5.2.1 Introduction

The need for IP address translation arises when a network's internal IP addresses cannot be used outside the network either for privacy reasons or because they are invalid for use outside the network. Network topology outside a local domain can change in many ways. Customers may change providers, company backbones may be reorganized, or providers may merge or split. Whenever external topology changes with time, address assignment for nodes within the local domain must also change to reflect the external changes. Changes of this type can be hidden from users within the domain by centralizing changes to a single address translation router.

Basic address translation would (in many cases, except as noted in RFC 2663 and section 6 of RFC 3022) allow hosts in a private network to transparently access the external network and enable access to selective local hosts from the outside. Organizations with a network setup predominantly for internal use, with a need for occasional external access, are good candidates for this scheme.

Many Small Office and Home Office (SOHO) users as well as telecommuting employees have multiple network nodes in their office running TCP/UDP applications, but have a single IP address assigned to their remote access router by their service provider to access remote networks. This community of remote access users typically employs NAPT, which permits multiple nodes in a local network to simultaneously access remote networks using the single IP address assigned to their router.

There are limitations to using the translation method. It is mandatory that all requests and responses pertaining to a session be routed via the same NAT router. One way to ascertain this would be to have NAT based on a border router that is unique to a stub domain, where all IP packets are either originated from the domain or destined to the domain. There are other ways to ensure this with multiple NAT devices. For example, a private domain could have two distinct exit points to different providers and the session flow from the hosts in a private network could traverse through whichever NAT device has the best metric for an external host. When one of the NAT routers fails, the other could route traffic for all the connections. There is however a caveat with this approach, in that rerouted flows could fail at the time of switchover to the new NAT router. A way to overcome this potential problem is to have the routers share the same NAT configuration and exchange state information to ensure a fail-safe backup for each other.

Address translation is application-independent and often accompanied by Application Level Gateways (ALGs) to perform payload monitoring and alterations. FTP is the most popular ALG resident on NAT devices. Applications requiring ALG intervention must not have their payload encoded, as doing that effectively disables the ALG, unless the ALG has the key to decrypt the payload.

This solution has the disadvantage of taking away the end-to-end significance of an IP address, and making up for it with increased state in the network. As a result, end-to-end IP network level security assured by IPSec cannot be assumed to end hosts, with a NAT device enroute. The advantage of this approach, however, is that it can be installed without changes to hosts or routers.

Definition of terms such as "Address Realm," "Transparent Routing," "TU Ports," "ALG," and others may be found in RFC 2663.

5.2.2 Overview of Traditional NAT

The Address Translation operation presented in this RFC is referred to as "Traditional NAT." There are other variations of NAT that are explored in this RFC. Traditional NAT would allow hosts within a private network, in most cases, to transparently access hosts in the external network. In a traditional NAT, sessions are unidirectional, outbound from the private network. Sessions in the opposite direction may be allowed on an exceptional basis using static address maps for pre-selected hosts. Basic NAT and NAPT are two variations of traditional NAT, in that translation in Basic NAT is limited to IP addresses alone, whereas translation in NAPT is extended to include IP address and Transport identifier (such as a TCP/UDP port or ICMP query ID).

Unless mentioned otherwise, Address Translation or NAT throughout this section will pertain to traditional NAT—namely Basic NAT—as well as NAPT. Only the stub border routers as described in Figure 5.9 may be configured to perform address translation.

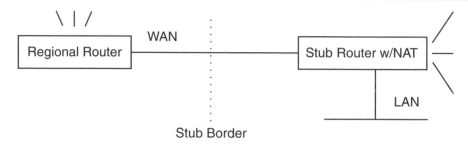

Figure 5.9: Traditional NAT configuration.

5.2.2.1 Overview of Basic NAT

Basic NAT operation is as follows. A stub domain with a set of private network addresses could be enabled to communicate with an external network by dynamically mapping the set of private addresses to a set of globally valid network addresses. If the number of local nodes is less than or equal to addresses in the global set, each local address is guaranteed a global address to map to. Otherwise, nodes allowed to have simultaneous access to external network are limited by the number of addresses in global set. Individual local addresses may be statically mapped to specific global addresses to ensure guaranteed access to the outside or to allow access to the local host from external hosts via a fixed public address. Multiple simultaneous sessions may be initiated from a local node using the same address mapping.

Addresses inside a stub domain are local to that domain and not valid outside the domain. Thus, addresses inside a stub domain can be reused by any other stub domain. For instance, a single Class A address could be used by many stub domains. At each exit point between a stub domain and backbone, NAT is installed. If there is more than one exit point, it is of great importance that each NAT have the same translation table.

For instance, in the example of Figure 5.10, both stubs A and B internally use class A private address block 10.0.0.0/8 (see RFC 1918). Stub A's NAT is assigned the class C address block 198.76.29.0/24, and Stub B's NAT is assigned the class C address block 198.76.28.0/24. The class C addresses are globally unique—no other NAT boxes can use them.

When stub A host 10.33.96.5 wishes to send a packet to stub B host 10.81.13.22, it uses the globally unique address 198.76.28.4 as destination, and sends the packet to its primary router. The stub router has a static route for net 198.76.0.0 so the packet is forwarded to the WAN-link. However, NAT translates the source address 10.33.96.5 of the IP header to the globally unique 198.76.29.7 before the packet is forwarded. Likewise, IP packets on the return path go through similar address translations.

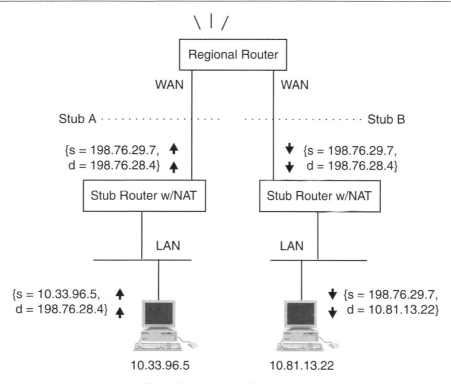

Figure 5.10: Basic NAT operation.

Notice that this requires no changes to hosts or routers. For instance, as far as the stub A host is concerned, 198.76.28.4 is the address used by the host in stub B. The address translations are transparent to end hosts in most cases. Of course, this is just a simple example. There are numerous issues to be explored.

5.2.2.2 Overview of NAPT

Say, an organization has a private IP network and a WAN link to a service provider. The private network's stub router is assigned a globally valid address on the WAN link and the remaining nodes in the organization have IP addresses that have only local significance. In such a case, nodes on the private network could be allowed simultaneous access to the external network, using the single registered IP address with the aid of NAPT. NAPT would allow mapping of tuples of the type (local IP addresses, local TU port number) to tuples of the type (registered IP address, assigned TU port number).

This model fits the requirements of most Small Office/Home Office (SOHO) groups to access external network using a single service provider assigned IP address. This model could be extended to allow inbound access by statically mapping a local node per each service TU port of the registered IP address.

In the example of Figure 5.11, stub A internally uses class A address block 10.0.0.0/8. The stub router's WAN interface is assigned an IP address 138.76.28.4 by the service provider.

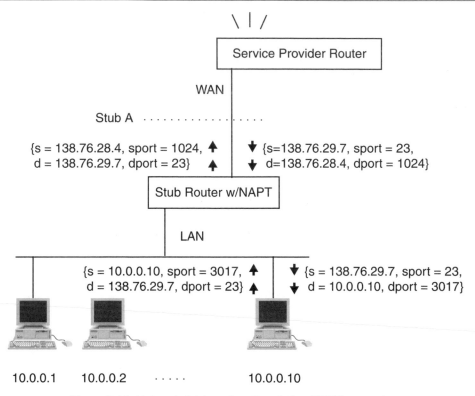

Figure 5.11: Network Address Port Translation (NAPT) operation.

When stub A host 10.0.0.10 sends a telnet packet to host 138.76.29.7, it uses the globally unique address 138.76.29.7 as destination, and sends the packet to it's primary router. The stub router has a static route for the subnet 138.76.0.0/16 so the packet is forwarded to the WAN-link. However, NAPT translates the tuple of source address 10.0.0.10 and source TCP port 3017 in the IP and TCP headers into the globally unique 138.76.28.4 and a uniquely assigned TCP port, say 1024, before the packet is forwarded. Packets on the return path go through similar address and TCP port translations for the target IP address and target TCP port. Notice that this requires no changes to hosts or routers. The translation is completely transparent.

In this setup, only TCP/UDP sessions are allowed and must originate from the local network. However, there are services such as DNS that demand inbound access. There may be other services for which an organization wishes to allow inbound session access. It is possible to statically configure a well known TU port service (RFC 1700) on the stub router to be directed to a specific node in the private network.

In addition to TCP/UDP sessions, ICMP messages, with the exception of REDIRECT message types, may also be monitored by a NAPT router. ICMP query type packets are translated in a manner similar to the way TCP/UDP packets are translated in that the identifier field in an ICMP message header will be uniquely mapped to a query identifier of the registered IP address. The identifier field in ICMP query messages is set by Query sender and returned unchanged in a response message from the Query responder. So, the tuple of (`Local IP address, local ICMP query identifier`) is mapped to a tuple of (`registered IP address, assigned ICMP query Identifier`) by the NAPT router to uniquely identify ICMP queries of all types from any of the local hosts. Modifications to ICMP error messages are discussed in a later section as that involves modifications to the ICMP payload as well as the IP and ICMP headers.

In NAPT setup, where the registered IP address is the same as the IP address of the stub router WAN interface, the router has to be sure to make distinction between TCP, UDP, or ICMP query sessions originated from itself versus those originated from the nodes on a local network. All inbound sessions (including TCP, UDP, and ICMP query sessions) are assumed to be directed to the NAT router as the end node, unless the target service port is statically mapped to a different node in the local network.

Sessions other than TCP, UDP and ICMP query type are simply not permitted from local nodes serviced by a NAPT router.

5.2.3 Translation Phases of a Session

The translation phases with traditional NAT are the same as those described in RFC 2663. The following subsections identify items that are specific to traditional NAT.

5.2.3.1 Address Binding

With Basic NAT, a private address is bound to an external address when the first outgoing session is initiated from the private host. Subsequent to that, all other outgoing sessions originating from the same private address will use the same address binding for packet translation.

In the case of NAPT, where many private addresses are mapped to a single globally unique address, the binding would be from the tuple of (private address, private TU port) to the tuple of (assigned address, assigned TU port). As with Basic NAT, this binding is determined when the first outgoing session is initiated by the tuple of (private address, private TU port) on the private host. While not a common practice, it is possible to have an application on private host establish multiple simultaneous sessions originating from the same tuple of (private address, private TU port). In such a case, a single binding for the tuple of (private address, private TU port) may be used for translation of packets pertaining to all sessions originating from the same tuple on a host.

5.2.3.2 Address Lookup and Translation

After an address binding or (address, TU port) tuple binding in case of NAPT is established, a soft state may be maintained for each of the connections using the binding. Packets belonging to the same session will be subject to session lookup for translation purposes. The exact nature of translation is discussed in the follow-on section.

5.2.3.3 Address Unbinding

When the last session based on an address or (address, TU port) tuple binding is terminated, the binding itself may be terminated.

5.2.4 Packet Translations

Packets pertaining to NAT-managed sessions undergo translation in either direction. Individual packet translation issues are covered in detail in the following subsections.

5.2.4.1 IP, TCP, UDP, and ICMP Header Manipulations

In Basic NAT model, the IP header of every packet must be modified. This modification includes IP address (source IP address for outbound packets and destination IP address for inbound packets) and the IP checksum.

For TCP and UDP sessions, modifications must include update of checksum in the TCP/UDP headers. This is because TCP/UDP checksum also covers a pseudo header which contains the source and destination IP addresses. As an exception, UDP headers with 0 checksum should not be modified. As for ICMP Query packets ([ICMP]), no further changes in ICMP header are required as the checksum in ICMP header does not cover IP addresses.

In a NAPT model, modifications to an IP header are similar to that of Basic NAT. For TCP/UDP sessions, modifications must be extended to include translation of TU port (source TU port for outbound packets and destination TU port for inbound packets) in the TCP/UDP header. The ICMP header in ICMP Query packets must also be modified to replace the query ID and ICMP header checksum. Private host query ID must be translated into assigned ID on the outbound and the exact reverse on the inbound. ICMP header checksum must be corrected to account for Query ID translation.

5.2.4.2 Checksum Adjustment

NAT modifications are applied on a packet-by-packet basis and can be very compute intensive, as they involve one or more checksum modifications in addition to simple field translations. Luckily, we have an algorithm below, which makes checksum adjustment to IP, TCP, UDP and ICMP headers very simple and efficient. Since all these headers use a one's complement sum, it is sufficient to calculate the arithmetic difference between the before-translation and after-translation addresses and add this to the checksum. The algorithm below is applicable only for even offsets (i.e., optr below must be at an even offset from start of header) and even lengths (i.e., olen and nlen below must be even). Sample code (in C) for this is as follows.

```
void checksumadjust(unsigned char *chksum, unsigned char *optr,
int olen, unsigned char *nptr, int nlen)
/* assuming: unsigned char is 8 bits, long is 32 bits.
  - chksum points to the chksum in the packet
  - optr points to the old data in the packet
  - nptr points to the new data in the packet
*/
{
  long x, old, new;
  x=chksum[0]*256+chksum[1];
  x=~x & 0xFFFF;
  while (olen)
  {
      old=optr[0]*256+optr[1]; optr+=2;
      x-=old & 0xffff;
      if (x<=0) { x--; x&=0xffff; }
      olen-=2;
  }
  while (nlen)
  {
      new=nptr[0]*256+nptr[1]; nptr+=2;
      x+=new & 0xffff;
      if (x & 0x10000) { x++; x&=0xffff; }
      nlen-=2;
  }
  x=~x & 0xFFFF;
  chksum[0]=x/256; chksum[1]=x & 0xff;
}
```

5.2.4.3 ICMP Error Packet Modifications

Changes to ICMP error message will include changes to IP and ICMP headers on the outer layer as well as changes to headers of the packet embedded within the ICMP-error message payload.

In order for NAT to be transparent to end-host, the IP address of the IP header embedded within the payload of ICMP-Error message must be modified, the checksum field of the embedded IP header must be modified, and lastly, the ICMP header checksum must also be modified to reflect changes to payload.

In a NAPT setup, if the IP message embedded within ICMP happens to be a TCP, UDP, or ICMP Query packet, you will also need to modify the appropriate TU port number within the TCP/UDP header or the Query Identifier field in the ICMP Query header.

Lastly, the IP header of the ICMP packet must also be modified.

5.2.4.4 FTP Support

One of the most popular applications, "FTP," would require an ALG to monitor the control session payload to determine the ensuing data session parameters. FTP ALG is an integral part of most NAT implementations.

The FTP ALG requires a special table to correct the TCP sequence and acknowledge numbers with source port FTP or destination port FTP. The table entries should have source address, destination address, source port, destination port, delta for sequence numbers and a timestamp. New entries are created only when FTP PORT commands or PASV responses are seen. The sequence number delta may be increased or decreased for every FTP PORT command or PASV response. Sequence numbers are incremented on the outbound and acknowledge numbers are decremented on the inbound by this delta.

FTP payload translations are limited to private addresses and their assigned external addresses (encoded as individual octets in ASCII) for Basic NAT. For NAPT setup, however, the translations must be extended to include the TCP port octets (in ASCII) following the address octets.

5.2.4.5 DNS Support

Considering that sessions in a traditional NAT are predominantly outbound from a private domain, DNS ALG may be obviated from use in conjunction with traditional NAT as follows. DNS server(s) internal to the private domain maintain mapping of names to IP addresses for internal hosts and possibly some external hosts. External DNS servers maintain name mapping for external hosts alone and not for any of the internal hosts. If the private network does not have an internal DNS server, all DNS requests may be directed to the external DNS server to find address mapping for the external hosts.

5.2.4.6 IP Option Handling

An IP datagram with any of the IP options Record Route, Strict Source Route, or Loose Source Route would involve recording or using IP addresses of intermediate routers. A NAT intermediate router may choose not to support these options or leave the addresses untranslated while processing the options. The result of leaving the addresses untranslated would be that private addresses along the source route are exposed end-to-end. This should not jeopardize the traversal path of the packet, per se, as each router is supposed to look at the next hop router only.

5.2.5 Miscellaneous Issues

5.2.5.1 Partitioning of Local and Global Addresses

For NAT to operate as described in this RFC, it is necessary to partition the IP address space into two parts—the private addresses used internal to stub domain and the globally unique addresses. Any given address must either be a private address or a global address. There is no overlap.

The problem with overlap is the following. Say a host in stub A wished to send packets to a host in stub B, but the global addresses of stub B overlapped the private addressees of stub A. In this case, the routers in stub A would not be able to distinguish the global address of stub B from its own private addresses.

5.2.5.2 Private Address Space Recommendation

RFC 1918 has recommendations on address space allocation for private networks. Internet Assigned Numbers Authority (IANA) has three blocks of IP address space, namely 10.0.0.0/8, 172.16.0.0/12, and 192.168.0.0/16 for private internets. In pre-CIDR notation, the first block is nothing but a single class A network number, while the second block is a set of 16 contiguous class B networks, and the third block is a set of 256 contiguous class C networks.

An organization that decides to use IP addresses in the address space defined above can do so without any coordination with IANA or an Internet registry. The address space can thus be used privately by many independent organizations at the same time, with NAT operation enabled on their border routers.

5.2.5.3 Routing Across NAT

The router running NAT should not advertise the private networks to the backbone. Only the networks with global addresses may be known outside the stub. However, global information that NAT receives from the stub border router can be advertised in the stub the usual way.

Typically, the NAT stub router will have a static route configured to forward all external traffic to service provider router over WAN link, and the service provider router will have a static route configured to forward NAT packets (i.e., those whose destination IP address fall within the range of NAT managed global address list) to NAT router over WAN link.

5.2.5.4 Switch-Over from Basic NAT to NAPT

In Basic NAT setup, when private network nodes outnumber global addresses available for mapping (say, a class B private network mapped to a class C global address block), external network access to some of the local nodes is abruptly cut off after the last global address from the address list is used up. This is very inconvenient and constraining. Such an incident can be safely avoided by optionally allowing the Basic NAT router to switch over to NAPT setup for the last global address in the address list. Doing this will ensure that hosts on private network will have continued, uninterrupted access to the external nodes and services for most applications. Note, however, it could be confusing if some of the applications that used to work with Basic NAT suddenly break due to the switch-over to NAPT.

5.2.6 NAT Limitations

RFC 2663 covers the limitations of all flavors of NAT, broadly speaking. The following subsections identify limitations specific to traditional NAT.

5.2.6.1 Privacy and Security

Traditional NAT can be viewed as providing a privacy mechanism since sessions are unidirectional from private hosts, and the actual addresses of the private hosts are not visible to external hosts. The same characteristic that enhances privacy potentially makes debugging problems (including security violations) more difficult. If a host in a private network is abusing the Internet in some way (such as trying to attack another machine or even sending large amounts of spam) it is more difficult to track the actual source of trouble because the IP address of the host is hidden in a NAT router.

5.2.6.2 ARP responses to NAT Mapped Global Addresses on a LAN Interface

NAT must be enabled only on border routers of a stub domain. The examples provided in the document to illustrate Basic NAT and NAPT have maintained a WAN link for connection to external router (i.e., service provider router) from NAT router. However, if the WAN link were to be replaced by a LAN connection and if part or all of the global address space used for NAT mapping belongs to the same IP subnet as the LAN segment, the NAT router would be expected to provide ARP support for the address range that belongs to the same subnet. Responding to ARP requests for the NAT mapped global addresses with its own MAC address is a must in such a situation with Basic NAT setup. If the NAT router did not respond to these requests, there is no other node in the network that has ownership of these addresses and hence will go unresponded.

This scenario is unlikely with NAPT setup except when the single address used in NAPT mapping is not the interface address of the NAT router (as in the case of a switch-over from Basic NAT to NAPT explained in 5.2.5.4 above, for example).

Using an address range from a directly connected subnet for NAT address mapping would obviate static route configuration on the service provider router.

It is the opinion of the authors that a LAN link to a service provider router is not very common. However, vendors may be interested to optionally support proxy ARP just in case.

5.2.6.3 Translation of Outbound TCP/UDP Fragmented Packets in NAPT Setup

Translation of outbound TCP/UDP fragments (i.e., those originating from private hosts) in NAPT setup are doomed to fail. This is because only the first fragment contains the TCP/UDP header that would be necessary to associate the packet to a session for translation purposes. Subsequent fragments do not contain TCP/UDP port information, but simply carry the same fragmentation identifier specified in the first fragment. Say, two private hosts originated fragmented TCP/UDP packets to the same destination host. And, they happened to use the same fragmentation identifier. When the target host receives the two unrelated datagrams, carrying the same fragmentation ID, and from the same assigned host address, it is unable to determine which of the two sessions the datagrams belong to. Consequently, both sessions will be corrupted.

5.3 STUN—Simple Traversal of User Datagram Protocol (UDP) Through Network Address Translators (NATs)

STUN is a lightweight protocol described in RFC 3489 that allows applications to discover the presence and types of NATs and firewalls between them and the public Internet. It also provides the ability for applications to determine the public IP addresses allocated to them by the NAT. STUN works with many existing NATs and does not require any special behavior from them. As a result, it allows a variety of applications to work through existing NAT infrastructure [ROS200301] (however, up to now it has not experienced major acceptance/deployment). The STUN operation described in this section is based on IETF RFC 3489 [ROS200301]. Developers should refer to the original RFP for any normative guidance.

5.3.1 Applicability Statement

It is recognized that STUN is not a cure-all for the problems associated with NAT. It does not enable incoming TCP connections through NAT. It allows incoming UDP packets through NAT, but only through a subset of existing NAT types. In particular, STUN does not enable incoming UDP packets through symmetric NATs, which are common in large enterprises. STUN's discovery procedures are based on assumptions on NAT treatment of UDP; such assumptions may prove invalid down the road as new NAT devices are deployed. STUN does not work when it is used to obtain an address to communicate with a peer that happens to be behind the same NAT. STUN does not work when the STUN server is not in a common shared address realm.

5.3.2 Introduction

NATs, while providing many benefits, also come with many drawbacks. The most troublesome of those drawbacks is the fact that they break many existing IP applications, and make it difficult to deploy new ones. Guidelines have been developed that describe how to build "NAT friendly" protocols, but many protocols simply cannot be constructed according to those guidelines. Examples of such protocols include almost all peer-to-peer protocols, such as multimedia communications, file sharing, and games.

To combat this problem, Application Layer Gateways (ALGs) have been embedded in NATs. ALGs perform the application layer functions required for a particular protocol to traverse a NAT. Typically, this involves rewriting application layer messages to contain translated addresses, rather than the ones inserted by the sender of the message. ALGs have serious limitations, including scalability, reliability, and speed of deploying new applications. To resolve these problems, the Middlebox Communications (MIDCOM) protocol has been developed (see RFC 3303). MIDCOM allows an application entity, such as an end client or network server of some sort (like a SIP proxy discussed in Chapter 3 in the context of RFC 3261) to control a NAT (or firewall) in order to obtain NAT bindings and open or close pinholes. In this way, NATs and applications can be separated once more, eliminating the need for embedding ALGs in NATs and resolving the limitations imposed by current architectures. MIDCOM is covered in Section 5.4 of this chapter.

Unfortunately, MIDCOM requires upgrades to existing NATs and firewalls in addition to application components. Complete upgrades of these NAT and firewall products will take a long time, potentially years. This is due, in part, to the fact that the deployers of NATs and firewalls are not the same people who are deploying and using applications. As a result, the incentive to upgrade these devices will be low in many cases. Consider, for example, an airport Internet lounge that provides access with a NAT. A user connecting to the NATed network may wish to use a peer-to-peer service, but cannot, because the NAT does not support it. Since the administrators of the lounge are not the ones providing the service, they are not motivated to upgrade their NAT equipment to support it, using either an ALG or MIDCOM.

Another problem is that the MIDCOM protocol requires that the agent controlling the middleboxes know the identity of those middleboxes, and have a relationship with them which permits control. In many configurations, this will not be possible. For example, many cable access providers use NAT in front of their entire access network. This NAT could be in addition to a residential NAT purchased and operated by the end user. The end user will probably not have a control relationship with the NAT in the cable access network, and may not even know of its existence.

Many existing proprietary protocols, such as those for online games and VoIP, have developed "tricks" that allow them to operate through NATs without changing those NATs. RFC 3489 is an attempt to take some of those ideas, and codify them into an interoperable protocol that can meet the needs of many applications. STUN allows entities behind a NAT to first discover the presence of a NAT and the type of NAT, and then to learn the address bindings allocated by the NAT. STUN requires no changes to NATs and works with an arbitrary number of NATs in tandem between the application entity and the public Internet.

5.3.3 Applicability to VoIP

The primary usage STUN has found is in the area of VoIP, facilitating allocation of addresses for receiving RTP traffic. In that application, the periodic keepalives are provided by the RTP traffic itself. However, several practical problems arise for RTP. First, RTP assumes that RTCP traffic is on a port one higher than the RTP traffic. This pairing property cannot be guaranteed through NATs that are not directly controllable. As a result, RTCP traffic may not be properly received. Protocol extensions to SDP have been proposed which mitigate this by allowing the client to signal a different port for RTCP. However, there will be interoperability problems for some time. For VoIP, silence suppression can cause a gap in the transmission of RTP packets. This could result in the loss of a binding in the middle of a call, if that silence period exceeds the binding timeout. This can be mitigated by sending occasional silence packets to keep the binding alive. However, the result is additional brittleness; proper operation depends on the silence suppression algorithm in use, the usage of a comfort noise codec, the duration of the silence period, and the binding lifetime in the NAT.

5.3.4 Definitions

STUN Client: A STUN client (also just referred to as a client) is an entity that generates STUN requests. A STUN client can execute on an end system, such as a user's PC, or can run in a network element, such as a conferencing server.

STUN Server: A STUN Server (also just referred to as a server) is an entity that receives STUN requests, and sends STUN responses. STUN servers are generally attached to the public Internet.

5.3.5 NAT Variations

It has been observed that NAT treatment of UDP varies among implementations. The four treatments observed in implementations are:

Full Cone: A full cone NAT is one where all requests from the same internal IP address and port are mapped to the same external IP address and port. Furthermore, any external host can send a packet to the internal host, by sending a packet to the mapped external address.

Restricted Cone: A restricted cone NAT is one where all requests from the same internal IP address and port are mapped to the same external IP address and port. Unlike a full cone NAT, an external host (with IP address X) can send a packet to the internal host only if the internal host had previously sent a packet to IP address X.

Port Restricted Cone: A port restricted cone NAT is like a restricted cone NAT, but the restriction includes port numbers. Specifically, an external host can send a packet, with source IP address X and source port P, to the internal host only if the internal host had previously sent a packet to IP address X and port P.

Symmetric: A symmetric NAT is one where all requests from the same internal IP address and port, to a specific destination IP address and port, are mapped to the same external IP address and port. If the same host sends a packet with the same source address and port, but to a different destination, a different mapping is used. Furthermore, only the external host that receives a packet can send a UDP packet back to the internal host.

Determining the type of NAT is important in many cases. Depending on what the application wants to do, it may need to take the particular behavior into account.

5.3.6 Overview of Operation

This section is descriptive only (normative behavior is described in Sections 5.3.8 and 5.3.9.) The typical STUN configuration is shown in Figure 5.12. A STUN client is connected to private network 1. This network connects to private network 2 through NAT 1. Private network 2 connects to the public Internet through NAT 2. The STUN server resides on the public Internet.

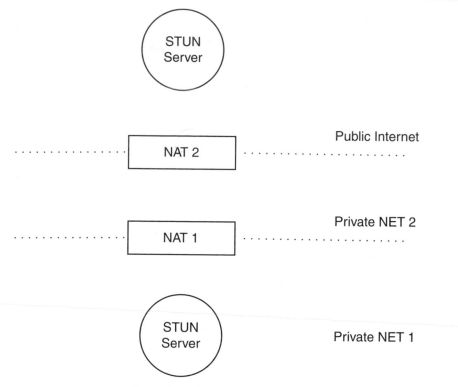

Figure 5.12: STUN configuration.

STUN is a simple client-server protocol. A client sends a request to a server, and the server returns a response. There are two types of requests—Binding Requests, sent over UDP, and Shared Secret Requests, sent over TLS over TCP. Shared Secret Requests ask the server to return a temporary username and password. This username and password are used in a subsequent Binding Request and Binding Response, for the purposes of authentication and message integrity.

Binding requests are used to determine the bindings allocated by NATs. The client sends a Binding Request to the server, over UDP. The server examines the source IP address and port of the request, and copies them into a response that is sent back to the client. There are some parameters in the request that allow the client to ask that the response be sent elsewhere, or that the server send the response from a different address and port. There are attributes for providing message integrity and authentication.

The trick is using STUN to discover the presence of NAT, and to learn and use the bindings they allocate.

The STUN client is typically embedded in an application which needs to obtain a public IP address and port that can be used to receive data. For example, it might need to obtain an IP address and port to receive Real Time Transport Protocol (RTP) traffic. When the application starts, the STUN client within the application sends a STUN Shared Secret Request to its server, obtains a username and password, and then sends it a Binding Request. STUN servers can be discovered through DNS SRV records, and it is generally assumed that the client is configured with the domain it needs to use to find the STUN server. Generally, this will be the domain of the provider of the service the application is using (such a provider is incented to deploy

STUN servers in order to allow its customers to use its application through NAT). Of course, a client can determine the address or domain name of a STUN server through other means. A STUN server can even be embedded within an end system.

The STUN Binding Request is used to discover the presence of a NAT, and to discover the public IP address and port mappings generated by the NAT. Binding Requests are sent to the STUN server using UDP. When a Binding Request arrives at the STUN server, it may have passed through one or more NATs between the STUN client and the STUN server. As a result, the source address of the request received by the server will be the mapped address created by the NAT closest to the server. The STUN server copies that source IP address and port into a STUN Binding Response, and sends it back to the source IP address and port of the STUN request. For all of the NAT types above, this response will arrive at the STUN client.

When the STUN client receives the STUN Binding Response, it compares the IP address and port in the packet with the local IP address and port it bound to when the request was sent. If these do not match, the STUN client is behind one or more NATs. In the case of a full-cone NAT, the IP address and port in the body of the STUN response are public, and can be used by any host on the public Internet to send packets to the application that sent the STUN request. An application need only listen in on the IP address and port from which the STUN request was sent. Any packets sent by a host on the public Internet to the public address and port learned by STUN will be received by the application.

Of course, the host may not be behind a full-cone NAT. Indeed, it does not yet know what type of NAT it is behind. To determine that, the client uses additional STUN Binding Requests. The exact procedure is flexible, but would generally work as follows. The client would send a second STUN Binding Request, this time to a different IP address, but from the same source IP address and port. If the IP address and port in the response are different from those in the first response, the client knows it is behind a symmetric NAT. To determine if it is behind a full-cone NAT, the client can send a STUN Binding Request with flags that tell the STUN server to send a response from a different IP address and port than the request was received on. In other words, if the client sent a Binding Request to IP address/port A/B using a source IP address/port of X/Y, the STUN server would send the Binding Response to X/Y using source IP address/port C/D. If the client receives this response, it knows it is behind a full cone NAT.

STUN also allows the client to ask the server to send the Binding Response from the same IP address the request was received on, but with a different port. This can be used to detect whether the client is behind a port restricted cone NAT or just a restricted cone NAT.

It should be noted that the configuration in Figure 5.12 is not the only permissible configuration. The STUN server can be located anywhere, including within another client. The only requirement is that the STUN server is reachable by the client, and if the client is trying to obtain a publicly routable address, that the server reside on the public Internet.

5.3.7 Message Overview

STUN messages are TLV (type-length-value) encoded using big endian (network ordered) binary. All STUN messages start with a STUN header, followed by a STUN payload. The payload is a series of STUN attributes, the set of which depends on the message type. The STUN header contains a STUN message type, transaction ID, and length. The message type can be Binding Request, Binding Response, Binding Error Response, Shared Secret Request, Shared Secret Response, or Shared Secret Error Response. The transaction ID is used to correlate requests and responses. The length indicates the total length of the STUN payload, not including the header. This allows STUN to run over TCP. Shared Secret Requests are always sent over TCP (indeed, using TLS over TCP).

Several STUN attributes are defined. The first is a MAPPED-ADDRESS attribute, which is an IP address and port. It is always placed in the Binding Response, and it indicates the source IP address and port the server saw in the Binding Request. There is also a RESPONSE-ADDRESS attribute, which contains an IP address and port. The RESPONSE-ADDRESS attribute can be present in the Binding Request, and indicates where the Binding Response is to be sent. It's optional, and when not present, the Binding Response is sent to the source IP address and port of the Binding Request.

The third attribute is the CHANGE-REQUEST attribute, and it contains two flags to control the IP address and port used to send the response. These flags are called *change IP* and *change port flags*. The CHANGE-REQUEST attribute is allowed only in the Binding Request. The "change IP" and "change port" flags are useful for determining whether the client is behind a restricted cone NAT or restricted port cone NAT. They instruct the server to send the Binding Responses from a different source IP address and port. The CHANGE-REQUEST attribute is optional in the Binding Request.

The fourth attribute is the CHANGED-ADDRESS attribute. It is present in Binding Responses. It informs the client of the source IP address and port that would be used if the client requested the "change IP" and "change port" behavior.

The fifth attribute is the SOURCE-ADDRESS attribute. It is only present in Binding Responses. It indicates the source IP address and port where the response was sent from. It is useful for detecting twice NAT configurations.

The sixth attribute is the USERNAME attribute. It is present in a Shared Secret Response, which provides the client with a temporary username and password (encoded in the PASSWORD attribute). The USER-NAME is also present in Binding Requests, serving as an index to the shared secret used for the integrity protection of the Binding Request. The seventh attribute, PASSWORD, is only found in Shared Secret Response messages. The eighth attribute is the MESSAGE-INTEGRITY attribute, which contains a message integrity check over the Binding Request or Binding Response.

The ninth attribute is the ERROR-CODE attribute. This is present in the Binding Error Response and Shared Secret Error Response. It indicates the error that has occurred. The tenth attribute is the UNKNOWN-ATTRIBUTES attribute which is present in either the Binding Error Response or Shared Secret Error Response. It indicates the mandatory attributes from the request which were unknown. The eleventh attribute is the REFLECTED-FROM attribute which is present in Binding Responses. It indicates the IP address and port of the sender of a Binding Request used for traceability purposes to prevent certain denial-of-service attacks.

5.3.8 Server Behavior

The server behavior depends on whether the request is a Binding Request or a Shared Secret Request.

5.3.8.1 Binding Requests

A STUN server must be prepared to receive Binding Requests on four address/port combinations—(A1, P1), (A2, P1), (A1, P2), and (A2, P2). (A1, P1) represent the primary address and port, and these are the ones obtained through the client discovery procedures below. Typically, P1 will be port 3478, the default STUN port. A2 and P2 are arbitrary. A2 and P2 are advertised by the server through the CHANGED-ADDRESS attribute, as described below.

It is recommended that the server check the Binding Request for a MESSAGE-INTEGRITY attribute. If not present, and the server requires integrity checks on the request, it generates a Binding Error Response with an ERROR-CODE attribute with response code 401. If the MESSAGE-INTEGRITY attribute was present, the server computes the HMAC over the request as described in Section 5.3.11.2. The key to use depends on the shared secret mechanism. If the STUN Shared Secret Request was used, the key must be the one associated with the USERNAME attribute present in the request. If the USERNAME attribute was not present, the server must generate a Binding Error Response. The Binding Error Response must include an ERROR-CODE attribute with response code 432. If the USERNAME is present, but the server does not remember the shared secret for that USERNAME (because it timed out, for example), the server must generate a Binding Error Response. The Binding Error Response must include an ERROR-CODE attribute with response code 430. If the server does know the shared secret, but the computed HMAC differs from the one in the request, the server must generate a Binding Error Response with an ERROR-CODE attribute with response code 431. The Binding Error Response is sent to the IP address and port the Binding Request came from, and sent from the IP address and port the Binding Request was sent to.

Assuming the message integrity check passed, processing continues. The server must check for any attributes in the request with values less than or equal to 0x7fff which it does not understand. If it encounters any, the server must generate a Binding Error Response, and it MUST include an ERROR-CODE attribute with a 420 response code.

That response must contain an UNKNOWN-ATTRIBUTES attribute listing the attributes with values less than or equal to 0x7fff which were not understood. The Binding Error Response is sent to the IP address and port the Binding Request came from, and sent from the IP address and port the Binding Request was sent to.

Assuming the request was correctly formed, the server must generate a single Binding Response. The Binding Response must contain the same transaction ID contained in the Binding Request. The length in the message header must contain the total length of the message in bytes, excluding the header. The Binding Response must have a message type of "Binding Response."

The server must add a MAPPED-ADDRESS attribute to the Binding Response. The IP address component of this attribute must be set to the source IP address observed in the Binding Request. The port component of this attribute must be set to the source port observed in the Binding Request.

If the RESPONSE-ADDRESS attribute was absent from the Binding Request, the destination address and port of the Binding Response must be the same as the source address and port of the Binding Request. Otherwise, the destination address and port of the Binding Response must be the value of the IP address and port in the RESPONSE-ADDRESS attribute.

The source address and port of the Binding Response depend on the value of the CHANGE-REQUEST attribute and on the address and port the Binding Request was received on, and are summarized in Table 5.2.

Let Da represent the destination IP address of the Binding Request (which will be either A1 or A2), and Dp represent the destination port of the Binding Request (which will be either P1 or P2). Let Ca represent the other address, so that if Da is A1, Ca is A2. If Da is A2, Ca is A1. Similarly, let Cp represent the other port, so that if Dp is P1, Cp is P2. If Dp is P2, Cp is P1. If the "change port" flag was set in the CHANGE-REQUEST attribute of the Binding Request, and the "change IP" flag was not set, the source IP address of the Binding Response must be Da and the source port of the Binding Response must be Cp. If the "change IP" flag was set in the Binding Request, and the "change port" flag was not set, the source IP address of the Binding Response must be Ca and the source port of the Binding Response MUST be Dp. When both flags are set, the source IP address of the Binding Response MUST be Ca and the source port of the Binding Response MUST be Cp. If neither flag is set, or if the CHANGE-REQUEST attribute is absent entirely, the source IP address of the Binding Response MUST be Da and the source port of the Binding Response must be Dp.

Table 5.2: Impact of flags on packet source and CHANGED-ADDRESS.

Flags	Source Address	Source Port	CHANGED-ADDRESS
none	Da	Dp	Ca:Cp
Change IP	Ca	Dp	Ca:Cp
Change port	Da	Cp	Ca:Cp
Change IP and Change port	Ca	Cp	Ca:Cp

The server must add a SOURCE-ADDRESS attribute to the Binding Response, containing the source address and port used to send the Binding Response.

The server must add a CHANGED-ADDRESS attribute to the Binding Response. This contains the source IP address and port that would be used if the client had set the "change IP" and "change port" flags in the Binding Request. As summarized in Table 5.2, these are Ca and Cp, respectively, regardless of the value of the CHANGE-REQUEST flags.

If the Binding Request contained both the USERNAME and MESSAGE-INTEGRITY attributes, the server must add a MESSAGE-INTEGRITY attribute to the Binding Response. The attribute contains an HMAC over the response, as described in Section 5.3.11.2. The key to use depends on the shared secret mechanism. If the STUN Shared Secret Request was used, the key must be the one associated with the USERNAME attribute present in the Binding Request.

If the Binding Request contained a RESPONSE-ADDRESS attribute, the server MUST add a REFLECTED-FROM attribute to the response. If the Binding Request was authenticated using a username obtained from a Shared Secret Request, the REFLECTED-FROM attribute MUST contain the source IP address and port where that Shared Secret Request came from. If the username present in the request was not allocated using a Shared Secret Request, the REFLECTED-FROM attribute must contain the source address and port of the entity which obtained the username, as best can be verified with the mechanism used to allocate the username. If the username was not present in the request, and the server was willing to process the request, the REFLECTED-FROM attribute should contain the source IP address and port where the request came from.

The server should not retransmit the response. Reliability is achieved by having the client periodically resend the request, each of which triggers a response from the server.

5.3.8.2 Shared Secret Requests

Shared Secret Requests are always received on TLS connections. When the server receives a request from the client to establish a TLS connection, it must proceed with TLS, and should present a site certificate. The TLS ciphersuite TLS_RSA_WITH_AES_128_CBC_SHA should be used. Client TLS authentication must not be done, since the server is not allocating any resources to clients, and the computational burden can be a source of attacks.

If the server receives a Shared Secret Request, it must verify that the request arrived on a TLS connection. If it did not receive the request over TLS, it must generate a Shared Secret Error Response, and it must include an ERROR-CODE attribute with a 433 response code. The destination for the error response depends on the transport on which the request was received. If the Shared Secret Request was received over TCP, the Shared Secret Error Response is sent over the same connection the request was received on. If the Shared Secret Request was receive over UDP, the Shared Secret Error Response is sent to the source IP address and port that the request came from.

The server must check for any attributes in the request with values less than or equal to 0x7fff which it does not understand. If it encounters any, the server must generate a Shared Secret Error Response, and it must include an ERROR-CODE attribute with a 420 response code. That response must contain an UN-KNOWN-ATTRIBUTES attribute listing the attributes with values less than or equal to 0x7fff which were not understood. The Shared Secret Error Response is sent over the TLS connection.

All Shared Secret Error Responses must contain the same transaction ID contained in the Shared Secret Request. The length in the message header must contain the total length of the message in bytes, excluding the header. The Shared Secret Error Response must have a message type of "Shared Secret Error Response" (0x0112).

Assuming the request was properly constructed, the server creates a Shared Secret Response. The Shared Secret Response must contain the same transaction ID contained in the Shared Secret Request. The length in the message header must contain the total length of the message in bytes, excluding the header. The Shared Secret Response must have a message type of "Shared Secret Response." The Shared Secret Response must contain a USERNAME attribute and a PASSWORD attribute. The USERNAME attribute serves as an index to the password, which is contained in the PASSWORD attribute. The server can use any mechanism it chooses to generate the username. However, the username must be valid for a period of at least 10 minutes. Validity means that the server can compute the password for that username. There MUST be a single password for each username. In other words, the server cannot, 10 minutes later, assign a different password to the same username. The server must hand out a different username for each distinct Shared Secret Request. Distinct, in this case, implies a different transaction ID. It is recommended that the server explicitly invalidate the username after ten minutes. It must invalidate the username after 30 minutes. The PASSWORD contains the password bound to that username. The password must have at least 128 bits. The likelihood that the server assigns the same password for two different usernames must be vanishingly small, and the passwords must be unguessable. In other words, they must be a cryptographically random function of the username.

These requirements can still be met using a stateless server, by intelligently computing the USERNAME and PASSWORD. One approach is to construct the USERNAME as:

```
USERNAME = <prefix,rounded-time,clientIP,hmac>
```

Where prefix is some random text string (different for each shared secret request), rounded-time is the current time modulo 20 minutes, clientIP is the source IP address where the Shared Secret Request came from, and hmac is an HMAC over the prefix, rounded-time, and client IP, using a server private key. The password is then computed as:

```
password = <hmac(USERNAME,anotherprivatekey)>
```

With this structure, the username itself, which will be present in the Binding Request, contains the source IP address where the Shared Secret Request came from. That allows the server to meet the requirements specified in Section 5.3.8.1 for constructing the REFLECTED-FROM attribute. The server can verify that the username was not tampered with, using the hmac present in the username.

The Shared Secret Response is sent over the same TLS connection the request was received on. The server should keep the connection open, and let the client close it.

5.3.9 Client Behavior

The behavior of the client is very straightforward. Its task is to discover the STUN server, obtain a shared secret, formulate the Binding Request, handle request reliability, and process the Binding Responses.

5.3.9.1 Discovery

Generally, the client will be configured with a domain name of the provider of the STUN servers. This domain name is resolved to an IP address and port using the SRV procedures specified in RFC 2782.

Specifically, the service name is "stun." The protocol is "udp" for sending Binding Requests or "tcp" for sending Shared Secret Requests. The procedures of RFC 2782 are followed to determine the server to contact. RFC 2782 spells out the details of how a set of SRV records are sorted and then tried. However, it only states that the client should "try to connect to the (protocol, address, service)" without giving any details on what happens in the event of failure. Those details are described here for STUN.

For STUN requests, failure occurs if there is a transport failure of some sort (generally, due to fatal ICMP errors in UDP or connection failures in TCP). Failure also occurs if the transaction fails due to timeout. This occurs 9.5 seconds after the first request is sent, for both Shared Secret Requests and Binding Requests. See Section 5.3.9.3 for details on transaction timeouts for Binding Requests. If a failure occurs, the client should create a new request, which is identical to the previous, but has a different transaction ID and MESSAGE INTEGRITY attribute (the HMAC will change because the transaction ID has changed). That request is sent to the next element in the list as specified by RFC 2782.

The default port for STUN requests is 3478, for both TCP and UDP. Administrators should use this port in their SRV records, but may use others.

If no SRV records were found, the client performs an A record lookup of the domain name. The result will be a list of IP addresses, each of which can be contacted at the default port.

This would allow a firewall admin to open the STUN port, so hosts within the enterprise could access new applications. Whether they will or will not do this is a relevant question.

5.3.9.2 Obtaining a Shared Secret

There are several attacks possible on STUN systems. Many of these are prevented through integrity of requests and responses. To provide that integrity, STUN makes use of a shared secret between client and server, used as the keying material for an HMAC in both the Binding Request and Binding Response. STUN allows for the shared secret to be obtained in any way (for example, Kerberos). However, it must have at least 128 bits of randomness. In order to ensure interoperability, this specification describes a TLS-based mechanism. This mechanism, described in this section, must be implemented by clients and servers.

First, the client determines the IP address and port that it will open a TCP connection to. This is done using the discovery procedures in Section 5.3.9.1. The client opens up the connection to that address and port, and immediately begins TLS negotiation. The client must verify the identity of the server. To do that, it follows the identification procedures defined in Section 3.1 of RFC 2818. Those procedures assume the client is dereferencing a URI. For purposes of usage with this specification, the client treats the domain name or IP address used in Section 5.3.9.1 as the host portion of the URI that has been dereferenced.

Once the connection is opened, the client sends a Shared Secret request. This request has no attributes, just the header. The transaction ID in the header must meet the requirements outlined for the transaction ID in a binding request, described in Section 5.3.9.3 below. The server generates a response, which can either be a Shared Secret Response or a Shared Secret Error Response.

If the response was a Shared Secret Error Response, the client checks the response code in the ERROR-CODE attribute. Interpretation of those response codes is identical to the processing of Section 5.3.9.4 for the Binding Error Response.

If a client receives a Shared Secret Response with an attribute whose type is greater than 0x7fff, the attribute must be ignored. If the client receives a Shared Secret Response with an attribute whose type is less than or equal to 0x7fff, the response is ignored.

If the response was a Shared Secret Response, it will contain a short-lived username and password encoded in the USERNAME and PASSWORD attributes, respectively.

The client may generate multiple Shared Secret Requests on the connection, and it may do so before receiving Shared Secret Responses to previous Shared Secret Requests. The client should close the connection as soon as it has finished obtaining usernames and passwords.

Section 5.3.9.3 describes how these passwords are used to provide integrity protection over Binding Requests, and Section 5.3.8.1 describes how it is used in Binding Responses.

5.3.9.3 Formulating the Binding Request

A Binding Request formulated by the client follows the syntax rules defined in Section 5.3.11. Any two requests that are not bit-wise identical, and not sent to the same server from the same IP address and port, must carry different transaction IDs. The transaction ID must be uniformly and randomly distributed between 0 and $2^{**}128 - 1$. The large range is needed because the transaction ID serves as a form of randomization, helping to prevent replays of previously signed responses from the server. The message type of the request must be "Binding Request."

The RESPONSE-ADDRESS attribute is optional in the Binding Request. It is used if the client wishes the response to be sent to a different IP address and port than the one the request was sent from. This is useful for determining whether the client is behind a firewall, and for applications that have separated control and data components. See Section 5.3.10.3 for more details. The CHANGE-REQUEST attribute is also optional. Whether it is present depends on what the application is trying to accomplish. See Section 5.3.10 for some example uses.

The client should add MESSAGE-INTEGRITY and USERNAME attributes to the Binding Request. This MESSAGE-INTEGRITY attribute contains an HMAC. The value of the username, and the key to use in the MESSAGE-INTEGRITY attribute depend on the shared secret mechanism. If the STUN Shared Secret Request was used, the USERNAME must be a valid username obtained from a Shared Secret Response within the last nine minutes. The shared secret for the HMAC is the value of the PASSWORD attribute obtained from the same Shared Secret Response.

Once formulated, the client sends the Binding Request. Reliability is accomplished through client retransmissions. Clients should retransmit the request starting with an interval of 100ms, doubling every retransmit until the interval reaches 1.6s. Retransmissions continue with intervals of 1.6s until a response is received, or a total of nine requests have been sent. If no response is received by 1.6 seconds after the last request has been sent, the client should consider the transaction to have failed. In other words, requests would be sent at times 0ms, 100ms, 300ms, 700ms, 1500ms, 3100ms, 4700ms, 6300ms, and 7900ms. At 9500ms, the client considers the transaction to have failed if no response has been received.

5.3.9.4 Processing Binding Responses

The response can either be a Binding Response or Binding Error Response. Binding Error Responses are always received on the source address and port the request was sent from. A Binding Response will be received on the address and port placed in the RESPONSE-ADDRESS attribute of the request. If none was present, the Binding Responses will be received on the source address and port the request was sent from.

If the response is a Binding Error Response, the client checks the response code from the ERROR-CODE attribute of the response. For a 400 response code, the client should display the reason phrase to the user. For a 420 response code, the client should retry the request, this time omitting any attributes listed in the UNKNOWN-ATTRIBUTES attribute of the response. For a 430 response code, the client should obtain a new shared secret, and retry the Binding Request with a new transaction. For 401 and 432 response codes, if the client had omitted the USERNAME or MESSAGE-INTEGRITY attribute as indicated by the error, it should try again with those attributes. For a 431 response code, the client should alert the user, and may try the request again after obtaining a new username and password. For a 500 response code, the client may wait several seconds and then retry the request. For a 600 response code, the client must not retry the request, and should display the reason phrase to the user. Unknown attributes between 400 and 499 are treated like a 400, unknown attributes between 500 and 599 are treated like a 500, and unknown attributes between 600 and 699 are treated like a 600. Any response between 100 and 399 must result in the cessation of request retransmissions, but otherwise is discarded.

If a client receives a response with an attribute whose type is greater than 0x7fff, the attribute MUST be ignored. If the client receives a response with an attribute whose type is less than or equal to 0x7fff, request retransmissions must cease, but the entire response is otherwise ignored. If the response is a Binding Response, the client should check the response for a MESSAGE-INTEGRITY attribute. If not present, and the client placed a MESSAGE-INTEGRITY attribute into the request, it must discard the response. If present, the client computes the HMAC over the response as described in Section 5.3.11.2. The key to use depends on the shared secret mechanism. If the STUN Shared Secret Request was used, the key must be the same as that used to compute the MESSAGE-INTEGRITY attribute in the request. If the computed HMAC differs from the one in the response, the client must discard the response, and should alert the user about a possible attack. If the computed HMAC matches the one from the response, processing continues.

Reception of a response (either a Binding Error Response or Binding Response) to a Binding Request will terminate retransmissions of that request. However, clients must continue to listen for responses to a Binding Request for 10 seconds after the first response. If it receives any responses in this interval with different message types (Binding Responses and Binding Error Responses, for example) or different MAPPED-ADDRESSes, it is an indication of a possible attack. The client must not use the MAPPED-ADDRESS from any of the responses it received (either the first or the additional ones), and should alert the user.

Furthermore, if a client receives more than twice as many Binding Responses as the number of Binding Requests it sent, it must not use the MAPPED-ADDRESS from any of those responses, and should alert the user about a potential attack.

If the Binding Response is authenticated, and the MAPPED-ADDRESS was not discarded because of a potential attack, the CLIENT may use the MAPPED-ADDRESS and SOURCE-ADDRESS attributes.

5.3.10 Use Cases

The rules of Sections 8 and 9 describe exactly how a client and server interact to send requests and get responses. However, they do not dictate how the STUN protocol is used to accomplish useful tasks. That is at the discretion of the client. Here, we provide some useful scenarios for applying STUN.

5.3.10.1 Discovery Process

In this scenario, a user is running a multimedia application and needs to determine which of the following scenarios applies to it:

1. On the open Internet;
2. Firewall that blocks UDP;
3. Firewall that allows UDP out, and responses have to come back to the source of the request (like a symmetric NAT, but no translation; this is called a *symmetric UDP firewall*);
4. Full-cone NAT;

5. Symmetric NAT;
6. Restricted cone or restricted port cone NAT.

The determination of which of the six scenarios applies can be achieved through the flow chart shown in Figure 5.13. The chart refers only to the sequence of Binding Requests; Shared Secret Requests will, of course, be needed to authenticate each Binding Request used in the sequence. The flow makes use of three tests. In test I, the client sends a STUN Binding Request to a server, without any flags set in the CHANGE-REQUEST attribute, and without the RESPONSE-ADDRESS attribute. This causes the server to send the response back to the address and port that the request came from. In test II, the client sends a Binding Request with both the "change IP" and "change port" flags from the CHANGE-REQUEST attribute set. In test III, the client sends a Binding Request with only the "change port" flag set.

The client begins by initiating test I. If this test yields no response, the client knows right away that it is not capable of UDP connectivity. If the test produces a response, the client examines the MAPPED-ADDRESS attribute. If this address and port are the same as the local IP address and port of the socket used to send the request, the client knows that it is not NATed. It executes test II.

If a response is received, the client knows that it has open access to the Internet (or, at least, it's behind a firewall that behaves like a full-cone NAT, but without the translation). If no response is received, the client knows it is behind a symmetric UDP firewall.

In the event that the IP address and port of the socket did not match the MAPPED-ADDRESS attribute in the response to test I, the client knows that it is behind a NAT. It performs test II. If a response is received, the client knows that it is behind a full-cone NAT. If no response is received, it performs test I again, but this time, does so to the address and port from the CHANGED-ADDRESS attribute from the response to test I. If the IP address and port returned in the MAPPED-ADDRESS attribute are not the same as the ones from the first test I, the client knows it's behind a symmetric NAT. If the address and port are the same, the client is either behind a restricted or port restricted NAT. To make a determination about which one it is behind, the client initiates test III. If a response is received, it is behind a restricted NAT, and if no response is received, it is behind a port-restricted NAT.

This procedure yields substantial information about the operating condition of the client application. In the event of multiple NATs between the client and the Internet, the type that is discovered will be the type of the most restrictive NAT between the client and the Internet. The types of NAT, in order of restrictiveness, from most to least, are: symmetric, port-restricted cone, restricted cone, and full cone.

Typically, a client will redo this discovery process periodically to detect changes, or look for inconsistent results. It is important to note that when the discovery process is redone, it should not generally be done from the same local address and port used in the previous discovery process. If the same local address and port are reused, bindings from the previous test may still be in existence, and these will invalidate the results of the test. Using a different local address and port for subsequent tests resolves this problem. An alternative is to wait sufficiently long to be confident that the old bindings have expired (half an hour should more than suffice).

5.3.10.2 Binding Lifetime Discovery

STUN can also be used to discover the lifetimes of the bindings created by the NAT. In many cases, the client will need to refresh the binding, either through a new STUN request, or an application packet, in order for the application to continue to use the binding. By discovering the binding lifetime, the client can determine how frequently it needs to refresh.

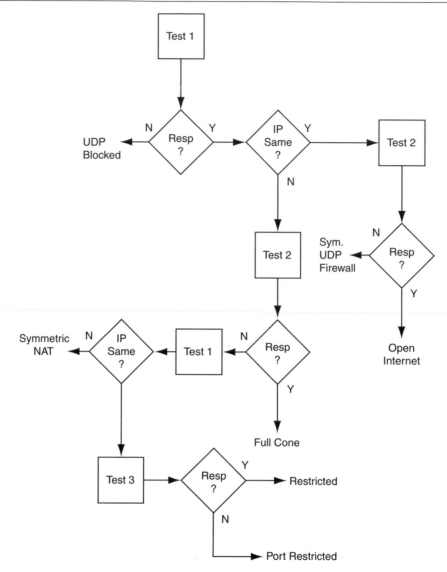

Figure 5.13: Flow for type discovery process.

To determine the binding lifetime, the client first sends a Binding Request to the server from a particular socket, X. This creates a binding in the NAT. The response from the server contains a MAPPED-ADDRESS attribute, providing the public address and port on the NAT. Call this Pa and Pp, respectively. The client then starts a timer with a value of T seconds. When this timer fires, the client sends another Binding Request to the server, using the same destination address and port, but from a different socket, Y. This request contains a RESPONSE-ADDRESS address attribute, set to (Pa,Pp). This will create a new binding on the NAT, and cause the STUN server to send a Binding Response that would match the old binding, if it still exists. If the client receives the Binding Response on socket X, it knows that the binding has not expired. If the client receives the Binding Response on socket Y (which is possible if the old binding expired, and the NAT allocated the same public address and port to the new binding), or receives no response at all, it knows that the binding has expired.

The client can find the value of the binding lifetime by doing a binary search through T, arriving eventually at the value where the response is not received for any timer greater than T, but is received for any timer less than T.

This discovery process takes quite a bit of time, and is something that will typically be run in the background on a device once it boots.

It is possible that the client can get inconsistent results each time this process is run. For example, if the NAT should reboot, or be reset for some reason, the process may discover a lifetime than is shorter than the actual one. For this reason, implementations are encouraged to run the test numerous times, and be prepared to get inconsistent results.

5.3.10.3 Binding Acquisition

Consider once more the case of a VoIP phone. It used the discovery process above when it started up to discover its environment. Now, it wants to make a call. As part of the discovery process, it determined that it was behind a full-cone NAT.

Consider further that this phone consists of two logically separated components—a control component that handles signaling, and a media component that handles the audio, video, and RTP. Both are behind the same NAT. Because of this separation of control and media, we wish to minimize the communication required between them. In fact, they may not even run on the same host.

In order to make a voice call, the phone needs to obtain an IP address and port that it can place in the call setup message as the destination for receiving audio.

To obtain an address, the control component sends a Shared Secret Request to the server, obtains a shared secret, and then sends a Binding Request to the server. No CHANGE-REQUEST attribute is present in the Binding Request, and neither is the RESPONSE-ADDRESS attribute. The Binding Response contains a mapped address. The control component then formulates a second Binding Request. This request contains a RESPONSE-ADDRESS which is set to the mapped address learned from the previous Binding Response. This Binding Request is passed to the media component, along with the IP address and port of the STUN server. The media component sends the Binding Request. The request goes to the STUN server which sends the Binding Response back to the control component. The control component receives this, and now has learned an IP address and port that will be routed back to the media component that sent the request.

The client will be able to receive media from anywhere on this mapped address.

In the case of silence suppression, there may be periods where the client receives no media. In this case, the UDP bindings could timeout (UDP bindings in NATs are typically short; 30 seconds is common). To deal with this, the application can periodically retransmit the query in order to keep the binding fresh.

It is possible that both participants in the multimedia session are behind the same NAT. In that case, both will repeat this procedure above, and both will obtain public address bindings. When one sends media to the other, the media is routed to the NAT, and then turns right back around to come back into the enterprise, where it is translated to the private address of the recipient. This is not particularly efficient, and unfortunately, does not work in many commercial NATs. In such cases, the clients may need to retry using private addresses.

5.3.11 Protocol Details

This section presents the detailed encoding of a STUN message. As noted, STUN is a request-response protocol. Clients send a request, and the server sends a response. There are two requests, Binding Request, and Shared Secret Request. The response to a Binding Request can either be the Binding Response or Binding Error Response. The response to a Shared Secret Request can either be a Shared Secret Response or a Shared Secret Error Response.

STUN messages are encoded using binary fields. All integer fields are carried in network byte order, that is, most significant byte (octet) first. This byte order is commonly known as big-endian. The transmission order is described in detail in Appendix B of RFC 791. Unless otherwise noted, numeric constants are in decimal (base 10).

5.3.11.1 Message Header

All STUN messages consist of a 20 byte header:

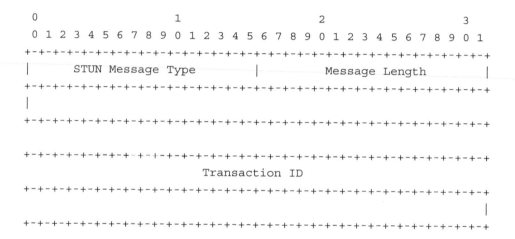

The Message Types can take on the following values:

    0x0001 : Binding Request

    0x0101 : Binding Response

    0x0111 : Binding Error Response

    0x0002 : Shared Secret Request

    0x0102 : Shared Secret Response

    0x0112 : Shared Secret Error Response

The Message Length is the count, in bytes, of the size of the message, not including the 20 byte header.

The Transaction ID is a 128 bit identifier. It also serves as salt to randomize the request and the response. All responses carry the same identifier as the request they correspond to.

5.3.11.2 Message Attributes

After the header are 0 or more attributes. Each attribute is TLV encoded, with a 16-bit type, 16-bit length, and variable value:

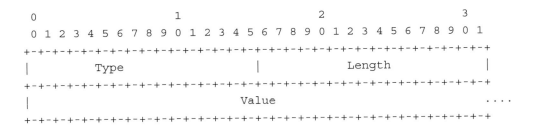

```
 0                   1                   2                   3
 0 1 2 3 4 5 6 7 8 9 0 1 2 3 4 5 6 7 8 9 0 1 2 3 4 5 6 7 8 9 0 1
+-+-+-+-+-+-+-+-+-+-+-+-+-+-+-+-+-+-+-+-+-+-+-+-+-+-+-+-+-+-+-+-+
|         Type             |            Length                 |
+-+-+-+-+-+-+-+-+-+-+-+-+-+-+-+-+-+-+-+-+-+-+-+-+-+-+-+-+-+-+-+-+
|                         Value                           ....
+-+-+-+-+-+-+-+-+-+-+-+-+-+-+-+-+-+-+-+-+-+-+-+-+-+-+-+-+-+-+-+-+
```

The following types are defined:

0x0001: MAPPED-ADDRESS

0x0002: RESPONSE-ADDRESS

0x0003: CHANGE-REQUEST

0x0004: SOURCE-ADDRESS

0x0005: CHANGED-ADDRESS

0x0006: USERNAME

0x0007: PASSWORD

0x0008: MESSAGE-INTEGRITY

0x0009: ERROR-CODE

0x000a: UNKNOWN-ATTRIBUTES

0x000b: REFLECTED-FROM

To allow future revisions of the specification to add new attributes if needed, the attribute space is divided into optional and mandatory ones. Attributes with values greater than 0x7fff are optional, which means that the message can be processed by the client or server even though the attribute is not understood. Attributes with values less than or equal to 0x7fff are mandatory to understand, which means that the client or server cannot process the message unless it understands the attribute.

The MESSAGE-INTEGRITY attribute must be the last attribute within a message. Any attributes that are known, but are not supposed to be present in a message (MAPPED-ADDRESS in a request, for example) must be ignored.

Table 5.3 indicates which attributes are present in which messages. An M indicates that inclusion of the attribute in the message is mandatory, O means its optional, C means it is conditional based on some other aspect of the message, and N/A means that the attribute is not applicable to that message type.

Table 5.3: Summary of Attributes

Att.	Binding Req.	Binding Resp.	Binding Error Resp.	Shared Secret Req.	Shared Secret Resp.	Secret Shared Error Resp.
MAPPED-ADDRESS	N/A	M	N/A	N/A	N/A	N/A
RESPONSE-ADDRESS	O	N/A	N/A	N/A	N/A	N/A
CHANGE-REQUEST	O	N/A	N/A	N/A	N/A	N/A
SOURCE-ADDRESS	N/A	M	N/A	N/A	N/A	N/A
CHANGED-ADDRESS	N/A	M	N/A	N/A	N/A	N/A
USERNAME	O	N/A	N/A	N/A	M	N/A
PASSWORD	N/A	N/A	N/A	N/A	M	N/A
MESSAGE-INTEGRITY	O	O	N/A	N/A	N/A	N/A
ERROR-CODE	N/A	N/A	M	N/A	N/A	M
UNKNOWN-ATTRIBUTES	N/A	N/A	C	N/A	N/A	C
REFLECTED-FROM	N/A	C	N/A	N/A	N/A	N/A

The length refers to the length of the value element, expressed as an unsigned integral number of bytes.

MAPPED-ADDRESS

The MAPPED-ADDRESS attribute indicates the mapped IP address and port. It consists of an eight-bit address family, and a sixteen bit port, followed by a fixed length value representing the IP address.

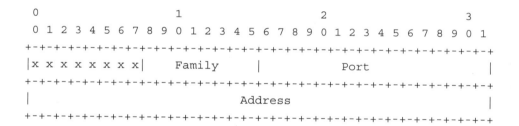

```
 0                   1                   2                   3
 0 1 2 3 4 5 6 7 8 9 0 1 2 3 4 5 6 7 8 9 0 1 2 3 4 5 6 7 8 9 0 1
+-+-+-+-+-+-+-+-+-+-+-+-+-+-+-+-+-+-+-+-+-+-+-+-+-+-+-+-+-+-+-+-+
|x x x x x x x x|     Family    |           Port                |
+-+-+-+-+-+-+-+-+-+-+-+-+-+-+-+-+-+-+-+-+-+-+-+-+-+-+-+-+-+-+-+-+
|                            Address                            |
+-+-+-+-+-+-+-+-+-+-+-+-+-+-+-+-+-+-+-+-+-+-+-+-+-+-+-+-+-+-+-+-+
```

The port is a network byte-ordered representation of the mapped port. The address family is always 0x01, corresponding to IPv4. The first 8 bits of the MAPPED-ADDRESS are ignored, for the purposes of aligning parameters on natural boundaries. The IPv4 address is 32 bits.

RESPONSE-ADDRESS

The RESPONSE-ADDRESS attribute indicates where the response to a Binding Request should be sent. Its syntax is identical to MAPPED-ADDRESS.

CHANGED-ADDRESS

The CHANGED-ADDRESS attribute indicates the IP address and port where responses would have been sent from if the "change IP" and "change port" flags had been set in the CHANGE-REQUEST attribute of the Binding Request. The attribute is always present in a Binding Response, independent of the value of the flags. Its syntax is identical to MAPPED-ADDRESS.

CHANGE-REQUEST

The CHANGE-REQUEST attribute is used by the client to request that the server use a different address and/or port when sending the response. The attribute is 32 bits long, although only two bits (A and B) are used:

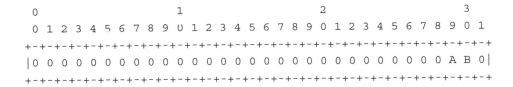

The meaning of the flags is:

A: This is the "change IP" flag. If true, it requests the server to send the Binding Response with a different IP address than the one the Binding Request was received on.

B: This is the "change port" flag. If true, it requests the server to send the Binding Response with a different port than the one the Binding Request was received on.

SOURCE-ADDRESS

The SOURCE-ADDRESS attribute is present in Binding Responses. It indicates the source IP address and port that the server is sending the response from. Its syntax is identical to that of MAPPED-ADDRESS.

USERNAME

The USERNAME attribute is used for message integrity. It serves as a means to identify the shared secret used in the message integrity check. The USERNAME is always present in a Shared Secret Response, along with the PASSWORD. It is optionally present in a Binding Request when message integrity is used.

The value of USERNAME is a variable length opaque value. Its length MUST be a multiple of 4 (measured in bytes) in order to guarantee alignment of attributes on word boundaries.

PASSWORD

The PASSWORD attribute is used in Shared Secret Responses. It is always present in a Shared Secret Response, along with the USERNAME.

The value of PASSWORD is a variable length value that is to be used as a shared secret. Its length MUST be a multiple of 4 (measured in bytes) in order to guarantee alignment of attributes on word boundaries.

MESSAGE-INTEGRITY

The MESSAGE-INTEGRITY attribute contains an HMAC-SHA1 of the STUN message. It can be present in Binding Requests or Binding Responses. Since it uses the SHA1 hash, the HMAC will be 20 bytes. The text used as input to HMAC is the STUN message, including the header, up to and including the attribute preceding the MESSAGE-INTEGRITY attribute. That text is then padded with zeroes so as to be a multiple of 64 bytes. As a result, the MESSAGE-INTEGRITY attribute must be the last attribute in any STUN message. The key used as input to HMAC depends on the context.

ERROR-CODE

The ERROR-CODE attribute is present in the Binding Error Response and Shared Secret Error Response. It is a numeric value in the range of 100 to 699 plus a textual reason phrase encoded in UTF-8, and is consistent in its code assignments and semantics with SIP and HTTP. The reason phrase is meant for user consumption, and can be anything appropriate for the response code. The lengths of the reason phrases must be a multiple of 4 (measured in bytes). This can be accomplished by added spaces to the end of the text, if necessary. Recommended reason phrases for the defined response codes are presented below.

To facilitate processing, the class of the error code (the hundreds digit) is encoded separately from the rest of the code.

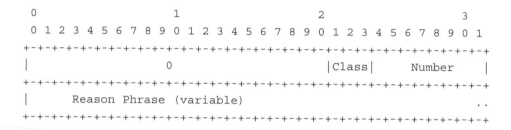

The class represents the hundreds digit of the response code. The value must be between 1 and 6. The number represents the response code modulo 100, and its value must be between 0 and 99.

The following response codes, along with their recommended reason phrases (in brackets) are defined at this time.

400 (Bad Request): The request was malformed. The client should not retry the request without modification from the previous attempt.

401 (Unauthorized): The Binding Request did not contain a MESSAGE-INTEGRITY attribute.

420 (Unknown Attribute): The server did not understand a mandatory attribute in the request.

430 (Stale Credentials): The Binding Request did contain a MESSAGE-INTEGRITY attribute, but it used a shared secret that has expired. The client should obtain a new shared secret and try again.

431 (Integrity Check Failure): The Binding Request contained a MESSAGE-INTEGRITY attribute, but the HMAC failed verification. This could be a sign of a potential attack or client implementation error.

432 (Missing Username): The Binding Request contained a MESSAGE-INTEGRITY attribute, but not a USERNAME attribute. Both must be present for integrity checks.

433 (Use TLS): The Shared Secret request has to be sent over TLS, but was not received over TLS.

500 (Server Error): The server has suffered a temporary error. The client should try again.

600 (Global Failure): The server is refusing to fulfill the request. The client should not retry.

UNKNOWN-ATTRIBUTES

The UNKNOWN-ATTRIBUTES attribute is present only in a Binding Error Response or Shared Secret Error Response when the response code in the ERROR-CODE attribute is 420.

The attribute contains a list of 16 bit values, each of which represents an attribute type that was not understood by the server. If the number of unknown attributes is an odd number, one of the attributes must be repeated in the list, so that the total length of the list is a multiple of 4 bytes.

```
 0                   1                   2                   3
 0 1 2 3 4 5 6 7 8 9 0 1 2 3 4 5 6 7 8 9 0 1 2 3 4 5 6 7 8 9 0 1
+-+-+-+-+-+-+-+-+-+-+-+-+-+-+-+-+-+-+-+-+-+-+-+-+-+-+-+-+-+-+-+-+
|          Attribute 1 Type         |         Attribute 2 Type          |
+-+-+-+-+-+-+-+-+-+-+-+-+-+-+-+-+-+-+-+-+-+-+-+-+-+-+-+-+-+-+-+-+
|          Attribute 3 Type         |         Attribute 4 Type     ...
+-+-+-+-+-+-+-+-+-+-+-+-+-+-+-+-+-+-+-+-+-+-+-+-+-+-+-+-+-+-+-+-+
```

REFLECTED-FROM

The REFLECTED-FROM attribute is present only in Binding Responses, when the Binding Request contained a RESPONSE-ADDRESS attribute. The attribute contains the identity (in terms of IP address) of the source where the request came from. Its purpose is to provide traceability, so that a STUN server cannot be used as a reflector for denial-of-service attacks. Its syntax is identical to the MAPPED-ADDRESS attribute.

5.4 Overview of MIDCOM Approaches

This section looks at the newly-defined topic of Middlebox Communications (MIDCOM), which was alluded to above in the context of STUN. A principal objective of RFC 3303 is to describe the underlying framework of MIDCOM to enable complex applications through the middleboxes, seamlessly using a trusted third party. This discussion is based on RFC 3303 [SRI200201]. Developers should refer to the original RFC and all supportive extensions, updates, etc., for normative development guidance.

5.4.1 Background

There are a variety of intermediate devices in the Internet today that require application intelligence for their operation. Datagrams pertaining to real-time streaming applications, such as SIP and H.323, and peer-to-peer applications, such as Napster and NetMeeting, cannot be identified by merely examining packet headers. Middleboxes implementing Firewall and Network Address Translator services typically embed application intelligence within the device for their operation. The document specifies an architecture and framework in which trusted third parties can be delegated to assist the middleboxes to perform their operation, without resorting to embedding application intelligence. Doing this will allow a middlebox to continue to provide the services while keeping the middlebox application agnostic.

Intermediate devices requiring application intelligence are the subject of RFC 3303. These devices are referred to as middleboxes throughout the document. Many of these devices enforce application-specific policy-based functions such as packet filtering, VPN (Virtual Private Network) tunneling, Intrusion detection, security, and so forth. Network Address Translator service, on the other hand, provides routing transparency across address realms (within IPv4 routing network or across V4 and V6 routing realms) independent of applications. Application Level Gateways (ALGs) are used in conjunction with NAT to examine and optionally modify application payload so the end-to-end application behavior remains unchanged for many of the applications traversing NAT middleboxes. There may be other types of services requiring embedding application intelligence in middleboxes for their operation. The discussion scope of this RFC is however limited to Firewall and NAT services. Nonetheless, the MIDCOM framework is designed to be extensible to support the deployment of new services.

Tight coupling of application intelligence with middleboxes makes maintenance of middleboxes hard with the advent of new applications. Built-in application awareness typically requires updates of operating systems with new applications or newer versions of existing applications. Operators requiring support for newer applications will not be able to use third party software/hardware specific to the application and are at the mercy of their middlebox vendor to make the necessary upgrade. Further, embedding intelligence for a large number of application protocols within the same middlebox increases complexity of the middlebox and is likely to be error prone and degrade in performance.

RFC 3303 describes a framework in which application intelligence can be moved from middleboxes into external MIDCOM agents. The premise of the framework is to devise a MIDCOM protocol that is application independent so the middleboxes can stay focused on services such as firewall and NAT. The framework document includes some explicit and implied requirements for the MIDCOM protocol. However, it must be noted that these requirements are only a subset. A separate requirements document lists the requirements in detail.

MIDCOM agents with application intelligence can assist the middleboxes through the MIDCOM protocol in permitting applications such as FTP, SIP and H.323. The communication between a MIDCOM agent and a middlebox will not be noticeable to the end-hosts that take part in the application, unless one of the end-hosts assumes the role of a MIDCOM agent. Discovery of middleboxes or MIDCOM agents in the path of an application instance is outside the scope of this RFC. Further, any communication amongst middleboxes is also outside the scope of RFC 3303.

RFC 3303 describes the framework in which middlebox communication takes place and the various elements that constitute the framework. Section 5.4.2 describes the terms used in the document. Section 5.4.3 defines the architectural framework of a middlebox for communication with MIDCOM agents. The remaining sections cover the components of the framework, illustration using sample flows, and operational considerations with the MIDCOM architecture. Section 5.4.4 describes the nature of MIDCOM protocol. Section 5.4.5 identifies entities that could potentially host the MIDCOM agent function. Section 5.4.6 considers the role of Policy server and its function with regard to communicating MIDCOM agent authorization policies. Section 5.4.7 is an illustration of SIP flows using a MIDCOM framework in which the MIDCOM agent is co-resident on a SIP proxy server. Section 5.4.8 addresses operational considerations in deploying a protocol adhering to the framework described here. Section 5.4.9 is an applicability statement, scoping the location of middleboxes.

5.4.2   Terminology

Below are the definitions for the terms used in RFC 3303.

5.4.2.1 Middlebox Function/Service

A middlebox function or a middlebox service is an operation or method performed by a network intermediary that may require application-specific intelligence for its operation. Policy-based packet filtering (a.k.a. firewall), Network Address Translation (NAT), Intrusion detection, Load balancing, Policy-based tunneling, and IPsec security are all examples of a middlebox function (or service).

5.4.2.2   Middlebox

A middlebox is a network intermediate device that implements one or more of the middlebox services. A NAT middlebox is a middlebox implementing NAT service. A firewall middlebox is a middlebox implementing firewall service.

Traditional middleboxes embed application intelligence within the device to support specific application traversal. Middleboxes supporting the MIDCOM protocol will be able to externalize application intelligence into MIDCOM agents. In reality, some of the middleboxes may continue to embed application intelligence for certain applications and depend on MIDCOM protocol and MIDCOM agents for the support of remaining applications.

5.4.2.3   Firewall

Firewall is a policy-based packet-filtering middlebox function, typically used for restricting access to/from specific devices and applications. The policies are often termed Access Control Lists (ACLs).

5.4.2.4   NAT

Network Address Translation is a method by which IP addresses are mapped from one address realm to another, providing transparent routing to end-hosts. Transparent routing here refers to modifying end-node addresses en route and maintaining state for these updates so that when a datagram leaves one realm and enters another, datagrams pertaining to a session are forwarded to the right end-host in either realm. Refer to RFC 2663 for the definition of Transparent routing, various NAT types, and the associated terms in use. Two types of NAT are most common. Basic-NAT, where only an IP address (and the related IP, TCP/UDP checksums) of packets is altered and NAPT (Network Address Port Translation), where both an IP address and a transport layer identifier, such as a TCP/UDP port (and the related IP, TCP/UDP checksums), are altered.

The term NAT here is very similar to the IPv4 NAT described in RFC 2663, but is extended beyond IPv4 networks to include the IPv4-v6 NAT-PT described in RFC 2766. While the IPv4 NAT translates one IPv4 address into another IPv4 address to provide routing between private v4 and external V4 address realms, IPv4-v6 NAT-PT (RFC 2766) translates an IPv4 address into an IPv6 address, and vice versa, to provide routing between a v6 address realm and an external v4 address realm. Unless specified otherwise, NAT is a middlebox function referring to both IPv4 NAT, as well as IPv4-v6 NAT-PT.

5.4.2.5   Proxy

A proxy is an intermediate relay agent between clients and servers of an application, relaying application messages between the two. Proxies use special protocol mechanisms to communicate with proxy clients and relay client data to servers and vice versa. A proxy terminates sessions with both the client and the server, acting as server to the end-host client and as client to the end-host server.

Applications such as FTP, SIP, and RTSP use a control session to establish data sessions. These control and data sessions can take divergent paths. While a proxy can intercept both the control and data sessions, it might intercept only the control session. This is often the case with real-time streaming applications such as SIP and RTSP.

5.4.2.6   ALG

Application Level Gateways are entities that possess the application-specific intelligence and knowledge of an associated middlebox function. They examine application traffic in transit and assist the middlebox in carrying out its function.

An ALG may be a co-resident with a middlebox or reside externally, communicating through a middlebox communication protocol. It interacts with a middlebox to set up state, access control filters, use middlebox state information, modify application specific payload, or perform whatever else is necessary to enable the application to run through the middlebox.

ALGs are different from proxies in that they are not visible to end-hosts, unlike the proxies which are relay agents terminating sessions with both end-hosts. They do not terminate sessions with either end-host. Instead, they examine, and optionally modify, application payload content to facilitate the flow of application traffic through a middlebox. ALGs are middlebox centric, in that they assist the middleboxes in carrying out their function, whereas, the proxies act as a focal point for application servers, relaying traffic between application clients and servers.

ALGs are similar to Proxies, in that both ALGs and proxies facilitate application-specific communication between clients and servers.

5.4.2.7 End-Hosts

End-hosts are entities that are party to a networked application instance. End-hosts referred to in this RFC, are specifically those terminating Real-time streaming Voice-over-IP applications such as SIP and H.323, and peer-to-peer applications such as Napster and NetMeeting.

5.4.2.8   MIDCOM Agents

MIDCOM agents are entities performing ALG functions, logically external to a middlebox. MIDCOM agents possess a combination of application awareness and knowledge of the middlebox function. This combination enables the agents to facilitate traversal of the middlebox by the application's packets. A MIDCOM agent may interact with one or more middleboxes.

Only "In-Path MIDCOM agents" are considered in this RFC. In-Path MIDCOM agents are agents which are within the path of those datagrams that the agent needs to examine and/or modify in fulfilling its role as a MIDCOM agent. "Within the path" here simply means that the packets in question flow through the node that hosts the agent. The packets may be addressed to the agent node at the IP layer. Alternatively, they may not be addressed to the agent node, but may be constrained by other factors to flow through it. In fact, it is immaterial to the MIDCOM protocol which of these is the case. Some examples of In-Path MIDCOM agents are application proxies, gateways, or even end-hosts that are party to the application.

Agents not resident on nodes that are within the path of their relevant application flows are referred to as "Out-of-Path (OOP) MIDCOM agents" and are out of the scope of this RFC.

5.4.2.9   MIDCOM PDP

MIDCOM Policy Decision Point (PDP) is primarily a Policy Decision Point (PDP) as defined in RFC 3198; and also acts as a policy repository, holding MIDCOM-related policy profiles in order to make authorization decisions. RFC 3198 defines a PDP as "a logical entity that makes policy decisions for itself or for other network elements that request such decisions"; and a policy repository as "a specific data store that holds policy rules, their conditions and actions, and related policy data."

A middlebox and a MIDCOM PDP may communicate further if the MIDCOM PDP's policy changes or if a middlebox needs further information. The MIDCOM PDP may, at any time, notify the middlebox to terminate authorization for an agent.

The protocol facilitating the communication between a middlebox and MIDCOM PDP need not be part of the MIDCOM protocol. Section 5.4.6 in the document addresses the MIDCOM PDP interface and protocol framework independent of the MIDCOM framework.

Application-specific policy data and policy interface between an agent or application endpoint and a MIDCOM PDP is out of bounds for this RFC. The MIDCOM PDP issues addressed in the document are focused at an aggregate domain level as befitting the middlebox. For example, a SIP MIDCOM agent may choose to query a MIDCOM PDP for the administrative (or corporate) domain to find whether a certain user is allowed to make an outgoing call. This type of application-specific policy data, as befitting an end user, is out of bounds for the MIDCOM PDP considered in this RFC. It is within bounds, however, for the MIDCOM PDP to specify the specific end-user applications (or tuples) for which an agent is permitted to be an ALG.

5.4.2.10 Middlebox Communication (MIDCOM) protocol

The protocol between a MIDCOM agent and a middlebox allows the MIDCOM agent to invoke services of the middlebox and allow the middlebox to delegate application specific processing to the MIDCOM agent. The MIDCOM protocol allows the middlebox to perform its operation with the aid of MIDCOM agents, without resorting to embedding application intelligence. The principal motivation behind architecting this protocol is to enable complex applications through middleboxes, seamlessly using a trusted third party, i.e., a MIDCOM agent.

This is a protocol yet to be devised.

5.4.2.11 MIDCOM Agent Registration

A MIDCOM agent registration is defined as the process of provisioning agent profile information with the middlebox or a MIDCOM PDP. MIDCOM agent registration is often a manual operation performed by an operator rather than the agent itself.

A MIDCOM agent profile may include agent authorization policy (i.e., session tuples for which the agent is authorized to act as ALG), agent-hosting-entity (e.g., Proxy, Gateway, or end-host which hosts the agent), agent accessibility profile (including any host level authentication information), and security profile (for the messages exchanged between the middlebox and the agent).

5.4.2.12 MIDCOM Session

A MIDCOM session is defined to be a lasting association between a MIDCOM agent and a middlebox. The MIDCOM session is not assumed to imply any specific transport layer protocol. Specifically, this should not be construed as referring to a connection-oriented TCP protocol.

5.4.2.13 Filter

A filter is packet matching information that identifies a set of packets to be treated a certain way by a middlebox. This definition is consistent with RFC 3198, which defines a filter as "A set of terms and/or criteria used for the purpose of separating or categorizing. This is accomplished via single- or multifield matching of traffic header and/or payload data."

5-Tuple specification of packets in the case of a firewall and 5-tuple specification of a session in the case of a NAT middlebox function are examples of a filter.

5.4.2.14 Policy action (or) Action

Policy action (or Action) is a description of the middlebox treatment/service to be applied to a set of packets. This definition is consistent with RFC 3198, which defines a policy action as "Definition of what is to be done to enforce a policy rule, when the conditions of the rule are met. Policy actions may result in the execution of one or more operations to affect and/or configure network traffic and network resources."

NAT Address-BIND (or Port-BIND in the case of NAPT) and firewall permit/deny action are examples of an Action.

5.4.2.15 Policy Rule(s)

The combination of one or more filters and one or more actions. Packets matching a filter are to be treated as specified by the associated action(s). The Policy rules may also contain auxiliary attributes such as individual rule type, timeout values, creating agent, etc.

Policy rules are communicated through the MIDCOM protocol.

5.4.3 Architectural Framework for Middleboxes

A middlebox may implement one or more of the middlebox functions selectively on multiple interfaces of the device. There can be a variety of MIDCOM agents interfacing with the middlebox to communicate with one or more of the middlebox functions on an interface. As such, the middlebox communication protocol must allow for selective communication between a specific MIDCOM agent and one or more middlebox functions on the interface. Figure 5.14 identifies a possible layering of the service supported by a middlebox and a list of MIDCOM agents that might interact with it.

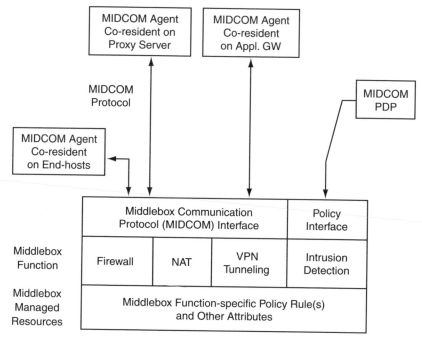

Figure 5.14: MIDCOM agents interfacing with a middlebox.

Firewall ACLs, NAT-BINDs, NAT address-maps, and Session-state are a few of the middlebox function-specific policy rules. A session state may include middlebox function-specific attributes, such as timeout values, NAT translation parameters (i.e., NAT-BINDS), and so forth. As Session-state may be shared across middlebox functions, a Session-state may be created by a function, and terminated by a different function. For example, a session-state may be created by the firewall function, but terminated by the NAT function, when a session timer expires.

Application specific MIDCOM agents (co-resident on the middlebox or external to the middlebox) would examine the IP datagrams and help identify the application the datagram belongs to, and assist the middlebox in performing functions unique to the application and the middlebox service. For example, a MIDCOM agent, assisting a NAT middlebox, might perform payload translations, whereas a MIDCOM agent assisting a firewall middlebox might request the firewall to permit access to application-specific, dynamically-generated session traffic.

5.4.4 MIDCOM Protocol

The MIDCOM protocol between a MIDCOM agent and a middlebox allows the MIDCOM agent to invoke services of the middlebox and allow the middlebox to delegate application-specific processing to the

MIDCOM agent. The protocol will allow MIDCOM agents to signal the middleboxes, to let complex applications using dynamic port-based sessions through them (i.e., middleboxes) seamlessly.

It is important to note that an agent and a middlebox can be on the same physical device. In such a case, they may communicate using a MIDCOM protocol message format (but using a non-IP based transport, such as IPC messaging), (or) they may communicate using well-defined API/DLL, (or) the application intelligence is fully embedded into the middlebox service (as it is done today in many stateful inspection firewall devices and NAT devices).

The MIDCOM protocol will consist of a session setup phase, run-time session phase, and a session termination phase.

Session setup must be preceded by registration of the MIDCOM agent with either the middlebox or the MIDCOM PDP. The MIDCOM agent access and authorization profile may either be preconfigured on the middlebox (or) listed on a MIDCOM PDP; the middlebox is configured to consult. MIDCOM shall be a client-server protocol initiated by the agent.

A MIDCOM session may be terminated by either of the parties. A MIDCOM session termination may also be triggered by (a) the middlebox or the agent going out of service and not being available for further MIDCOM operations, or (b) the MIDCOM PDP notifying the middlebox that a particular MIDCOM agent is no longer authorized.

The MIDCOM protocol data exchanged during runtime is governed principally by the middlebox services the protocol supports. Firewall and NAT middlebox services are considered in this RFC. Nonetheless, the MIDCOM framework is designed to be extensible to support the deployment of other services as well.

5.4.5 MIDCOM Agents

MIDCOM agents are logical entities which may reside physically on nodes external to a middlebox, possessing a combination of application awareness and knowledge of middlebox function. A MIDCOM agent may communicate with one or more middleboxes. The issues of middleboxes discovering agents, or vice versa, are outside the scope of this RFC. The focus of the document is the framework in which a MIDCOM agent communicates with a middlebox using MIDCOM protocol, which is yet to be devised. Specifically, the focus is restricted to just the In-Path agents.

In-Path MIDCOM agents are MIDCOM agents that are located naturally within the message path of the application(s) they are associated with. Bundled session applications, such as H.323, SIP, and RTSP which have separate control and data sessions, may have their sessions take divergent paths. In those scenarios, In-Path MIDCOM agents are those that find themselves in the control path. In a majority of cases, a middlebox will likely require the assistance of a single agent for an application in the control path alone. However, it is possible that a middlebox function, or a specific application traversing the middlebox might require the intervention of more than a single MIDCOM agent for the same application, one for each sub-session of the application.

Application Proxies and gateways are a good choice for In-Path MIDCOM agents as these entities, by definition, are in the path of an application between a client and server. In addition to hosting the MIDCOM agent function, these natively in-path application-specific entities may also enforce application-specific choices locally, such as dropping messages infected with known viruses or lacking user authentication. These entities can be interjecting both the control and data sessions. For example, FTP control and Data sessions are interjected by an FTP proxy server.

However, proxies may also be interjecting just the control session and not the data sessions, as is the case with real-time streaming applications such as SIP and RTSP. Note, applications may not always traverse a proxy and some applications may not have a proxy server available.

SIP proxies and H.323 gatekeepers may be used to host MIDCOM agent functions to control middleboxes implementing firewall and NAT functions. The advantage of using in-path entities, as opposed to creating an entirely new agent, is that the in-path entities already possess application intelligence. You will need to merely enable the use of the MIDCOM protocol to be an effective MIDCOM agent. Figure 5.15 illustrates a scenario where the in-path MIDCOM agents interface with the middlebox. Let us say, the MIDCOM PDP has preconfigured the in-path proxies as trusted MIDCOM agents on the middlebox and the packet filter implements a 'default-deny' packet filtering policy. Proxies use their application-awareness knowledge to control the firewall function and selectively permit a certain number of voice stream sessions dynamically using MIDCOM protocol.

In the illustration below, the proxies and the MIDCOM PDP are shown inside a private domain. The intent however, is not to imply that they be inside the private boundary alone. The proxies may also reside external to the domain. The only requirement is that there be a trust relationship with the middlebox.

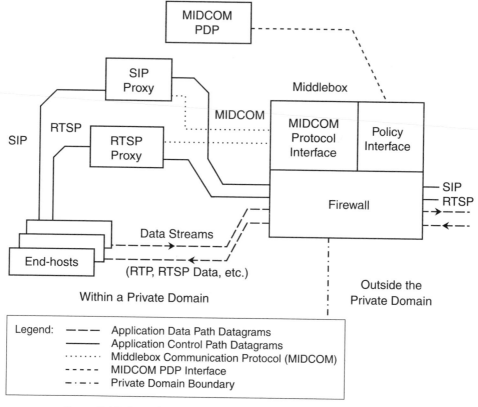

Figure 5.15: In-path MIDCOM agents for middlebox communication.

5.4.5.1 End-hosts as In-path MIDCOM Agents

End-hosts are another variation of In-Path MIDCOM agents. Unlike Proxies, End-hosts are a direct party to the application and possess all the end-to-end application intelligence there is to it. End-hosts presumably terminate both the control and data paths of an application. Unlike other entities hosting MIDCOM agents, end-host is able to process secure datagrams. However, the problem would be one of manageability—upgrading all the end-hosts running a specific application.

5.4.6 MIDCOM PDP Functions

The functional decomposition of the MIDCOM architecture assumes the existence of a logical entity, known as MIDCOM PDP, responsible for performing authorization and related provisioning services for the middlebox as depicted in Figure 5.14. The MIDCOM PDP is a logical entity which may reside physically on a middlebox or on a node external to the middlebox. The protocol employed for communication between the middlebox and the MIDCOM PDP is unrelated to the MIDCOM protocol.

Agents are registered with a MIDCOM PDP for authorization to invoke services of the middlebox. The MIDCOM PDP maintains a list of agents that are authorized to connect to each of the middleboxes the MIDCOM PDP supports. In the context of the MIDCOM Framework, the MIDCOM PDP does not assist a middlebox in the implementation of the services it provides.

The MIDCOM PDP acts in an advisory capacity to a middlebox, to authorize or terminate authorization for an agent attempting connectivity to the middlebox. The primary objective of a MIDCOM PDP is to communicate agent authorization information so as to ensure that the security and integrity of a middlebox is not jeopardized. Specifically, the MIDCOM PDP should associate a trust level with each agent attempting to connect to a middlebox and provide a security profile. The MIDCOM PDP should be capable of addressing cases when end-hosts are agents to the middlebox.

5.4.6.1 Authentication, Integrity and Confidentiality

Host authenticity and individual message security are two distinct types of security considerations. Host authentication refers to credentials required of a MIDCOM agent to authenticate itself to the middlebox and vice versa. When authentication fails, the middlebox must not process signaling requests received from the agent that failed authentication. Two-way authentication should be supported. In some cases, the two-way authentication may be tightly linked to the establishment of keys to protect subsequent traffic. Two-way authentication is often required to prevent various active attacks on the MIDCOM protocol and secure establishment of keying material.

Security services such as authentication, data integrity, confidentiality and replay protection may be adapted to secure MIDCOM messages in an untrusted domain. Message authentication is the same as data origin authentication and is an affirmation that the sender of the message is who it claims to be. Data integrity refers to the ability to ensure that a message has not been accidentally (maliciously or otherwise) altered or destroyed. Confidentiality is the encryption of a message with a key, so that only those in possession of the key can decipher the message content. Lastly, replay protection is a form of sequence integrity, so when an intruder plays back a previously-recorded sequence of messages, the receiver of the replay messages will simply drop the replay messages into bit-bucket. Certain applications of the MIDCOM protocol might require support for nonrepudiation as an option of the data integrity service. Typically, support for nonrepudiation is required for billing, service level agreements, payment orders, and rec eipts for delivery of service.

IPsec IP Authentication Header (AH) offers data-origin authentication, data integrity and protection from message replay. IPsec Encapsulating Security Payload (ESP) provides data-origin authentication to a lesser degree (same as IPsec AH if the MIDCOM transport protocol turns out to be TCP or UDP), message confidentiality, data integrity, and protection from replay. Besides the IPsec based protocols, there are other security options as well. TLS based transport layer security is one option. There are also many application-layer security mechanisms available. Simple Source-address based security is a minimal form of security and should be relied on only in the most trusted environments, where those hosts will not be spoofed.

The MIDCOM message security shall use existing standards, whenever the existing standards satisfy the requirements. Security shall be specified to minimize the impact on sessions that do not use the security option. Security should be designed to avoid introducing, and to minimize the impact of, denial of service attacks. Some security mechanisms and algorithms require substantial processing or storage, in which case the security protocols should protect themselves as well against possible flooding attacks that overwhelm the endpoint (i.e., the middlebox or the agent) with such processing. For connection-oriented protocols (such as TCP) using security services, the security protocol should detect premature closure or truncation attacks.

5.4.6.2 Registration and Deregistration of MIDCOM Agents

Prior to allowing MIDCOM agents to invoke services of the middlebox, a registration process must take place. Registration is a different process than establishing a MIDCOM session. The former requires provisioning agent profile information with the middlebox or a MIDCOM PDP. Agent registration is often a manual operation performed by an operator rather than the agent itself. Setting up a MIDCOM session refers to establishing a MIDCOM transport session and exchanging security credentials between an agent and a middlebox. The transport session uses the registered information for session establishment.

Profile of a MIDCOM agent includes agent authorization policy (i.e., session tuples for which the agent is authorized to act as ALG), agent-hosting-entity (e.g., Proxy, Gateway or end-host which hosts the agent), agent accessibility profile (including any host level authentication information), and security profile (i.e., security requirements for messages exchanged between the middlebox and the agent).

MIDCOM agent profile may be preconfigured on a middlebox. Subsequent to that, the agent may choose to initiate a MIDCOM session prior to any data traffic. For example, MIDCOM agent authorization policy for a middlebox service may be preconfigured by specifying the agent in conjunction with a filter. In the case of a firewall, for example, the ACL tuple may be altered to reflect the optional Agent presence. The revised ACL may look something like the following.

```
(<Session-Direction>, <Source-Address>, <Destination-Address>, <IP-
Protocol>, <Source-Port>, <Destination-Port>, <Agent>)
```

The reader should note that this is an illustrative example and not necessarily the actual definition of an ACL tuple. The formal description of the ACL is yet to be devised. Agent accessibility information should also be provisioned. For a MIDCOM agent, accessibility information includes the IP address, trust level, host authentication parameters, and message authentication parameters. Once a session is established between a middlebox and a MIDCOM agent, that session should be usable with multiple instances of the application(s), as appropriate. Note, all of this could be captured in an agent profile for ease of management.

The technique described above is necessary for the pre-registration of MIDCOM agents with the middlebox. The middlebox provisioning may remain unchanged, if the middlebox learns of the registered agents through a MIDCOM PDP. In either case, the MIDCOM agent should initiate the session prior to the start of the application. If the agent session is delayed until after the application has started, the agent might be unable to process the control stream to permit the data sessions. When a middlebox notices an incoming MIDCOM session, and the middlebox has no prior profile of the MIDCOM agent, the middlebox will consult its MIDCOM PDP for authenticity, authorization, and trust guidelines for the session.

5.4.7 MIDCOM Framework Illustration Using an In-Path Agent

In Figure 5.16, one considers SIP applications to illustrate the operation of the MIDCOM protocol. Specifically, the application assumes that a caller, external to a private domain, initiates the call. The middlebox is assumed to be located at the edge of the private domain. A SIP phone (SIP User Agent Client/Server) inside the private domain is capable of receiving calls from external SIP phones. The caller uses a SIP Proxy, node located external to the private domain, as its outbound proxy. No interior proxy is assumed for the callee. Lastly, the external SIP proxy node is designated to host the MIDCOM agent function.

Arrows 1 and 8 in the figure below refer to a SIP call setup exchange between the external SIP phone and the SIP proxy. Arrows 4 and 5 refer to a SIP call setup exchange between the SIP proxy and the interior SIP phone, and are assumed to be traversing the middlebox. Arrows 2, 3, 6 and 7 below, between the SIP proxy and the middlebox, refer to MIDCOM communication. Na and Nb represent RTP/RTCP media traffic path in the external network. Nc and Nd represent media traffic inside the private domain.

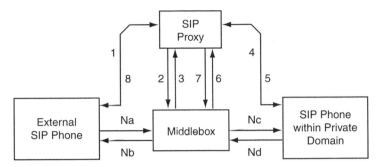

Figure 5.16: MIDCOM framework illustration with in-path SIP proxy.

As for the SIP application, we make the assumption that the middlebox is preconfigured to accept SIP calls into the private SIP phone. Specifically, this would imply that the middlebox implementing firewall service is preconfigured to permit SIP calls (destination TCP or UDP port number set to 5060) into the private phone. Likewise, middlebox implementing NAPT service would have been preconfigured to provide a port binding, to permit incoming SIP calls to be redirected to the specific private SIP phone. In other words, the INVITE from the external caller is not made to the private IP address but to the NAPT external address.

The objective of the MIDCOM agent in the following illustration is to merely permit the RTP/RTCP media stream through the middlebox, when using the MIDCOM protocol architecture outlined in the document. A SIP session typically establishes two RTP/RTCP media streams—one from the callee to the caller and another from the caller to the callee. These media sessions are UDP based and will use dynamic ports. The dynamic ports used for the media stream are specified in the SDP section of the SIP payload message. The MIDCOM agent will parse the SDP section and use the MIDCOM protocol to (a) open pinholes (i.e., permit RTP/RTCP session tuples) in a middlebox implementing firewall service, or (b) create PORT bindings and appropriately modify the SDP content to permit the RTP/RTCP streams through a middlebox implementing NAT service. The MIDCOM protocol should be sufficiently rich and expressive to support the operations described under the timelines. The examples do not show the timers maintained by the agent to keep the middlebox policy rule(s) from timing out.

MIDCOM agent Registration and connectivity between the MIDCOM agent and the middlebox are not shown in the interest of restricting the focus of the MIDCOM transactions to enabling the middlebox to let the media stream through. MIDCOM PDP is also not shown in the diagram below or on the timelines for the same reason.

The following subsections illustrate a typical timeline sequence of operations that transpire with the various elements involved in a SIP telephony application path. Each subsection is devoted to a specific instantiation of a middlebox service: NAPT, firewall, and a combination of both NAPT and firewall are considered.

5.4.7.1 Timeline Flow—Middlebox Implementing Firewall Service

Figure 5.17 assumes a middlebox implementing a firewall service. One further assumes that the middlebox is preconfigured to permit SIP calls (destination TCP or UDP port number set to 5060) into the private phone. The following timeline illustrates the operations performed by the MIDCOM agent, to permit RTP/ RTCP media stream through the middlebox.

The INVITE from the caller (external) is assumed to include the SDP payload. You will note that the MID- COM agent requests the middlebox to permit the Private-to-external RTP/RTCP flows before the INVITE is relayed to the callee. This is because, in SIP, the calling party must be ready to receive the media when it sends the INVITE with a session description. If the called party (private phone) assumes this and sends "early media" before sending the 200 OK response, the firewall will have blocked these packets without this initial MIDCOM signaling from the agent.

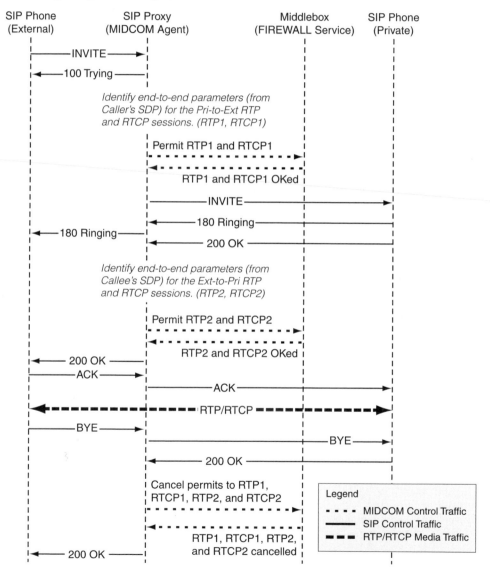

Figure 5.17: Timeline flow—Middlebox implementing firewall service.

5.4.7.2 Timeline Tow—Middlebox Implementing NAPT Service

Figure 5.18 assumes a middlebox implementing NAPT service. One makes the assumption that the middlebox is preconfigured to redirect SIP calls to the specific private SIP phone application. i.e., the INVITE from the external caller is not made to the private IP address, but to the NAPT external address. Let us say, the external phone's IP address is Ea, NAPT middlebox external Address is Ma, and the internal SIP phone's private address is Pa. SIP calls to the private SIP phone will arrive as TCP/UDP sessions, with the destination address and port set to Ma and 5060 respectively. The middlebox will redirect these datagrams to the internal SIP phone. The following timeline will illustrate the operations necessary to be performed by the MIDCOM agent to permit the RTP/RTCP media stream through the middlebox.

As with the previous example (Section 5.4.7.1), the INVITE from the caller (external) is assumed to include the SDP payload. You will note that the MIDCOM agent requests the middlebox to create NAT session descriptors for the private-to-external RTP/RTCP flows before the INVITE is relayed to the private SIP phone (for the same reasons as described in Section 5.4.7.1). If the called party (private phone) sends "early media" before sending the 200 OK response, the NAPT middlebox will have blocked these packets without the initial MIDCOM signaling from the agent. Also, note that after the 200 OK is received by the proxy from the private phone, the agent requests the middlebox to allocate NAT session descriptors for the external-to-private RTP2 and RTCP2 flows, such that the ports assigned on the Ma for RTP2 and RTCP2 are contiguous. The RTCP stream does not happen with a noncontiguous port. Lastly, you will note that even though each media stream (RTP1, RTCP1, RTP2 and RTCP2) is independent, they are all tied to the single SIP control session, while their NAT session descriptors were being created. Finally, when the agent issues a terminate session bundle command for the SIP session, the middlebox is assumed to delete all associated media stream sessions automatically.

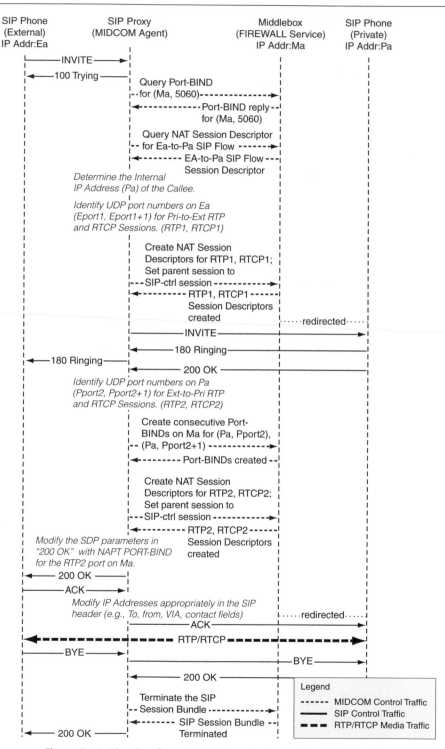

Figure 5.18: Timeline flow—Middlebox implementing NAPT service.

5.4.7.3 Timeline flow—Middlebox implementing NAPT and firewall.

Figure 5.19 assumes a middlebox implementing a combination of a firewall and a stateful NAPT service. One makes the assumption that the NAPT function is configured to translate the IP and TCP headers of the initial SIP session into the private SIP phone, and the firewall function is configured to permit the initial SIP session.

In the following timeline, it may be noted that the firewall description is based on packet fields on the wire (for example, as seen on the external interface of the middlebox). In order to ensure correct behavior of the individual services, you will notice that NAT specific MIDCOM operations precede firewall specific operations on the MIDCOM agent. This is noticeable in the timeline below when the MIDCOM agent processes the "200 OK" from the private SIP phone. The MIDCOM agent initially requests the NAT service on the middlebox to set up port-BIND and session-descriptors for the media stream in both directions. Subsequent to that, the MIDCOM agent determines the session parameters (i.e., the dynamic UDP ports) for the media stream, as viewed by the external interface, and requests the firewall service on the middlebox to permit those sessions through.

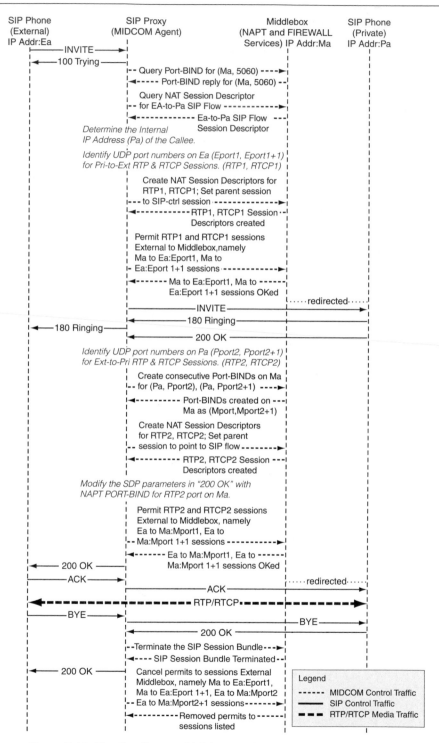

Figure 5.19: Timeline flow—Middlebox implementing NAPT and firewall.

5.4.8 Operational Considerations

5.4.8.1 Multiple MIDCOM sessions between agents and middlebox

A middlebox cannot be assumed to be a simple device implementing just one middlebox function and no more than a couple of interfaces. Middleboxes often combine multiple intermediate functions into the same device and have the ability to provision individual interfaces of the same device with different sets of functions and varied provisioning for the same function across the interfaces.

As such, a MIDCOM agent ought to be able to have a single MIDCOM session with a middlebox and use the MIDCOM interface on the middlebox to interface with different services on the same middlebox.

5.4.8.2 Asynchronous Notification to MIDCOM Agents

Asynchronous notification by the middlebox to a MIDCOM agent can be useful for events such as Session creation, Session termination, MIDCOM protocol failure, middlebox function failure or any other significant event. Independently, ICMP error codes can also be useful to notify transport layer failures to the agents.

In addition, periodic notification of various forms of data, such as statistics update, would also be a useful function that would be beneficial to certain types of agents.

5.4.8.3 Timers on Middlebox Considered Useful

When supporting the MIDCOM protocol, the middlebox is required to allocate dynamic resources, as specified in policy rule(s), upon request from agents. Explicit release of dynamically allocated resources happens when the application session is ended or when a MIDCOM agent requests the middlebox to release the resource.

However, the middlebox should be able to recover the dynamically allocated resources, even as the agent that was responsible for the allocation is not alive. Associating a lifetime for these dynamic resources and using a timer to track the lifetime can be a good way to accomplish this.

5.4.8.4 Middleboxes Supporting Multiple Services

A middlebox could be implementing a variety of services (e.g. NAT and firewall) in the same box. Some of these services might have interdependency on shared resources and sequence of operation. Others may be independent of each other. Generally speaking, the sequence in which these function operations may be performed on datagrams is not within the scope of this RFC.

In the case of a middlebox implementing NAT and firewall services, it is safe to state that the NAT operation on an interface will precede a firewall on the egress and will follow a firewall on the ingress. Further, firewall access control lists used by a firewall are assumed to be based on session parameters, as seen on the interface supporting firewall service.

5.4.8.5 Signaling and Data Traffic

The class of applications the MIDCOM architecture addresses focus around applications that have a combination of one or more signaling and data traffic sessions. The signaling may be done out-of-band, using a dedicated stand-alone session or may be done in-band, within a data session. Alternately, signaling may also be done as a combination of both stand-alone and in-band sessions.

SIP is an example of an application based on distinct signaling and data sessions. A SIP signaling session is used for call setup between a caller and a callee. A MIDCOM agent may be required to examine/modify SIP payload content to administer the middlebox so as to let the media streams (RTP/RTCP based) through. A MIDCOM agent is not required to intervene in the data traffic.

Signaling and context-specific Header information is sent in-band, within the same data stream for applications such as HTTP embedded applications, Sun-RPC (embedding a variety of NFS apps), Oracle transactions (embedding Oracle SQL+, MS ODBC, Peoplesoft) etc.

H.323 is an example of an application that sends signaling in both dedicated stand-alone sessions, as well as in conjunction with data. H.225.0 call signaling traffic traverses middleboxes by virtue of static policy, no MIDCOM control needed. H.225.0 call signaling also negotiates ports for an H.245 TCP stream. A MIDCOM agent is required to examine/modify the contents of the H.245 so that H.245 can traverse it.

H.245 traverses the middlebox and also carries Open Logical Channel information for media data. So, the MIDCOM agent is once again required to examine/modify the payload content needs to let the media traffic flow.

The MIDCOM architecture takes into consideration, supporting applications with independent signaling and data sessions as well as applications that have signaling and data communicated over the same session.

In the cases where signaling is done on a single stand-alone session, it is desirable to have a MIDCOM agent interpret the signaling stream and program the middlebox (that transits the data stream) so as to let the data traffic through uninterrupted.

5.4.9 Applicability Statement

Middleboxes may be stationed in a number of topologies. However, the signaling framework outlined in this RFC may be limited to only those middleboxes that are located in a DMZ (Demilitarized Zone) at the edge of a private domain, connecting to the Internet. Specifically, the assumption is that you have a single middlebox (running NAT or firewall) along the application route. Discovery of a middlebox along an application route is outside the scope of this RFC. It is conceivable to have middleboxes located between departments within the same domain or inside the service provider's domain and so forth. However, care must be taken to review each individual scenario and determine the applicability on a case-by-case basis.

The applicability may also be illustrated as follows. Real-time and streaming applications, such as Voice-Over-IP, and peer-to-peer applications, such as Napster and Netmeeting, require administering firewalls and NAT middleboxes to let their media streams reach hosts inside a private domain. The requirements are in the form of establishing a "pin-hole" to permit a TCP/UDP session (the port parameters of which are dynamically determined) through a firewall or retain an address/port bind in the NAT device to permit sessions to a port. These requirements are met by current generation middleboxes using adhoc methods, such as embedding application intelligence within a middlebox to identify the dynamic session parameters and administering the middlebox internally as appropriate. The objective of the MIDCOM architecture is to create a unified, standard way to exercise this functionality, currently existing in an ad-hoc fashion, in some of the middleboxes.

By adopting MIDCOM architecture, middleboxes will be able to support newer applications they have not been able to support thus far. MIDCOM architecture does not, and must not in anyway, change the fundamental characteristic of the services supported on the middlebox.

Typically, organizations shield a majority of their corporate resources (such as end-hosts) from visibility to the external network by the use of a DMZ at the domain edge. Only a portion of these hosts are allowed to be accessed by the external world. The remaining hosts and their names are unique to the private domain. Hosts visible to the external world and the authoritative name server that maps their names to network addresses are often configured within a DMZ in front of a firewall. Hosts and middleboxes within DMZ are referred to as DMZ nodes.

Figure 5.20 illustrates the configuration of a private domain with a DMZ at its edge. Actual configurations may vary. Internal hosts are accessed only by users inside the domain. Middleboxes, located in the DMZ may be accessed by agents inside or outside the domain.

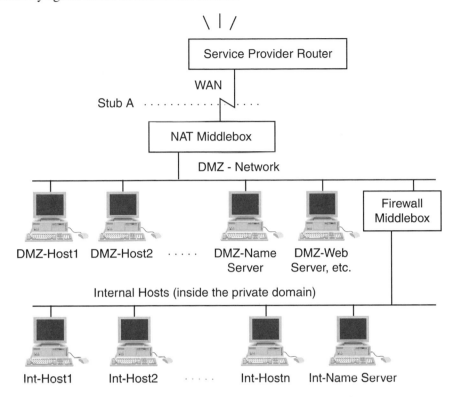

Figure 5.20: DMZ network configuration of a private domain.

5.5 Pragmatic Approaches using SIP Border Gateways

The previous sections of this chapter discussed some of the issues involved in supporting VoIP on a large scale due to addressing problems, and some current approaches (e.g., STUN, MIDCOM) to address these concerns. As an outgrowth of these limitations, Session Border Controllers (SBC) have emerged of late to assist service providers support VoIP and real-time interactive IP-based video/multimedia sessions in five areas: security, service reach maximization (end-to-end feasibility), SLA assurance, revenue and profit protection, and regulatory and law enforcement [OUE200501]. The interest in this context is on the first two items in this list.

A session border controller is a piece of network equipment or a collection of functions that control real-time session traffic at the signaling, call-control, and packet layers as they cross a notional packet-to-packet network border between networks or between network segments. SBCs are critical to the deployment of VoIP networks, because they address the inability of real-time session traffic to cross NAT device or firewall boundaries. Signaling protocols such as H.323, MGCP, and SIP transfer information including media session endpoint IP addresses and UDP port numbers in different layers above OSI Layer 4 (IETF TCP/UDP). This information cannot be seen by a normal firewall or NAT device, so the subsequent sessions set up are not recognized, do not pass through firewalls, and have incompatible IP addresses across NAT boundaries. SBCs allow NAT and firewall traversal, normally by incorporating those elements with signaling controllers for the required signaling protocols [LIG200502].

SBCs are devices used in VoIP to deal with a number of interworking issues. The controller refreshes NAT bindings for SIP registrations. The SBC compresses SIP packets to less than 1492 bytes (UDP fragmentation). It hides routing information to the outside world. The SBC ensures that an end-to-end media path is established. Also it helps the teardown process after the call is completed.

SBCs address the requirements at the boundary where different service provider networks interconnect or "peer." In general, session border controllers integrate signaling and media control, encompassing the following three functional subelements: (a) Interconnect Border Control Function, (b) Interworking Function, and (c) Interconnect Border Gateway Function.

One example of SBC usage is in the IP Multimedia Subsystem (IMS). IMS is an architecture defined by the Third Generation Partnership Project (3GPP) for the delivery of real-time voice, video and multimedia services using SIP over packet-switched networks with a focus on mobile wireless access networks. This architecture has been extended by ETSI to more completely satisfy the service delivery requirements in fixed-wireline access networks. Some of these additional requirements include [OUE200501]:

- Premise-based NAT traversal;
- Overlapping private address space and enterprise MPLS VPN bridging;
- IPv4 to IPv6 interworking for signaling and media;
- SIP interworking for H.323 IP PBXs and gatekeeper trunking/termination networks;
- Media-based DTMF (RFC 2833) to signaling-based DTMF translations.

Within the extended IMS architecture, two different types of session border controllers that integrate signaling and media control play very important roles: the Access SBC and the Interconnect SBC. The integration of signaling and media control provides several architectural benefits:

Security: SBC prevents DoS attacks on core (IMS) elements by dynamically discovering and blocking malicious signaling and media attacks or nonmalicious overloads (e.g., endpoint re-registering very frequently). Advanced SBCs using hardware-based features, can protect themselves against attack without loss of service.

Scalability: SBC provides distributed edge processing function for signaling and media offloading core (IMS) elements for connection and encryption management (e.g., TCP, TLS, IPSec), NAT traversal processing and other processor-intensive tasks.

Manageability: SBC incorporates multiple (IMS) functions resulting in fewer network elements, fewer networking protocols, and more robust fault and performance management (e.g., media QoS monitoring incorporated with session layer accounting).

The Interconnect Session Border Controller addresses the requirements at the boundary where different service provider networks interconnect or "peer." The controller integrates three functional elements from the ETSI TISPAN architecture [OUE200501].

1. Interconnect Border Control Function (IBCF): Provides overall control of the boundary between different service provider networks. It provides security for the IMS core in terms of signaling information by implementing a Topology Hiding Internetwork Gateway (THIG) subfunction. This subfunction performs signaling-based topology hiding, IPv4-IPv6 interworking, and session screening based upon source and destination signaling addresses. The IBCF also invokes the interworking function (described below) when connecting non-SIP or non-IPv6 networks, and performs admission control and bandwidth allocation using local policies or via interface to ETSI TISPAN Resource and

Admission Control Subsystem (RACS). Lastly, the IBCF interacts with I-BGF (described below) for control of the boundary at the transport layers including pinhole firewall, NAPT, and numerous other features.

2. Interworking Function (IWF): Provides signaling protocol interworking between the SIP-based IMS network and other service provider networks using H.323 or different SIP profiles.

3. Interconnect Border Gateway Function (I-BGF): Controls the transport boundary at layers 3 and 4 between service provider networks. This function acts as a pinhole firewall and NAT device protecting the service provider's IMS core. It controls access by packet filtering on IP address/port and opening/closing gates (pinholes) into the network. It uses NAPT to hide the IP addresses/ports of the service elements in the IMS core. QoS packet marking, bandwidth and signaling rate policing, usage metering and QoS measurements for the media flows are additional features supported by the I-BGF.

The Access Session Border Controller satisfies the requirements at the border where subscribers access the IMS core. It integrates two functional elements from the IMS and ETSI TISPAN architectures [OUE200501].

1. Proxy-Call Session Control Function (P-CSCF): Is the SIP signaling contact point, the outbound/inbound "proxy," for subscribers within IMS as defined by 3GPP. However, the term "proxy" is deceiving since to fulfill its complete set of responsibilities it must be able to proactively initiate SIP requests. This requires implementation as a SIP Back-to-Back User Agent (SIP B2BUA), not a simple SIP proxy. The P-CSCF is responsible for forwarding SIP registration messages from the subscriber's endpoint, the User Element (UE), in a visited network to the Interrogating-CSCF (I-CSCF) and subsequent call set-up requests and responses to the Serving-CSCF (S-CSCF). The P-CSCF maintains the mapping between logical subscriber SIP URI address and physical UE IP address and a security association, for both authentication and confidentiality, with the UE using TLS for example. It supports emergency call (E911) local routing within the visited network, accounting, session timers, and admission control. Admission control requires an interface to an external IMS Policy Decision Function (PDF)/ESTI TISPAN RACS. The P-CSCF interacts with an Access Border Gateway Function (A-BGF) for control of the boundary at the transport layers including pinhole firewall, NAPT and numerous other features. In addition, for wireline networks, ETSI's RACS is responsible for network-based NAT traversal.

2. Access Border Gateway Function (A-BGF): Controls the transport boundary at layers 3 and 4 between subscribers and the service provider's network. It performs all of the functions and features of the I-BGF. In addition, in wireline networks, it provides network-based NAT traversal for the media flows.

Session border controller product typically integrate signaling and media control in a single platform. Alternatively, session border control may be implemented using a distributed architecture using separate physical signaling and media control products for the three functional elements described above.

Figure 5.21 depicts, for illustrative purposes, an example of a commercial Session Border Controller; in this case the functionality runs parallel to the Gatekeeper [SYS200501]. In this example the Gateway/ Session Border Controller can operate in three ways:

1. Direct/static Mode to allow call resolution without Registration, Admission, or Status (RAS) message control (RAS is a protocol used in the H.323 protocol suite for discovering and interacting with a Gatekeeper). This mode will allow number translation and dynamic call control given that the participating gateways support the canMapAlias attribute.

2. Routed Mode to allow direct control of RAS messages with very low level of bandwidth utilization. This mode allows number translation and dynamic call control for gateways that do not support the canMapAlias attribute.

3. Proxy Mode to allow full RAS and RTP data transfer for gateways behind NAT or gateways that want to keep their identity. This bandwidth-intensive mode fully controls the RAS and Q.932 data streams and supports number translation and dynamic call control.

Figure 5.22 depicts, for illustrative purposes, a more complete enterprise SIP VoIP application to illustrate how the various elements interplay (ETH Zurich's PolyPhone environment) [LOR200501].

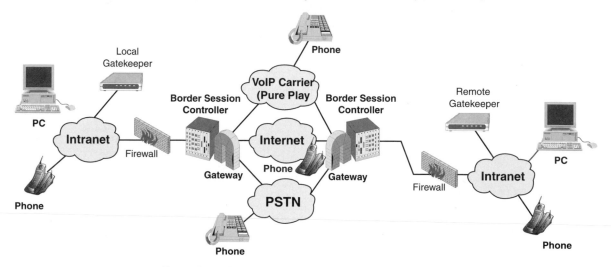

Figure 5.21: Gateway/session border controller example.

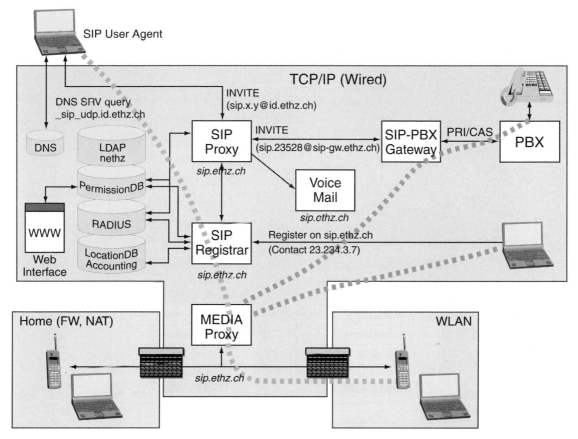

SIP Sessions

From the SDP specification (RFC 2327): "A multimedia session is a set of multimedia senders and receivers and the data streams flowing from senders to receivers. A multimedia conference is an example of a multimedia session." (A session as defined for SDP can comprise one or more RTP sessions.) As defined, a callee can be invited several times, by different calls, to the same session. If SDP is used, a session is defined by the concatenation of the SDP user name, session id, network type, address type, and address elements in the origin field. SIP supports stateless and stateful connections. A stateless proxy establishes the connection and then "gets out of the way." A stateful proxy stores all signaling events for the duration of the call (some SIP proxy servers deposit cookies in the IP phone/terminal as a method of providing state information).

SIP Proxy

An intermediate device that receives SIP requests from a client and then initiates requests on the client's behalf. The SIP proxy server provides similar functionality to a gatekeeper in an H.323 environment or a softswitch in an MGCP/MEGACO environment.

SIP Registrar

The default SER registrar where all active SIP clients are registered.

PeerPoint

A third party Border Gateway Controller is integrated to support NAT/Firewalled user agents and hide internal topology of SIP environment (Proxies, Gateways, Servers).

(Figure 5.22 continued on next page)

LDAP nethz

Stores usernames, passwords, E-Mail (primary and aliases) and internal phone numbers.

Lightweight Directory Access Protocol

An emerging software protocol for enabling anyone to locate organizations, individuals, and other resources such as files and devices in a network, whether on the Internet or on a corporate intranet. LDAP is a "lightweight" (smaller amount of code) version of DAP (Directory Access Protocol), which is part of X.500, a standard for directory services in a network.

PermissionDB

Stores phonenumbers, settings and permissions for users. In the near future, the PermissionDB will be integrated into the LDAP infrastructure.

Web-Interface

Instead of serweb, this implementation uses a custom website with interface to the SER proxy. In the future, the web interface may be integrated in the web interface of the existing LDAP services (n.ethz.ch). The web interface also gives useful information about this project and monitors the status of the environment components.

Radius

The following operations are authorized using the existing RADIUS server infrastructure:
- *Registration of SIP users (REGISTER)*
- *Establishing calls using an n.ethz.ch digest header (INVITE)*

Location DB, Accounting

The default SER Location DB. Accounting is only used for statistical purposes.

DNS

Domain Name Server of ethz.ch (any former e-mail could be resolved to a SIP account or an internal PSTN phone number.)

Gateways

Existing PSTN infrastructure has been integrated in the environment. Authorized SIP users can reach every internal and external phone.

Voice Mail

A voice mail-box system is available to the user.

TCP/IP (Wired)

The wired TCP/IP network of organization. Directly addressable IP numbers are used (no firewalls or number translation).

HOME (FW, NAT)

Infrastructure used by employees when at home or en route.

WLAN

The Wireless LAN infrastructure of the organization. For public users the WLAN allows only access to selected IP addresses inside the organization (e.g., the home page www.ethz.ch). Other addresses can only be reached after VPN validation. Since this is not currently possible with WLAN SIP phones, exceptional access to the SIP Server/ MEDIA Proxy has been granted. Phone registered on the SIP server will be able to establish connections to any other SIP phone on the Internet.

Figure 5.22: Example of an institutional SIP environment.

CHAPTER **6**

Basic IPv6 Concepts

This chapter focuses on the description of Internet protocol version 6 (IPv6). We introduced basic IPv6 concepts in Chapter 1, and we have alluded to in the chapters that followed the advantages and motivations for considering IPv6 in the VoIP context. The discussion is based on IETF RFC 2460 [DEE199801]. There is an extensive body of technical research literature on this topic (as documented in Appendix B of Chapter 1) and only the most basic concepts are covered in this chapter.

6.1 Introduction

IP version 6 (IPv6) is a new version of the Internet protocol, designed as the successor to IP version 4 (IPv4) described in RFC 791. The changes from IPv4 to IPv6 fall primarily into the following categories:

Expanded addressing capabilities: IPv6 increases the IP address size from 32 bits to 128 bits, to support more levels of addressing hierarchy, a much greater number of addressable nodes, and simpler auto-configuration of addresses. The scalability of multicast routing is improved by adding a "scope" field to multicast addresses; and a new type of address, called an *anycast address* is defined and used to send a packet to any one of a group of nodes.

Header format simplification: Some IPv4 header fields have been dropped or made optional, to reduce the common-case processing cost of packet handling and to limit the bandwidth cost of the IPv6 header.

Improved support for extensions and options. Changes in the way IP header options are encoded allows for more efficient forwarding, less stringent limits on the length of options, and greater flexibility for introducing new options in the future.

Flow labeling capability: A new capability is added to enable the labeling of packets belonging to particular traffic "flows" for which the sender requests special handling, such as nondefault quality of service or "real-time" service.

Authentication and privacy capabilities: Extensions to support authentication, data integrity, and data confidentiality (optional) are specified for IPv6.

RFC 2460 specifies the basic IPv6 header and the initially-defined IPv6 extension headers and options. It also discusses packet size issues, the semantics of flow labels and traffic classes, and the effects of IPv6 on upper-layer protocols. The format and semantics of IPv6 addresses are specified separately in RFC 2373 (now obsoleted by RFC 3513). The IPv6 version of ICMP, which all IPv6 implementations are required to include, is specified in ICMPv6 (RFC 2483). Developers should refer directly to all relevant IETF RFCs for normative guidelines.

6.2 Terminology

The following nomenclature is used in the standard:

Node: A device that implements IPv6.

Router: A node that forwards IPv6 packets and is not explicitly addressed to itself. (See note below).

Host: Any node that is not a router. (See note below).

Upper layer: A protocol layer immediately above IPv6. Examples are transport protocols such as TCP and UDP, control protocols such as ICMP, routing protocols such as OSPF, and Internet or lower-layer protocols being "tunneled" over (i.e., encapsulated in) IPv6 such as IPX, AppleTalk, or IPv6 itself.

Link: A communication facility or medium over which nodes can communicate at the link layer, i.e., the layer immediately below IPv6. Examples are Ethernets (simple or bridged); PPP links; X.25, Frame Relay, or ATM networks; and Internet (or higher) layer "tunnels," such as tunnels over IPv4 or IPv6 itself.

Neighbors: Nodes attached to the same link.

Interface: A node's attachment to a link.

Address: An IPv6-layer identifier for an interface or a set of interfaces.

Packet: An IPv6 header plus payload.

Link MTU: The maximum transmission unit, that is, maximum packet size in octets, that can be conveyed over a link.

Path MTU: The minimum link MTU of all the links in a path between a source node and a destination node.

Note: It is possible, though unusual, for a device with multiple interfaces to be configured to forward non-self-destined packets arriving from some set (fewer than all) of its interfaces, and to discard nonself-destined packets arriving from its other interfaces. Such a device must obey the protocol requirements for routers when receiving packets from, and interacting with neighbors over, the former (forwarding) interfaces. It must obey the protocol requirements for hosts when receiving packets from, and interacting with neighbors over, the latter (nonforwarding) interfaces.

6.3   IPv6 Header Format

Figure 6.1 depicts the IPv6 Header format.

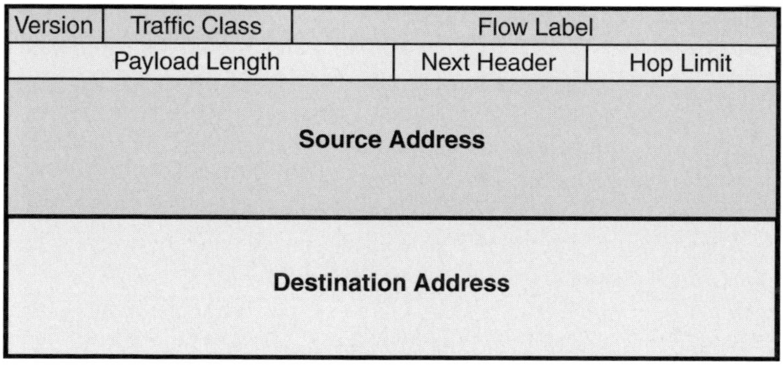

Figure 6.1: IPv6 header format.

The fields in the header have the following meanings:

Version: 4-bit Internet protocol version number = 6.

Traffic class: 8-bit traffic class field. See Section 6.7.

Flow label: 20-bit flow label. See Section 6.6.

Payload Length: 16-bit unsigned integer. Length of the IPv6 payload, i.e., the rest of the packet following this IPv6 header, in octets. (Note that any extension headers [Section 6.4] present are considered part of the payload, i.e., included in the length count.)

Next Header: 8-bit selector. Identifies the type of header immediately following the IPv6 header. Uses the same values as the IPv4 Protocol field.

Hop Limit: 8-bit unsigned integer. Decremented by one by each node that forwards the packet. The packet is discarded if Hop Limit is decremented to zero.

Source Address: 128-bit address of the originator of the packet. This is covered later in more detail.

Destination Address: 128-bit address of the intended recipient of the packet (possibly not the ultimate recipient, if a Routing header is present).

6.4 IPv6 Extension Headers

In IPv6, optional Internet-layer information is encoded in separate headers that may be placed between the IPv6 header and the upper-layer header in a packet. There are a small number of such extension headers, each identified by a distinct Next Header value. As illustrated in the examples in Figure 6.2, an IPv6 packet may carry zero, one, or more extension headers, each identified by the Next Header field of the preceding header:

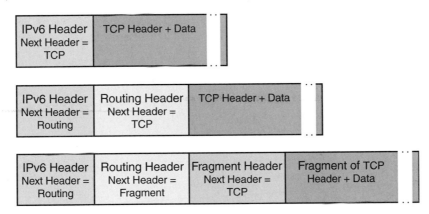

Figure 6.2: Examples of extension headers.

With one exception, extension headers are not examined or processed by any node along a packet's delivery path, until the packet reaches the node (or each of the set of nodes, in the case of multicast) identified in the Destination Address field of the IPv6 header. There, normal demultiplexing on the Next Header field of the IPv6 header invokes the module to process the first extension header, or the upper-layer header if no extension header is present. The contents and semantics of each extension header determine whether or not to proceed to the next header. Therefore, extension headers must be processed strictly in the order they appear in the packet; a receiver must not, for example, scan through a packet looking for a particular kind of extension header and process that header prior to processing all preceding ones.

The exception referred to in the preceding paragraph is the Hop-by-Hop Options header, which carries information that must be examined and processed by every node along a packet's delivery path, including the source and destination nodes. The Hop-by-Hop Options header, when present, must immediately follow the IPv6 header. Its presence is indicated by the value zero in the Next Header field of the IPv6 header.

If, as a result of processing a header, a node is required to proceed to the next header but the Next Header value in the current header is unrecognized by the node, it should discard the packet and send an ICMP Parameter Problem message to the source of the packet, with an ICMP Code value of 1 ("unrecognized Next Header type encountered") and the ICMP Pointer field containing the offset of the unrecognized value within the original packet. The same action should be taken if a node encounters a Next Header value of zero in any header other than an IPv6 header.

Each extension header is an integer multiple of 8 octets long, in order to retain 8-octet alignment for subsequent headers. Multioctet fields within each extension header are aligned on their natural boundaries, i.e., fields of width *n* octets are placed at an integer multiple of *n* octets from the start of the header, for $n = 1, 2, 4,$ or 8.

A full implementation of IPv6 includes implementation of the following extension headers:

- Hop-by-hop Options
- Routing (Type 0)
- Fragment
- Destination Options
- Authentication
- Encapsulating security payload

The first four are specified in this RFC; the last two are specified in RFC 2402 and RFC 2406, respectively.

6.4.1 Extension Header Order

When more than one extension header is used in the same packet, it is recommended that those headers appear in the following order:

1. IPv6 header
2. Hop-by-Hop Options header
3. Destination Options header (Note 1)
4. Routing header
5. Fragment header
6. Authentication header (Note 2)
7. Encapsulating Security Payload header (note 2)
8. Destination Options header (Note 3)
9. Upper-layer header

Note 1: For options to be processed by the first destination that appears in the IPv6 Destination Address field plus subsequent destinations listed in the Routing header.

Note 2: Additional recommendations regarding the relative order of the Authentication and Encapsulating Security Payload headers are given in RFC 2406.

Note 3: For options to be processed only by the final destination of the packet.

Each extension header should occur at most once, except for the Destination Options header which should occur at most twice (once before a Routing header and once before the upper-layer header).

If the upper-layer header is another IPv6 header (in the case of IPv6 being tunneled over or encapsulated in IPv6), it may be followed by its own extension headers, which are separately subject to the same ordering recommendations.

If and when other extension headers are defined, their ordering constraints relative to the above listed headers must be specified.

IPv6 nodes must accept and attempt to process extension headers in any order and occurring any number of times in the same packet, except for the Hop-by-Hop Options header which is restricted to appear immediately after an IPv6 header only. Nonetheless, it is strongly advised that sources of IPv6 packets adhere to the above recommended order until and unless subsequent specifications revise that recommendation.

6.4.2 Options

Two of the currently-defined extension headers—the Hop-by-Hop Options header and the Destination Options header—carry a variable number of type-length-value (TLV) encoded "options", of the format shown in Figure 6.3.

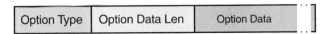

Figure 6.3: Extension headers options.

Option Type: 8-bit identifier of the type of option.
Opt Data Len: 8-bit unsigned integer. Length of the Option Data field of this option, in octets.
Option Data: Variable-length field. Option-Type-specific data.

The sequence of options within a header must be processed strictly in the order they appear in the header; a receiver must not, for example, scan through the header looking for a particular kind of option and process that option prior to processing all preceding ones.

The Option Type identifiers are internally encoded such that their highest-order two bits specify the action that must be taken if the processing IPv6 node does not recognize the Option Type:

00: Skip over this option and continue processing the header.
01: Discard the packet.
10: Discard the packet and, regardless of whether or not the packet's Destination Address was a multicast address, send an ICMP Parameter Problem, Code 2, message to the packet's Source Address, pointing to the unrecognized Option Type.
11: Discard the packet and, only if the packet's Destination Address was not a multicast address, send an ICMP Parameter Problem, Code 2, message to the packet's Source Address, pointing to the unrecognized Option Type.

The third-highest-order bit of the Option Type specifies whether or not the Option Data of that option can change en-route to the packet's final destination. When an Authentication header is present in the packet, for any option whose data may change en-route, its entire Option Data field must be treated as zero-valued octets when computing or verifying the packet's authenticating value.

0: Option Data does not change en-route.
1: Option Data may change en-route.

The three high-order bits described previously are to be treated as part of the Option Type, not independent of the Option Type. That is, a particular option is identified by a full 8-bit Option Type, not just the low-order 5 bits of an Option Type.

The same Option Type numbering space is used for both the Hop-by-Hop Options header and the Destination Options header; however, the specification of a particular option may restrict its use to only one of those two headers.

Individual options may have specific alignment requirements, to ensure that multioctet values within Option Data fields fall on natural boundaries. The alignment requirement of an option is specified using the notation $xn + y$, meaning the Option Type must appear at an integer multiple of x octets from the start of the header, plus y octets. For example:

2n: Means any 2-octet offset from the start of the header;
8n+2: Means any 8-octet offset from the start of the header, plus 2 octets.

There are two padding options that are used when necessary to align subsequent options and to pad out the containing header to a multiple of 8 octets in length. These padding options must be recognized by all IPv6 implementations:

Pad1 option (alignment requirement: none)

Note: the format of the Pad1 option is a special case—it does not have length and value fields.

The Pad1 option is used to insert one octet of padding into the Options area of a header. If more than one octet of padding is required, the PadN option, described next, should be used, rather than multiple Pad1 options.

PadN option (alignment requirement: none)

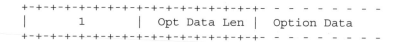

The PadN option is used to insert two or more octets of padding into the Options area of a header. For N octets of padding, the Opt Data Len field contains the value $N - 2$, and the Option Data consists of $N - 2$ zero-valued octets.

Section 6.10 contains formatting guidelines for designing new options.

6.4.3 Hop-by-Hop Options Header

The Hop-by-Hop Options header is used to carry optional information that must be examined by every node along a packet's delivery path. The Hop-by-Hop Options header is identified by a Next Header value of 0 in the IPv6 header, and has the format of Figure 6.4.

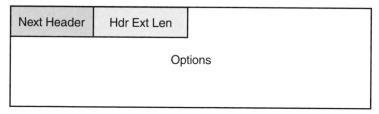

Figure 6.4: Hop-by-Hop options header.

The fields are as follows:

Next Header: 8-bit selector. Identifies the type of header immediately following the Hop-by-Hop Options header. Uses the same values as the IPv4 Protocol field (RFC 1700 et seq).

Hdr Ext Len: 8-bit unsigned integer. Length of the Hop-by-Hop Options header in 8-octet units, not including the first 8 octets.

Options: Variable-length field, of length such that the complete Hop-by-Hop Options header is an integer multiple of 8 octets long. Contains one or more TLV-encoded options, as described in Section 6.4.2.

The only hop-by-hop options defined in this RFC are the Pad1 and PadN options specified in Section 6.4.2.

6.4.4 Routing Header

The Routing header is used by an IPv6 source to list one or more intermediate nodes to be "visited" on the way to a packet's destination. This function is very similar to IPv4's Loose Source and Record Route option. The Routing header is identified by a Next Header value of 43 in the immediately preceding header, and has the format of Figure 6.5.

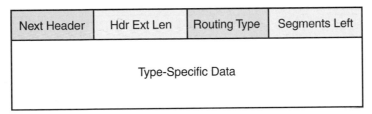

Figure 6.5: Routing header.

The fields are as follows:

Next Header: 8-bit selector. Identifies the type of header immediately following the Routing header. Uses the same values as the IPv4 Protocol field (RFC 1700).

Hdr Ext Len: 8-bit unsigned integer. Length of the Routing header in 8-octet units, not including the first 8 octets.

Routing Type: 8-bit identifier of a particular Routing header variant.

Segments Left: 8-bit unsigned integer. Number of route segments remaining, i.e., number of explicitly listed intermediate nodes still to be visited before reaching the final destination.

Type-specific data: Variable-length field, of format determined by the Routing Type, and of length such that the complete Routing header is an integer multiple of 8 octets long.

If, while processing a received packet, a node encounters a Routing header with an unrecognized Routing Type value, the required behavior of the node depends on the value of the Segments Left field, as follows:

If Segments Left is zero, the node must ignore the Routing header and proceed to process the next header in the packet, whose type is identified by the Next Header field in the Routing header.

If Segments Left is nonzero, the node must discard the packet and send an ICMP Parameter Problem, Code 0, message to the packet's Source Address, pointing to the unrecognized Routing Type.

If, after processing a Routing header of a received packet, an intermediate node determines that the packet is to be forwarded onto a link whose link MTU is less than the size of the packet, the node must discard the packet and send an ICMP Packet Too Big message to the packet's Source Address.

The Type 0 Routing header has the format shown in Figure 6.6.

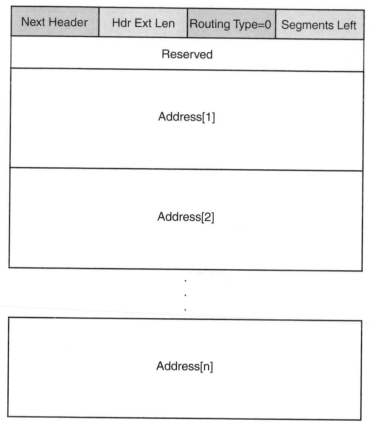

Next Header	Hdr Ext Len	Routing Type=0	Segments Left
Reserved			
Address[1]			
Address[2]			

.
.
.

Address[n]

Figure 6.6: Type 0 routing header.

The fields are as follows:

Next Header: 8-bit selector. Identifies the type of header immediately following the Routing header. Uses the same values as the IPv4 Protocol field (RFC 1700).

Hdr Ext Len: 8-bit unsigned integer. Length of the Routing header in 8-octet units, not including the first 8 octets. For the Type 0 Routing header, Hdr Ext Len is equal to two times the number ofaddresses in the header.

Routing Type: 0.

Segments Left: 8-bit unsigned integer. Number of route segments remaining, i.e., number of explicitly listed intermediate nodes still to be visited before reaching the final destination.

Reserved: 32-bit reserved field. Initialized to zero for transmission; ignored on reception.

Address[1, 2, ..., n]: Vector of 128-bit addresses, numbered 1 to n.

Multicast addresses must not appear in a Routing header of Type 0, or in the IPv6 Destination Address field of a packet carrying a Routing header of Type 0.

A Routing header is not examined or processed until it reaches the node identified in the Destination Address field of the IPv6 header. In that node, dispatching on the Next Header field of the immediately preceding header causes the Routing header module to be invoked, which, in the case of Routing Type 0, performs the following algorithm:

```
if Segments Left = 0 {
    proceed to process the next header in the packet, whose type is
    identified by the Next Header field in the Routing header
}
else if Hdr Ext Len is odd {
        send an ICMP Parameter Problem, Code 0, message to the Source
        Address, pointing to the Hdr Ext Len field, and discard the
        packet
}
else {
    compute n, the number of addresses in the Routing header, by
    dividing Hdr Ext Len by 2

    if Segments Left is greater than n {
        send an ICMP Parameter Problem, Code 0, message to the Source
        Address, pointing to the Segments Left field, and discard the
        packet
    }
    else {
        decrement Segments Left by 1;
        compute i, the index of the next address to be visited in
        the address vector, by subtracting Segments Left from n

        if Address [i] or the IPv6 Destination Address is multicast {
            discard the packet
        }
        else {
            swap the IPv6 Destination Address and Address[i]

            if the IPv6 Hop Limit is less than or equal to 1 {
                send an ICMP Time Exceeded—Hop Limit Exceeded in
                Transit message to the Source Address and discard the
                packet
            }
            else {
                decrement the Hop Limit by 1

                resubmit the packet to the IPv6 module for transmission
                to the new destination
            }
        }
    }
}
```

As an example of the effects of the above algorithm, consider the case of a source node S sending a packet to destination node D, using a Routing header to cause the packet to be routed via intermediate nodes I1, I2, and I3. The values of the relevant IPv6 header and Routing header fields on each segment of the delivery path would be as follows:

As the packet travels from S to I1:

```
        Source Address = S              Hdr Ext Len = 6
        Destination Address = I1        Segments Left = 3
                                        Address[1] = I2
                                        Address[2] = I3
                                        Address[3] = D
```

As the packet travels from I1 to I2:

```
        Source Address = S              Hdr Ext Len = 6
        Destination Address = I2        Segments Left = 2
                                        Address[1] = I1
                                        Address[2] = I3
                                        Address[3] = D
```

As the packet travels from I2 to I3:

```
        Source Address = S              Hdr Ext Len = 6
        Destination Address = I3        Segments Left = 1
                                        Address[1] = I1
                                        Address[2] = I2
                                        Address[3] = D
```

As the packet travels from I3 to D:

```
        Source Address = S              Hdr Ext Len = 6
        Destination Address = D         Segments Left = 0
                                        Address[1] = I1
                                        Address[2] = I2
                                        Address[3] = I3
```

6.4.5 Fragment Header

The Fragment header is used by an IPv6 source to send a packet larger than would fit in the path MTU to its destination. (Note: unlike IPv4, fragmentation in IPv6 is performed only by source nodes, not by routers along a packet's delivery path—see Section 6.5.) The Fragment header is identified by a Next Header value of 44 in the immediately preceding header, and has the format shown in Figure 6.7

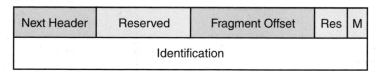

Figure 6.7: Fragment header.

The fields are as follows:

> *Next Header:* 8-bit selector. Identifies the initial header type of the Fragmentable Part of the original packet (defined below). Uses the same values as the IPv4 Protocol field (RFC 1700).
> *Reserved:* 8-bit reserved field. Initialized to zero for transmission; ignored on reception.

Fragment Offset: 13-bit unsigned integer. The offsct, in 8-octet units, of the data following this header, relative to the start of the Fragmentable Part of the original packet.

Res: 2-bit reserved field. Initialized to zero for transmission; ignored on reception.

M flag: 1 = more fragments; 0 = last fragment.

Identification: 32 bits. See description below.

In order to send a packet that is too large to fit in the MTU of the path to its destination, a source node may divide the packet into fragments and send each fragment as a separate packet, to be reassembled at the receiver.

For every packet that is to be fragmented, the source node generates an Identification value. The Identification must be different than that of any other fragmented packet sent recently* with the same Source Address and Destination Address. If a Routing header is present, the Destination Address of concern is that of the final destination.

The initial, large, unfragmented packet is referred to as the "original packet," and it is considered to consist of two parts, as see in Figure 6.8.

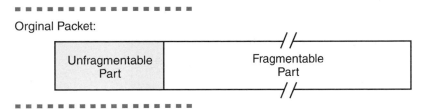

Orginal Packet:

Figure 6.8: Original packet.

The Unfragmentable Part consists of the IPv6 header plus any extension headers that must be processed by nodes en route to the destination, that is, all headers up to and including the Routing header if present, else the Hop-by-Hop Options header if present, else no extension headers.

The Fragmentable Part consists of the rest of the packet, that is, any extension headers that need be processed only by the final destination node(s), plus the upper-layer header and data.

The Fragmentable Part of the original packet is divided into fragments, each, except possibly the last ("right-most") one, being an integer multiple of 8 octets long. The fragments are transmitted in separate "fragment packets" as illustrated in Figure 6.9.

---

* "recently" means within the maximum likely lifetime of a packet, including transit time from source to destination and time spent awaiting reassembly with other fragments of the same packet. However, it is not required that a source node know the maximum packet lifetime. Rather, it is assumed that the requirement can be met by maintaining the Identification value as a simple, 32- bit, "wrap-around" counter, incremented each time a packet must be fragmented. It is an implementation choice whether to maintain a single counter for the node or multiple counters, e.g., one for each of the node's possible source addresses, or one for each active (source address, destination address) combination.

Orginal Packet:

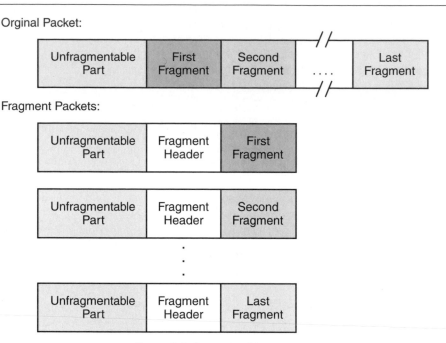

Figure 6.9: Fragmentable parts.

Each fragment packet is composed of:

1. The Unfragmentable Part of the original packet, with the Payload Length of the original IPv6 header changed to contain the length of this fragment packet only (excluding the length of the IPv6 header itself), and the Next Header field of the last header of the Unfragmentable Part changed to 44.
2. A Fragment header containing:
   • The Next Header value that identifies the first header of the Fragmentable Part of the original packet.
   • A Fragment Offset containing the offset of the fragment, in 8-octet units, relative to the start of the Fragmentable Part of the original packet. The Fragment Offset of the first ("leftmost") fragment is 0.
   • An M flag value of 0 if the fragment is the last ("rightmost") one, else an M flag value of 1.
   • The Identification value generated for the original packet.
3. The fragment itself.

The lengths of the fragments must be chosen such that the resulting fragment packets fit within the MTU of the path to the packets' destination(s).

At the destination, fragment packets are reassembled into their original, unfragmented form, as illustrated in Figure 6.10.

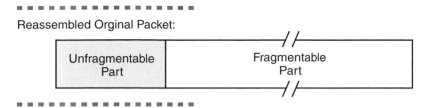

Figure 6.10: Reassembled original packet.

The following rules govern reassembly:

- An original packet is reassembled only from fragment packets that have the same Source Address, Destination Address, and Fragment Identification.
- The Unfragmentable Part of the reassembled packet consists of all headers up to, but not including, the Fragment header of the first fragment packet (that is, the packet whose Fragment Offset is zero), with the following two changes:
  1. The Next Header field of the last header of the Unfragmentable Part is obtained from the Next Header field of the first fragment's Fragment header.
  2. The Payload Length of the reassembled packet is computed from the length of the Unfragmentable Part and the length and offset of the last fragment. For example, a formula for computing the Payload Length of the reassembled original packet is:

     PL.orig = PL.first - FL.first – 8 + (8 * FO.last) + FL.last

     where,

     PL.orig = Payload Length field of reassembled packet.
     PL.first = Payload Length field of first fragment packet.
     FL.first = length of fragment following Fragment header of first fragment packet.
     FO.last = Fragment Offset field of Fragment header of last fragment packet.
     FL.last = length of fragment following Fragment header of last fragment packet.

The Fragmentable Part of the reassembled packet is constructed from the fragments following the Fragment headers in each of the fragment packets. The length of each fragment is computed by subtracting from the packet's Payload Length the length of the headers between the IPv6 header and fragment itself; its relative position in Fragmentable Part is computed from its Fragment Offset value.

The Fragment header is not present in the final, reassembled packet.

The following error conditions may arise when reassembling fragmented packets:

- If insufficient fragments are received to complete reassembly of a packet within 60 seconds of the reception of the first-arriving fragment of that packet, reassembly of that packet must be abandoned and all the fragments that have been received for that packet must be discarded. If the first fragment (i.e., the one with a Fragment Offset of zero) has been received, an ICMP Time Exceeded—Fragment Reassembly Time Exceeded message should be sent to the source of that fragment.
- If the length of a fragment, as derived from the fragment packet's Payload Length field, is not a multiple of 8 octets and the M flag of that fragment is 1, then that fragment must be discarded and an ICMP Parameter Problem, Code 0, message should be sent to the source of the fragment, pointing to the Payload Length field of the fragment packet.
- If the length and offset of a fragment are such that the Payload Length of the packet reassembled from that fragment would exceed 65,535 octets, then that fragment must be discarded and an ICMP Parameter Problem, Code 0, message should be sent to the source of the fragment, pointing to the Fragment Offset field of the fragment packet.

The following conditions are not expected to occur, but are not considered errors if they do:

- The number and content of the headers preceding the Fragment header of different fragments of the same original packet may differ. Whatever headers are present, preceding the Fragment header in each fragment packet, are processed when the packets arrive, prior to queueing the fragments for reassembly. Only those headers in the Offset zero fragment packet are retained in the reassembled packet.
- The Next Header values in the Fragment headers of different fragments of the same original packet may differ. Only the value from the Offset zero fragment packet is used for reassembly.

6.4.6 Destination Options Header

The Destination Options header is used to carry optional information that need be examined only by a packet's destination node(s). The Destination Options header is identified by a Next Header value of 60 in the immediately preceding header, and has the format shown in Figure 6.11.

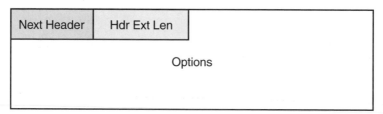

Figure 6.11: Destination Options Header.

Next Header: 8-bit selector. Identifies the type of header immediately following the Destination Options header. Uses the same values as the IPv4 Protocol field (RFC 1700)

Hdr Ext Len: 8-bit unsigned integer. Length of the Destination Options header in 8-octet units, not including the first 8 octets.

Options: Variable-length field, of length such that the complete Destination Options header is an integer multiple of 8 octets long. Contains one or more TLV-encoded options, as described in section 6.4.2.

The only destination options defined in this RFC are the Pad1 and PadN options specified in section 6.4.2.

Note that there are two possible ways to encode optional destination information in an IPv6 packet: either as an option in the Destination Options header, or as a separate extension header. The Fragment header and the Authentication header are examples of the latter approach. Which approach can be used depends on what action is desired of a destination node that does not understand the optional information:

- If the desired action is for the destination node to discard the packet and, only if the packet's Destination Address is not a multicast address, send an ICMP Unrecognized Type message to the packet's Source Address, then the information may be encoded either as a separate header or as an option in the Destination Options header whose Option Type has the value 11 in its highest-order two bits. The choice may depend on such factors as which takes fewer octets, or which yields better alignment or more efficient parsing.
- If any other action is desired, the information must be encoded as an option in the Destination Options header whose Option Type has the value 00, 01, or 10 in its highest-order two bits, specifying the desired action (see section 6.4.2).

6.4.7 No Next Header

The value 59 in the Next Header field of an IPv6 header or any extension header indicates that there is nothing following that header. If the Payload Length field of the IPv6 header indicates the presence of octets past the end of a header whose Next Header field contains 59, those octets must be ignored, and passed on unchanged if the packet is forwarded.

6.5 Packet Size Issues

IPv6 requires that every link in the Internet have an MTU of 1280 octets or greater. On any link that cannot convey a 1280-octet packet in one piece, link-specific fragmentation and reassembly must be provided at a layer below IPv6.

Links that have a configurable MTU (for example, PPP links defined in RFC 1661] must be configured to have an MTU of at least 1280 octets; it is recommended that they be configured with an MTU of 1500 octets or greater, to accommodate possible encapsulations (i.e., tunneling) without incurring IPv6-layer fragmentation.

From each link to which a node is directly attached, the node must be able to accept packets as large as that link's MTU.

It is strongly recommended that IPv6 nodes implement Path MTU Discovery (RFC 1981), in order to discover and take advantage of path MTUs greater than 1280 octets. However, a minimal IPv6 implementation (e.g., in a boot ROM) may simply restrict itself to sending packets no larger than 1280 octets, and omit implementation of Path MTU Discovery.

In order to send a packet larger than a path's MTU, a node may use the IPv6 Fragment header to fragment the packet at the source and have it reassembled at the destination(s). However, the use of such fragmentation is discouraged in any application that is able to adjust its packets to fit the measured path MTU (i.e., down to 1280 octets).

A node must be able to accept a fragmented packet that, after reassembly, is as large as 1500 octets. A node is permitted to accept fragmented packets that reassemble to more than 1500 octets. An upper-layer protocol or application that depends on IPv6 fragmentation to send packets larger than the MTU of a path should not send packets larger than 1500 octets unless it has assurance that the destination is capable of reassembling packets of that larger size.

In response to an IPv6 packet that is sent to an IPv4 destination (i.e., a packet that undergoes translation from IPv6 to IPv4), the originating IPv6 node may receive an ICMP Packet Too Big message reporting a Next-Hop MTU less than 1280. In that case, the IPv6 node is not required to reduce the size of subsequent packets to less than 1280, but must include a Fragment header in those packets so that the IPv6-to-IPv4 translating router can obtain a suitable Identification value to use in resulting IPv4 fragments. Note that this means the payload may have to be reduced to 1232 octets (1280 minus 40 for the IPv6 header and 8 for the Fragment header), and smaller still if additional extension headers are used.

6.6 Flow Labels

The 20-bit Flow Label field in the IPv6 header may be used by a source to label sequences of packets for which it requests special handling by the IPv6 routers, such as nondefault quality of service or "real-time" service. This aspect of IPv6 is still experimental to a large degree and subject to change as the requirements for flow support in the Internet become clearer (RFC 3697, March 2004 and RFC 3595 September 2003 provide some current thinking on the topic.) Hosts or routers that do not support the functions of the Flow Label field are required to set the field to zero when originating a packet, pass the field on unchanged when forwarding a packet, and ignore the field when receiving a packet.

Section 6.9 describes the current intended semantics and usage of the Flow Label field.

6.7 Traffic Classes

The 8-bit Traffic Class field in the IPv6 header is available for use by originating nodes and/or forwarding routers to identify and distinguish between different classes or priorities of IPv6 packets. At the point in time at which this specification is being written, there are a number of experiments underway in the use of the IPv4 Type of Service and/or Precedence bits to provide various forms of "differentiated service" for IP packets, other than through the use of explicit flow set-up. The Traffic Class field in the IPv6 header is intended to allow similar functionality to be supported in IPv6.

The expectation is that experimentation will eventually lead to agreement on what sorts of traffic classifications are most useful for IP packets. Detailed definitions of the syntax and semantics of all or some of the IPv6 Traffic Class bits, whether experimental or intended for eventual standardization, are to be provided in separate documents.

The following general requirements apply to the Traffic Class field:

- The service interface to the IPv6 service within a node must provide a means for an upper-layer protocol to supply the value of the Traffic Class bits in packets originated by that upper- layer protocol. The default value must be zero for all 8 bits.
- Nodes that support a specific (experimental or eventual standard) use of some or all of the Traffic Class bits are permitted to change the value of those bits in packets that they originate, forward, or receive, as required for that specific use. Nodes should ignore and leave unchanged any bits of the Traffic Class field for which they do not support a specific use.
- An upper-layer protocol must not assume that the value of the Traffic Class bits in a received packet are the same as the value sent by the packet's source.

6.8 Upper-Layer Protocol Issues

6.8.1 Upper-Layer Checksums

Any transport or other upper-layer protocol that includes the addresses from the IP header in its checksum computation must be modified for use over IPv6, to include the 128-bit IPv6 addresses instead of 32-bit IPv4 addresses. In particular, Figure 6.12 shows the TCP and UDP "pseudo-header" for IPv6:

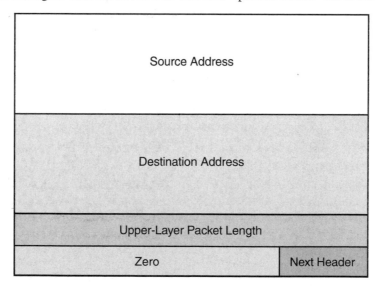

Figure 6.12: TCP and UDP "pseudo-header" for IPv6.

- If the IPv6 packet contains a Routing header, the Destination Address used in the pseudo-header is that of the final destination. At the originating node, that address will be in the last element of the Routing header; at the recipient(s), that address will be in the Destination Address field of the IPv6 header.
- The Next Header value in the pseudo-header identifies the upper-layer protocol (e.g., 6 for TCP, or 17 for UDP). It will differ from the Next Header value in the IPv6 header if there are extension headers between the IPv6 header and the upper- layer header.
- The Upper-Layer Packet Length in the pseudo-header is the length of the upper-layer header and data (e.g., TCP header plus TCP data). Some upper-layer protocols carry their own length information (e.g., the Length field in the UDP header); for such protocols, that is the length used in the pseudo- header. Other protocols (such as TCP) do not carry their own length information, in which case the length used in the pseudo-header is the Payload Length from the IPv6 header, minus the length of any extension headers present between the IPv6 header and the upper-layer header.
- Unlike IPv4, when UDP packets are originated by an IPv6 node, the UDP checksum is not optional. That is, whenever originating a UDP packet, an IPv6 node must compute a UDP checksum over the packet and the pseudo-header, and, if that computation yields a result of zero, it must be changed to hex FFFF for placement in the UDP header. IPv6 receivers must discard UDP packets containing a zero checksum, and should log the error.

The IPv6 version of ICMP includes the above pseudo-header in its checksum computation; this is a change from the IPv4 version of ICMP, which does not include a pseudo-header in its checksum. The reason for the change is to protect ICMP from misdelivery or corruption of those fields of the IPv6 header on which it depends, which, unlike IPv4, are not covered by an internet-layer checksum. The Next Header field in the pseudo-header for ICMP contains the value 58, which identifies the IPv6 version of ICMP.

6.8.2 Maximum Packet Lifetime

Unlike IPv4, IPv6 nodes are not required to enforce maximum packet lifetime. That is the reason the IPv4 "Time to Live" field was renamed "Hop Limit" in IPv6. In practice, very few, if any, IPv4 implementations conform to the requirement that they limit packet lifetime, so this is not a change in practice. Any upper-layer protocol that relies on the internet layer (whether IPv4 or IPv6) to limit packet lifetime ought to be upgraded to provide its own mechanisms for detecting and discarding obsolete packets.

6.8.3 Maximum Upper-Layer Payload Size

When computing the maximum payload size available for upper-layer data, an upper-layer protocol must take into account the larger size of the IPv6 header relative to the IPv4 header. For example, in IPv4, TCP's Maximum Segment Size (MSS) option is computed as the maximum packet size (a default value or a value learned through Path MTU Discovery) minus 40 octets (20 octets for the minimum-length IPv4 header and 20 octets for the minimum-length TCP header). When using TCP over IPv6, the MSS must be computed as the maximum packet size minus 60 octets, because the minimum-length IPv6 header (i.e., an IPv6 header with no extension headers) is 20 octets longer than a minimum-length IPv4 header.

6.8.4 Responding to Packets Carrying Routing Headers

When an upper-layer protocol sends one or more packets in response to a received packet that included a Routing header, the response packet(s) must not include a Routing header that was automatically derived by "reversing" the received Routing header UNLESS the integrity and authenticity of the received Source Address and Routing header have been verified (e.g., via the use of an Authentication header in the received packet). In other words, only the following kinds of packets are permitted in response to a received packet bearing a Routing header:

- Response packets that do not carry Routing headers.

- Response packets that carry Routing headers that were NOT derived by reversing the Routing header of the received packet (for example, a Routing header supplied by local configuration).
- Response packets that carry Routing headers that were derived by reversing the Routing header of the received packet IF AND ONLY IF the integrity and authenticity of the Source Address and Routing header from the received packet have been verified by the responder.

6.9   Semantics and Usage of the Flow Label Field

A flow is a sequence of packets sent from a particular source to a particular (unicast or multicast) destination for which the source desires special handling by the intervening routers. The nature of that special handling might be conveyed to the routers by a control protocol, such as a resource reservation protocol, or by information within the flow's packets themselves, e.g., in a hop-by-hop option. The details of such control protocols or options are beyond the scope of RFC 2460.

There may be multiple active flows from a source to a destination, as well as traffic that is not associated with any flow. A flow is uniquely identified by the combination of a source address and a nonzero flow label. Packets that do not belong to a flow carry a flow label of zero.

A flow label is assigned to a flow by the flow's source node. New flow labels must be chosen (pseudo-)randomly and uniformly from the range 1 to FFFFF hex. The purpose of the random allocation is to make any set of bits within the Flow Label field suitable for use as a hash key by routers, for looking up the state associated with the flow.

All packets belonging to the same flow must be sent with the same source address, destination address, and flow label. If any of those packets includes a Hop-by-Hop Options header, then they all must be originated with the same Hop-by-Hop Options header contents (excluding the Next Header field of the Hop-by-Hop Options header). If any of those packets includes a Routing header, then they all must be originated with the same contents in all extension headers up to and including the Routing header (excluding the Next Header field in the Routing header). The routers or destinations are permitted, but not required, to verify that these conditions are satisfied. If a violation is detected, it should be reported to the source by an ICMP Parameter Problem message, Code 0, pointing to the high-order octet of the Flow Label field (i.e., offset 1 within the IPv6 packet).

The maximum lifetime of any flow-handling state established along a flow's path must be specified as part of the description of the state-establishment mechanism, e.g., the resource reservation protocol or the flow-setup hop-by-hop option. A source must not re-use a flow label for a new flow within the maximum lifetime of any flow-handling state that might have been established for the prior use of that flow label.

When a node stops and restarts (e.g., as a result of a "crash"), it must be careful not to use a flow label that it might have used for an earlier flow whose lifetime may not have expired yet. This may be accomplished by recording flow label usage on stable storage so that it can be remembered across crashes, or by refraining from using any flow labels until the maximum lifetime of any possible previously established flows has expired. If the minimum time for rebooting the node is known, that time can be deducted from the necessary waiting period before starting to allocate flow labels.

There is no requirement that all, or even most, packets belong to flows, that is, carry nonzero flow labels. This observation is placed here to remind protocol designers and implementors not to assume otherwise. For example, it would be unwise to design a router whose performance would be adequate only if most packets belonged to flows, or to design a header compression scheme that only worked on packets that belonged to flows.

6.10 Formatting Guidelines for Options

This appendix gives some advice on how to lay out the fields when designing new options to be used in the Hop-by-Hop Options header or the Destination Options header, as described in Section 6.4.2. These guidelines are based on the following assumptions:

One desirable feature is that any multioctet fields within the Option Data area of an option be aligned on their natural boundaries, i.e., fields of width n octets should be placed at an integer multiple of n octets from the start of the Hop-by-Hop or Destination Options header, for $n = 1, 2, 4,$ or 8.

Another desirable feature is that the Hop-by-Hop or Destination Options header take up as little space as possible, subject to the requirement that the header be an integer multiple of 8 octets long.

It may be assumed that, when either of the option-bearing headers are present, they carry a very small number of options, usually only one.

These assumptions suggest the following approach to laying out the fields of an option: order the fields from smallest to largest, with no interior padding, then derive the alignment requirement for the entire option based on the alignment requirement of the largest field (up to a maximum alignment of 8 octets). This approach is illustrated in the following examples:

Example 1

If an option X required two data fields, one of length 8 octets and one of length 4 octets, it would be laid out as follows:

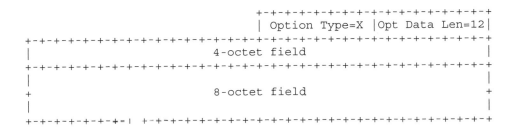

Its alignment requirement is $8n + 2$, to ensure that the 8-octet field starts at a multiple-of-8 offset from the start of the enclosing header. A complete Hop-by-Hop or Destination Options header containing this one option would look as follows:

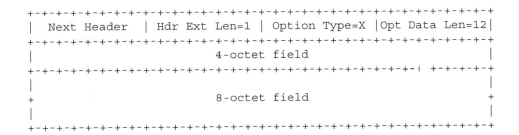

Example 2

If an option Y required three data fields, one of length 4 octets, one of length 2 octets, and one of length 1 octet, it would be laid out as follows:

```
                                              +-+-+-+-+-+-+-+-+
                                              | Option Type=Y |
 +-+-+-+-+-+-+-+-+-+-+-+-+-+-+-+-+-+-+-+-+-+-+-+-+-+-+-+-+-+-+-+-+
 |Opt Data Len=7 | 1-octet field |          2-octet field        |
 +-+-+-+-+-+-+-+-+-+-+-+-+-+-+-+-+-+-+-+-+-+-+-+-+-+-+-+-+-+-+-+-+
 |                          4-octet field                        |
 +-+-+-+-+-+-+-+-+-+-+-+-+-+-+-+-+-+-+-+-+-+-+-+-+-+-+-+-+-+-+-+-+
```

Its alignment requirement is $4n + 3$, to ensure that the 4-octet field starts at a multiple-of-4 offset from the start of the enclosing header. A complete Hop-by-Hop or Destination Options header containing this one option would look as follows:

```
 +-+-+-+-+-+-+-+-+-+-+-+-+-+-+-+-+-+-+-+-+-+-+-+-+-+-+-+-+-+-+-+-+
 |  Next Header  | Hdr Ext Len=1 | Pad1 Option=0 | Option Type=Y |
 +-+-+-+-+-+-+-+-+-+-+-+-+-+-+-+-+-+-+-+-+-+-+-+-+-+-+-+-+-+-+-+-+
 |Opt Data Len=7 | 1-octet field |          2-octet field        |
 +-+-+-+-+-+-+-+-+-+-+-+-+-+-+-+-+-+-+-+-+-+-+-+-+-+-+-+-+-+-+-+-+
 |                          4-octet field                        |
 +-+-+-+-+-+-+-+-+-+-+-+-+-+-+-+-+-+-+-+-+-+-+-+-+-+-+-+-+-+-+-+-+
 | PadN Option=1 |Opt Data Len=2 |       0       |       0       |
 +-+-+-+-+-+-+-+-+-+-+-+-+-+-+-+-+-+-+-+-+-+-+-+-+-+-+-+-+-+-+-+-+
```

Example 3

A Hop-by-Hop or Destination Options header containing both options X and Y from Examples 1 and 2 would have one of the two following formats, depending on which option appeared first:

```
 +-+-+-+-+-+-+-+-+-+-+-+-+-+-+-+-+-+-+-+-+-+-+-+-+-+-+-+-+-+-+-+-+
 |  Next Header  | Hdr Ext Len=3 | Option Type=X |Opt Data Len=12|
 +-+-+-+-+-+-+-+-+-+-+-+-+-+-+-+-+-+-+-+-+-+-+-+-+-+-+-+-+-+-+-+-+
 |                          4-octet field                        |
 +-+-+-+-+-+-+-+-+-+-+-+-+-+-+-+-+-+-+-+-+-+-+-+-+-+-+-+-+-+-+-+-+
 |                                                               |
 +                          8-octet field                        +
 |                                                               |
 +-+-+-+-+-+-+-+-+-+-+-+-+-+-+-+-+-+-+-+-+-+-+-+-+-+-+-+-+-+-+-+-+
 | PadN Option=1 |Opt Data Len=1 |       0       | Option Type=Y |
 +-+-+-+-+-+-+-+-+-+-+-+-+-+-+-+-+-+-+-+-+-+-+-+-+-+-+-+-+-+-+-+-+
 |Opt Data Len=7 | 1-octet field |          2-octet field        |
 +-+-+-+-+-+-+-+-+-+-+-+-+-+-+-+-+-+-+-+-+-+-+-+-+-+-+-+-+-+-+-+-+
 |                          4-octet field                        |
 +-+-+-+-+-+-+-+-+-+-+-+-+-+-+-+-+-+-+-+-+-+-+-+-+-+-+-+-+-+-+-+-+
 | PadN Option=1 |Opt Data Len=2 |       0       |       0       |
 +-+-+-+-+-+-+-+-+-+-+-+-+-+-+-+-+-+-+-+-+-+-+-+-+-+-+-+-+-+-+-+-+
```

```
+-+-+-+-+-+-+-+-+-+-+-+-+-+-+-+-+-+-+-+-+-+-+-+-+-+-+-+-+-+-+-+-+
|  Next Header   |  Hdr Ext Len=3 |  Pad1 Option=0 | Option Type=Y |
+-+-+-+-+-+-+-+-+-+-+-+-+-+-+-+-+-+-+-+-+-+-+-+-+-+-+-+-+-+-+-+-+
|Opt Data Len=7 | 1-octet field |          2-octet field          |
+-+-+-+-+-+-+-+-+-+-+-+-+-+-+-+-+-+-+-+-+-+-+-+-+-+-+-+-+-+-+-+-+
|                          4-octet field                          |
+-+-+-+-+-+-+-+-+-+-+-+-+-+-+-+-+-+-+-+-+-+-+-+-+-+-+-+-+-+-+-+-+
|  PadN Option=1 |Opt Data Len=4 |       0        |       0        |
+-+-+-+-+-+-+-+-+-+-+-+-+-+-+-+-+-+-+-+-+-+-+-+-+-+-+-+-+-+-+-+-+
|       0        |       0       | Option Type=X |Opt Data Len=12|
+-+-+-+-+-+-+-+-+-+-+-+-+-+-+-+-+-+-+-+-+-+-+-+-+-+-+-+-+-+-+-+-+
|                          4-octet field                          |
+-+-+-+-+-+-+-+-+-+-+-+-+-+-+-+-+-+-+-+-+-+-+-+-+-+-+-+-+-+-+-+-+
|                                                                 |
+                          8-octet field                          +
|                                                                 |
+-+-+-+-+-+-+-+-+-+-+-+-+-+-+-+-+-+-+-+-+-+-+-+-+-+-+-+-+-+-+-+-+
```

6.11 Introduction to Addressing

This section defines the addressing architecture of the IP Version 6 (IPv6) protocol. IETF RFC 3513 defines the addressing architecture of the IP Version 6 (IPv6) protocol; it includes the basic formats for the various types of IPv6 addresses (unicast, anycast, and multicast). This section covers the IPv6 addressing model, text representations of IPv6 addresses, definition of IPv6 unicast addresses, anycast addresses, and multicast addresses. The discussion is based on IETF RFC 3513 [HIN200301]. The discussion is for pedagogical purposes; developers should refer to the latest IETF documentation.

6.12 IPv6 Addressing

IPv6 addresses are 128-bit identifiers for interfaces and sets of interfaces (where "interface" is as defined in Section 2 of IPV6.) There are three types of addresses:

Unicast: An identifier for a single interface. A packet sent to a unicast address is delivered to the interface identified by that address.

Anycast: An identifier for a set of interfaces (typically belonging to different nodes). A packet sent to an anycast address is delivered to one of the interfaces identified by that address (the "nearest" one, according to the routing protocols' measure of distance).

Multicast: An identifier for a set of interfaces (typically belonging to different nodes). A packet sent to a multicast address is delivered to all interfaces identified by that address.

There are no broadcast addresses in IPv6, their function being superseded by multicast addresses.

In RFC 3513, fields in addresses are given a specific name, for example "subnet". When this name is used with the term "ID" for identifier after the name (for example, "subnet ID"), it refers to the contents of the named field. When it is used with the term "prefix" (for example, "subnet prefix") it refers to all of the address from the left up to and including this field. In IPv6, all zeros and all ones are legal values for any field, unless specifically excluded. Specifically, prefixes may contain, or end with, zero-valued fields.

6.12.1 Addressing Model

IPv6 addresses of all types are assigned to interfaces, not nodes. An IPv6 unicast address refers to a single interface. Since each interface belongs to a single node, any of that node's interfaces' unicast addresses may be used as an identifier for the node.

All interfaces are required to have at least one link-local unicast address (see section 6.12.8 for additional required addresses). A single interface may also have multiple IPv6 addresses of any type (unicast, anycast, and multicast) or scope. Unicast addresses with scope greater than link-scope are not needed for interfaces that are not used as the origin or destination of any IPv6 packets to or from nonneighbors. This is sometimes convenient for point-to-point interfaces. There is one exception to this addressing model:

A unicast address or a set of unicast addresses may be assigned to multiple physical interfaces if the implementation treats the multiple physical interfaces as one interface when presenting it to the internet layer. This is useful for load-sharing over multiple physical interfaces.

Currently IPv6 continues the IPv4 model that a subnet prefix is associated with one link. Multiple subnet prefixes may be assigned to the same link.

6.12.2 Text Representation of Addresses

There are three conventional forms for representing IPv6 addresses as text strings:

1. The preferred form is x:x:x:x:x:x:x:x, where the 'x's are the hexadecimal values of the eight 16-bit pieces of the address.

    Examples:

    ```
    FEDC:BA98:7654:3210:FEDC:BA98:7654:3210

    1080:0:0:0:8:800:200C:417A
    ```

    Note that it is not necessary to write the leading zeros in an individual field, but there must be at least one numeral in every field (except for the case described in 2.).

2. Due to some methods of allocating certain styles of IPv6 addresses, it will be common for addresses to contain long strings of zero bits. In order to make writing addresses containing zero bits easier a special syntax is available to compress the zeros. The use of "::" indicates one or more groups of 16 bits of zeros. The "::" can only appear once in an address. The "::" can also be used to compress leading or trailing zeros in an address.

    For example, the following addresses:

    ```
    1080:0:0:0:8:800:200C:417A    a unicast address
    FF01:0:0:0:0:0:101            a multicast address
    0:0:0:0:0:0:0:1               the loopback address
    0:0:0:0:0:0:0:0               the unspecified addresses
    ```

    may be represented as:

    ```
    1080::8:800:200C:417A         a unicast address
    FF01::101                     a multicast address
    ::1                           the loopback address
    ::                            the unspecified addresses
    ```

3. An alternative form that is sometimes more convenient when dealing with a mixed environment of IPv4 and IPv6 nodes is x:x:x:x:x:x:d.d.d.d, where the 'x's are the hexadecimal values of the six high-order 16-bit pieces of the address, and the 'd's are the decimal values of the four low-order 8-bit pieces of the address (standard IPv4 representation). Examples:

```
0:0:0:0:0:0:13.1.68.3

0:0:0:0:0:FFFF:129.144.52.38
```

or in compressed form:

```
::13.1.68.3

::FFFF:129.144.52.38
```

6.12.3 Text Representation of Address Prefixes

The text representation of IPv6 address prefixes is similar to the way IPv4 addresses prefixes are written in CIDR notation. An IPv6 address prefix is represented by the notation:

```
ipv6-address/prefix-length
```

where

> *ipv6-address:* is an IPv6 address in any of the notations listed in Section 6.12.2.
> *prefix-length:* is a decimal value specifying how many of the leftmost contiguous bits of the address comprise the prefix.

For example, the following are legal representations of the 60-bit prefix 12AB00000000CD3 (hexadecimal):

```
12AB:0000:0000:CD30:0000:0000:0000:0000/60
12AB::CD30:0:0:0:0/60
12AB:0:0:CD30::/60
```

The following are *not* legal representations of the above prefix:

```
12AB:0:0:CD3/60      may drop leading zeros, but not trailing zeros,
                     within any 16-bit chunk of the address

12AB::CD30/60        address to left of "/" expands to
                     12AB:0000:0000:0000:0000:000:0000:CD30

12AB::CD3/60         address to left of "/" expands to
                     12AB:0000:0000:0000:0000:000:0000:0CD3
```

When writing both a node address and a prefix of that node address (e.g., the node's subnet prefix), the two can combined as follows:

```
the node address      12AB:0:0:CD30:123:4567:89AB:CDEF
and its subnet number 12AB:0:0:CD30::/60
```

can be abbreviated as **12AB:0:0:CD30:123:4567:89AB:CDEF/60**

6.12.4 Address Type Identification

The type of an IPv6 address is identified by the high-order bits of the address, as follows:

```
Address type          Binary prefix          IPv6 notation    Section
------------          ------------           -------------    -------
Unspecified           00...0  (128 bits)     ::/128           6.12.5.2
Loopback              00...1  (128 bits)     ::1/128          6.12.5.3
Multicast             11111111               FF00::/8         6.12.7
Link-local unicast    1111111010             FE80::/10        6.12.5.6
Site-local unicast    1111111011             FEC0::/10        6.12.5.6
Global unicast        (everything else)
```

Anycast addresses are taken from the unicast address spaces (of any scope) and are not syntactically distinguishable from unicast addresses.

The general format of global unicast addresses is described in Section 6.12.5.4. Some special-purpose subtypes of global unicast addresses which contain embedded IPv4 addresses (for the purposes of IPv4-IPv6 interoperation) are described in Section 6.12.5.5.

Future specifications may redefine one or more sub-ranges of the global unicast space for other purposes, but unless and until that happens, implementations must treat all addresses that do not start with any of the above-listed prefixes as global unicast addresses.

6.12.5 Unicast Addresses

IPv6 unicast addresses are aggregable with prefixes of arbitrary bit-length similar to IPv4 addresses under Classless Interdomain Routing.

There are several types of unicast addresses in IPv6, in particular global unicast, site-local unicast, and link-local unicast. There are also some special-purpose subtypes of global unicast, such as IPv6 addresses with embedded IPv4 addresses or encoded NSAP addresses. Additional address types or subtypes can be defined in the future.

IPv6 nodes may have considerable or little knowledge of the internal structure of the IPv6 address, depending on the role the node plays (for instance, host versus router). At a minimum, a node may consider that unicast addresses (including its own) have no internal structure:

```
|                              128 bits                                 |
+-----------------------------------------------------------------------+
|                            node address                               |
+-----------------------------------------------------------------------+
```

A slightly sophisticated host (but still rather simple) may additionally be aware of subnet prefix(es) for the link(s) it is attached to, where different addresses may have different values for n:

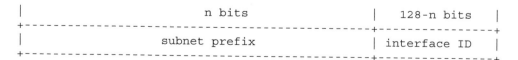

```
|                    n bits                    |    128-n bits    |
+----------------------------------------------+------------------+
|                  subnet prefix               |    interface ID  |
+----------------------------------------------+------------------+
```

Though a very simple router may have no knowledge of the internal structure of IPv6 unicast addresses, routers will more generally have knowledge of one or more of the hierarchical boundaries for the operation of routing protocols. The known boundaries will differ from router to router, depending on what positions the router holds in the routing hierarchy.

6.12.5.1 Interface Identifiers

Interface identifiers in IPv6 unicast addresses are used to identify interfaces on a link. They are required to be unique within a subnet prefix. It is recommended that the same interface identifier not be assigned to different nodes on a link. They may also be unique over a broader scope. In some cases an interface's identifier will be derived directly from that interface's link-layer address. The same interface identifier may be used on multiple interfaces on a single node, as long as they are attached to different subnets.

Note that the uniqueness of interface identifiers is independent of the uniqueness of IPv6 addresses. For example, a global unicast address may be created with a nonglobal scope interface identifier and a site-local address may be created with a global scope interface identifier.

For all unicast addresses, except those that start with binary value 000, Interface IDs are required to be 64 bits long and to be constructed in Modified EUI-64 format.

Modified EUI-64 format based Interface identifiers may have global scope when derived from a global token (e.g., IEEE 802 48-bit MAC or IEEE EUI-64 identifiers) or may have local scope where a global token is not available (e.g., serial links, tunnel end-points, etc.) or where global tokens are undesirable (e.g., temporary tokens for privacy.

Modified EUI-64 format interface identifiers are formed by inverting the "u" bit (universal/local bit in IEEE EUI-64 terminology) when forming the interface identifier from IEEE EUI-64 identifiers. In the resulting Modified EUI-64 format the "u" bit is set to one (1) to indicate global scope, and it is set to zero (0) to indicate local scope. The first three octets in binary of an IEEE EUI-64 identifier are as follows:

```
 0          0  0        1  1        2
|0          7  8        5  6        3|
 +----+----+----+----+----+----+
|cccc|ccug|cccc|cccc|cccc|cccc|
 +----+----+----+----+----+----+
```

written in Internet standard bit-order , where "u" is the universal/local bit, "g" is the individual/group bit, and "c" are the bits of the company_id.

The motivation for inverting the "u" bit when forming an interface identifier is to make it easy for system administrators to hand configure nonglobal identifiers when hardware tokens are not available. This is expected to be case for serial links, tunnel end- points, etc. The alternative would have been for these to be of the form 0200:0:0:1, 0200:0:0:2, etc., instead of the much simpler 1, 2, etc.

The use of the universal/local bit in the Modified EUI-64 format identifier is to allow development of future technology that can take advantage of interface identifiers with global scope.

The details of forming interface identifiers are defined in the appropriate "IPv6 over <link>" specification such as "IPv6 over Ethernet," "IPv6 over FDDI," and so on.

Chapter 6

6.12.5.2 The Unspecified Address

The address 0:0:0:0:0:0:0:0 is called the *unspecified address*. It must never be assigned to any node. It indicates the absence of an address. One example of its use is in the Source Address field of any IPv6 packets sent by an initializing host before it has learned its own address.

The unspecified address must not be used as the destination address of IPv6 packets or in IPv6 Routing Headers. An IPv6 packet with a source address of unspecified must never be forwarded by an IPv6 router.

6.12.5.3 The Loopback Address

The unicast address 0:0:0:0:0:0:0:1 is called the *loopback address*. It may be used by a node to send an IPv6 packet to itself. It may never be assigned to any physical interface. It is treated as having link-local scope, and may be thought of as the link-local unicast address of a virtual interface (typically called the *loopback interface*) to an imaginary link that goes nowhere.

The loopback address must not be used as the source address in IPv6 packets that are sent outside of a single node. An IPv6 packet with a destination address of loopback must never be sent outside of a single node and must never be forwarded by an IPv6 router. A packet received on an interface with destination address of loopback must be dropped.

6.12.5.4 Global Unicast Addresses

The general format for IPv6 global unicast addresses is as follows:

```
|          n bits          |  m bits  |        128-n-m bits       |
+--------------------------+----------+---------------------------+
| global routing prefix    | subnet ID|        interface ID       |
+--------------------------+----------+---------------------------+
```

where the global routing prefix is a (typically hierarchically- structured) value assigned to a site (a cluster of subnets/links), the subnet ID is an identifier of a link within the site, and the interface ID is as defined in section 6.12.5.1.

All global unicast addresses other than those that start with binary 000 have a 64-bit interface ID field (i.e., $n + m = 64$), formatted as described in Section 6.12.5.1. Global unicast addresses that start with binary 000 have no such constraint on the size or structure of the interface ID field.

Examples of global unicast addresses that start with binary 000 are the IPv6 address with embedded IPv4 addresses described in Section 6.12.5.5 and the IPv6 address containing encoded NSAP addresses

6.12.5.5 IPv6 Addresses with Embedded IPv4 Addresses

The IPv6 transition mechanisms include a technique for hosts and routers to dynamically tunnel IPv6 packets over IPv4 routing infrastructure. IPv6 nodes that use this technique are assigned special IPv6 unicast addresses that carry a global IPv4 address in the low-order 32 bits. This type of address is termed an "IPv4-compatible IPv6 address" and has the format:

```
|                 80 bits                  | 16 |      32 bits      |
+------------------------------------------+----+-------------------+
|0000..................................0000|0000|   IPv4 address    |
+------------------------------------------+----+-------------------+
```

Note: The IPv4 address used in the "IPv4-compatible IPv6 address" must be a globally-unique IPv4 unicast address.

A second type of IPv6 address which holds an embedded IPv4 address is also defined. This address type is used to represent the addresses of IPv4 nodes as IPv6 addresses. This type of address is termed an "IPv4-mapped IPv6 address" and has the format:

```
|               80 bits              | 16 |     32 bits      |
+-----------------------------------+----+------------------+
|0000...........................0000|FFFF|   IPv4 address   |
+-----------------------------------+----+------------------+
```

6.12.5.6 Local-Use IPv6 Unicast Addresses

There are two types of local-use unicast addresses defined. These are Link-Local and Site-Local. The Link-Local is for use on a single link and the Site-Local is for use in a single site. Link-Local addresses have the following format:

```
|   10     |                        |                          |
|  bits    |      54 bits           |         64 bits          |
+----------+------------------------+--------------------------+
|1111111010|          0             |       interface ID       |
+----------+------------------------+--------------------------+
```

Link-Local addresses are designed to be used for addressing on a single link for purposes such as automatic address configuration, neighbor discovery, or when no routers are present.

Routers must not forward any packets with link-local source or destination addresses to other links.

Site-Local addresses have the following format:

```
|   10     |                        |                          |
|  bits    |      54 bits           |         64 bits          |
+----------+------------------------+--------------------------+
|1111111011|       subnet ID        |       interface ID       |
+----------+------------------------+--------------------------+
```

Site-local addresses are designed to be used for addressing inside of a site without the need for a global prefix. Although a subnet ID may be up to 54-bits long, it is expected that globally-connected sites will use the same subnet IDs for site-local and global prefixes.

Routers must not forward any packets with site-local source or destination addresses outside of the site.

6.12.6 Anycast Addresses

An IPv6 anycast address is an address that is assigned to more than one interface (typically belonging to different nodes), with the property that a packet sent to an anycast address is routed to the "nearest" interface having that address, according to the routing protocols' measure of distance.

Anycast addresses are allocated from the unicast address space, using any of the defined unicast address formats. Thus, anycast addresses are syntactically indistinguishable from unicast addresses. When a unicast address is assigned to more than one interface, thus turning it into an anycast address, the nodes to which the address is assigned must be explicitly configured to know that it is an anycast address.

For any assigned anycast address, there is a longest prefix P of that address that identifies the topological region in which all interfaces belonging to that anycast address reside. Within the region identified by P, the anycast address must be maintained as a separate entry in the routing system (commonly referred to as a "host route"); outside the region identified by P, the anycast address may be aggregated into the routing entry for prefix P.

Note that in the worst case, the prefix P of an anycast set may be the null prefix, i.e., the members of the set may have no topological locality. In that case, the anycast address must be maintained as a separate routing entry throughout the entire internet, which presents a severe scaling limit on how many such "global" anycast sets may be supported. Therefore, it is expected that support for global anycast sets may be unavailable or very restricted.

One expected use of anycast addresses is to identify the set of routers belonging to an organization providing internet service. Such addresses could be used as intermediate addresses in an IPv6 Routing header, to cause a packet to be delivered via a particular service provider or sequence of service providers.

Some other possible uses are to identify the set of routers attached to a particular subnet, or the set of routers providing entry into a particular routing domain.

There is little experience with widespread, arbitrary use of internet anycast addresses, and some known complications and hazards when using them in their full generality. Until more experience has been gained and solutions are specified, the following restrictions are imposed on IPv6 anycast addresses:

An anycast address must not be used as the source address of an   IPv6 packet.

An anycast address must not be assigned to an IPv6 host, that is,   it may be assigned to an IPv6 router only.

6.12.6.1 Required Anycast Address

The Subnet-Router anycast address is predefined. Its format is as follows:

```
|                                            n bits              | 128-n bits     |
+------------------------------------------------+----------------+
|                    subnet prefix               | 00000000000000 |
+------------------------------------------------+----------------+
```

The "subnet prefix" in an anycast address is the prefix which identifies a specific link. This anycast address is syntactically the same as a unicast address for an interface on the link with the interface identifier set to zero.

Packets sent to the Subnet-Router anycast address will be delivered to one router on the subnet. All routers are required to support the Subnet-Router anycast addresses for the subnets to which they have interfaces.

The subnet-router anycast address is intended to be used for applications where a node needs to communicate with any one of the set of routers.

6.12.7 Multicast Addresses

An IPv6 multicast address is an identifier for a group of interfaces (typically on different nodes). An interface may belong to any number of multicast groups. Multicast addresses have the following format:

```
|   8    |  4 |  4 |                     112 bits                    |
+------- -+----+----+--------------------------------------------+
|11111111|flgs|scop|                     group ID                   |
+--------+----+----+--------------------------------------------+
```

- Binary 11111111 at the start of the address identifies the address as being a multicast address.
- Flgs is a set of 4 flags: |0|0|0|T|

The high-order 3 flags are reserved, and must be initialized to 0.

T = 0 indicates a permanently-assigned ("well-known") multicast address, assigned by the Internet Assigned Number Authority (IANA).

T = 1 indicates a nonpermanently-assigned ("transient") multicast address.

- Scop is a 4-bit multicast scope value used to limit the scope of the multicast group. The values are:

  0   reserved
  1   interface-local scope
  2   link-local scope
  3   reserved
  4   admin-local scope
  5   site-local scope
  6   (unassigned)
  7   (unassigned)
  8   organization-local scope
  9   (unassigned)
  A   (unassigned)
  B   (unassigned)
  C   (unassigned)
  D   (unassigned)
  E   global scope
  F   reserved

- Interface-local scope spans only a single interface on a node, and is useful only for loopback transmission of multicast.
- Link-local and site-local multicast scopes span the same topological regions as the corresponding unicast scopes.
- Admin-local scope is the smallest scope that must be administratively configured, i.e., not automatically derived from physical connectivity or other, nonmulticast-related configuration.
- Organization-local scope is intended to span multiple sites belonging to a single organization.
- Scopes labeled "(unassigned)" are available for administrators to define additional multicast regions.
- Group ID identifies the multicast group, either permanent or transient, within the given scope.

The "meaning" of a permanently-assigned multicast address is independent of the scope value. For example, if the "NTP servers group" is assigned a permanent multicast address with a group ID of 101 (hex), then:

FF01:0:0:0:0:0:0:101 means all NTP servers on the same interface (i.e., the same node) as the sender.
FF02:0:0:0:0:0:0:101 means all NTP servers on the same link as the sender.
FF05:0:0:0:0:0:0:101 means all NTP servers in the same site as the sender.
FF0E:0:0:0:0:0:0:101 means all NTP servers in the internet.

Nonpermanently-assigned multicast addresses are meaningful only within a given scope. For example, a group identified by the nonpermanent, site-local multicast address FF15:0:0:0:0:0:0:101 at one site bears no relationship to a group using the same address at a different site, nor to a nonpermanent group using the same group ID with different scope, nor to a permanent group with the same group ID.

Multicast addresses must not be used as source addresses in IPv6 packets or appear in any Routing header.

Routers must not forward any multicast packets beyond of the scope indicated by the scop field in the destination multicast address.

Nodes must not originate a packet to a multicast address whose scop field contains the reserved value 0; if such a packet is received, it must be silently dropped. Nodes should not originate a packet to a multicast address whose scop field contains the reserved value F; if such a packet is sent or received, it must be treated the same as packets destined to a global (scop E) multicast address.

6.12.7.1 Pre-Defined Multicast Addresses

The following well-known multicast addresses are pre-defined. The group ID's defined in this section are defined for explicit scope values.

Use of these group IDs for any other scope values, with the T flag equal to 0, is not allowed.

```
Reserved Multicast Addresses:     FF00:0:0:0:0:0:0:0
                                  FF01:0:0:0:0:0:0:0
                                  FF02:0:0:0:0:0:0:0
                                  FF03:0:0:0:0:0:0:0
                                  FF04:0:0:0:0:0:0:0
                                  FF05:0:0:0:0:0:0:0
                                  FF06:0:0:0:0:0:0:0
                                  FF07:0:0:0:0:0:0:0
                                  FF08:0:0:0:0:0:0:0
                                  FF09:0:0:0:0:0:0:0
                                  FF0A:0:0:0:0:0:0:0
                                  FF0B:0:0:0:0:0:0:0
                                  FF0C:0:0:0:0:0:0:0
                                  FF0D:0:0:0:0:0:0:0
                                  FF0E:0:0:0:0:0:0:0
                                  FF0F:0:0:0:0:0:0:0
```

The above multicast addresses are reserved and shall never be assigned to any multicast group.

```
All Nodes Addresses:     FF01:0:0:0:0:0:0:1
                         FF02:0:0:0:0:0:0:1
```

The above multicast addresses identify the group of all IPv6 nodes, within scope 1 (interface-local) or 2 (link-local).

```
All Routers Addresses:     FF01:0:0:0:0:0:0:2
                           FF02:0:0:0:0:0:0:2
                           FF05:0:0:0:0:0:0:2
```

The above multicast addresses identify the group of all IPv6 routers, within scope 1 (interface-local), 2 (link-local), or 5 (site-local).

```
Solicited-Node Address:   FF02:0:0:0:0:1:FFXX:XXXX
```

Solicited-node multicast address are computed as a function of a node's unicast and anycast addresses. A solicited-node multicast address is formed by taking the low-order 24 bits of an address (unicast or anycast) and appending those bits to the prefix FF02:0:0:0:0:1:FF00::/104 resulting in a multicast address in the range

```
FF02:0:0:0:0:1:FF00:0000
```
to
```
FF02:0:0:0:0:1:FFFF:FFFF
```

For example, the solicited node multicast address corresponding to the IPv6 address 4037::01:800:200E:8C6C is FF02::1:FF0E:8C6C. IPv6 addresses that differ only in the high-order bits, e.g., due to multiple high-order prefixes associated with different aggregations, will map to the same solicited-node address thereby, reducing the number of multicast addresses a node must join.

A node is required to compute and join (on the appropriate interface) the associated Solicited-Node multi cast addresses for every unicast and anycast address it is assigned.

6.12.8 A Node's Required Addresses

A host is required to recognize the following addresses as identifying itself:

- Its required Link-Local Address for each interface.
- Any additional Unicast and Anycast Addresses that have been configured for the node's interfaces (manually or automatically).
- The loopback address.
- The All-Nodes Multicast Addresses defined in Section 6.12.7.1.
- The Solicited-Node Multicast Address for each of its unicast and anycast addresses.
- Multicast Addresses of all other groups to which the node belongs.

A router is required to recognize all addresses that a host is required to recognize, plus the following addresses as identifying itself:

The Subnet-Router Anycast Addresses for all interfaces for which it is configured to act as a router.
All other Anycast Addresses with which the router has been configured.
The All-Routers Multicast Addresses defined in Section 6.12.7.1.

6.13 IANA Considerations

The initial assignment of IPv6 address space is as follows:

```
Allocation                      Prefix       Fraction of
                                (binary)     Address Space

-----------------------         --------     ------------
Unassigned (see Note 1 below)   0000 0000    1/256
Unassigned                      0000 0001    1/256
Reserved for NSAP Allocation    0000 001     1/128  [RFC1888]
Unassigned                      0000 01      1/64
Unassigned                      0000 1       1/32
Unassigned                      0001         1/16
Global Unicast                  001          1/8    (per RFC2374)
Unassigned                      010          1/8
Unassigned                      011          1/8
Unassigned                      100          1/8
Unassigned                      101          1/8
Unassigned                      110          1/8
Unassigned                      1110         1/16
Unassigned                      1111 0       1/32
Unassigned                      1111 10      1/64
Unassigned                      1111 110     1/128
Unassigned                      1111 1110 0  1/512
Link-Local Unicast Addresses    1111 1110 10 1/1024
Site-Local Unicast Addresses    1111 1110 11 1/1024
Multicast Addresses             1111 1111    1/256
```

Notes:

1. The "unspecified address", the "loopback address", and the IPv6 Addresses with Embedded IPv4 Addresses are assigned out of the 0000 0000 binary prefix space.

2. For now, IANA should limit its allocation of IPv6 unicast address space to the range of addresses that start with binary value 001. The rest of the global unicast address space (approximately 85% of the IPv6 address space) is reserved for future definition and use, and is not to be assigned by IANA at this time.

6.14 Creating Modified EUI-64 Format Interface Identifiers

Depending on the characteristics of a specific link or node there are a number of approaches for creating Modified EUI-64 format interface identifiers. This appendix describes some of these approaches.

EUI is defined in IEEE, "Guidelines for 64-bit Global Identifier (EUI-64) Registration Authority", March 1997 (see Section 6.15).

Links or Nodes with IEEE EUI-64 Identifiers

The only change needed to transform an IEEE EUI-64 identifier to an interface identifier is to invert the "u" (universal/local) bit. For example, a globally unique IEEE EUI-64 identifier of the form:

```
|0                   1|1                   3|3                   4|4                   6|
|0                   5|6                   1|2                   7|8                   3|
+-----------------+-----------------+-----------------+-----------------+
|cccccc0gcccccccc|ccccccccmmmmmmmm|mmmmmmmmmmmmmmmm|mmmmmmmmmmmmmmmm|
+-----------------+-----------------+-----------------+-----------------+
```

where "c" are the bits of the assigned company_id, "0" is the value of the universal/local bit to indicate global scope, "g" is individual/group bit, and "m" are the bits of the manufacturer-selected extension identifier. The IPv6 interface identifier would be of the form:

```
|0                   1|1                   3|3                   4|4                   6|
|0                   5|6                   1|2                   7|8                   3|
+-----------------+-----------------+-----------------+-----------------+
|cccccc1gcccccccc|ccccccccmmmmmmmm|mmmmmmmmmmmmmmmm|mmmmmmmmmmmmmmmm|
+-----------------+-----------------+-----------------+-----------------+
```

The only change is inverting the value of the universal/local bit.

Links or Nodes with IEEE 802 48 bit MAC's

EUI64 defines a method to create a IEEE EUI-64 identifier from an IEEE 48bit MAC identifier. This is to insert two octets, with hexadecimal values of 0xFF and 0xFE, in the middle of the 48 bit MAC (between the company_id and vendor supplied id). For example, the 48 bit IEEE MAC with global scope:

```
|0                   1|1                   3|3                   4|
|0                   5|6                   1|2                   7|
+-----------------+-----------------+-----------------+
|cccccc0gcccccccc|ccccccccmmmmmmmm|mmmmmmmmmmmmmmmm|
+-----------------+-----------------+-----------------+
```

where "c" are the bits of the assigned company_id, "0" is the valueof the universal/local bit to indicate global scope, "g" is individual/group bit, and "m" are the bits of the manufacturer-selected extension identifier. The interface identifier would be of the form:

```
| 0              1 | 1             3 | 3             4 | 4             6 |
| 0              5 | 6             1 | 2             7 | 8             3 |
+---------------+----------------+----------------+----------------+
|ccccccc1gcccccccc|cccccccc11111111|11111110mmmmmmmmm|mmmmmmmmmmmmmmmmm|
+---------------+----------------+----------------+----------------+
```

When IEEE 802 48bit MAC addresses are available (on an interface or a node), an implementation may use them to create interface identifiers due to their availability and uniqueness properties.

Links with Other Kinds of Identifiers

There are a number of types of links that have link-layer interface identifiers other than IEEE EIU-64 or IEEE 802 48-bit MACs. Examples include LocalTalk and Arcnet. The method to create an Modified EUI-64 format identifier is to take the link identifier (e.g., the LocalTalk 8 bit node identifier) and zero fill it to the left. For example, a LocalTalk 8 bit node identifier of hexadecimal value 0x4F results in the following interface identifier:

```
| 0              1 | 1             3 | 3             4 | 4             6 |
| 0              5 | 6             1 | 2             7 | 8             3 |
+---------------+----------------+----------------+----------------+
|0000000000000000|0000000000000000|0000000000000000|0000000001001111|
+---------------+----------------+----------------+----------------+
```

Note that this results in the universal/local bit set to "0" to indicate local scope.

Links without Identifiers

There are a number of links that do not have any type of built-in identifier. The most common of these are serial links and configured tunnels. Interface identifiers must be chosen that are unique within a subnet-prefix.

When no built-in identifier is available on a link the preferred approach is to use a global interface identifier from another interface or one which is assigned to the node itself. When using this approach no other interface connecting the same node to the same subnet-prefix may use the same identifier.

If there is no global interface identifier available for use on the link the implementation needs to create a local-scope interface identifier. The only requirement is that it be unique within a subnet prefix. There are many possible approaches to select a subnet-prefix-unique interface identifier. These include:

- Manual Configuration
- Node Serial Number
- Other node-specific token

The subnet-prefix-unique interface identifier should be generated in a manner that it does not change after a reboot of a node or if interfaces are added or deleted from the node.

The selection of the appropriate algorithm is link and implementation dependent. The details on forming interface identifiers are defined in the appropriate "IPv6 over <link>" specification. It is strongly recommended that a collision detection algorithm be implemented as part of any automatic algorithm.

6.15 64-Bit Global Identifier (EUI-64) Registration Authority

The IEEE-defined 64-bit extended unique identifier (EUI-64) is a concatenation of the 24-bit company_id value by the IEEE Registration Authority and a 40-bit extension identifier assigned by the organization with that company_id assignment. The IEEE administers the assignment of 24-bit company_id values. The assignments of these values are public, so that a user of an EUI 64 value can identify the manufacturer that provided any value. The IEEE/RAC has no control over the assignments of 40-bit extension identifiers and assumes no liability for assignments of duplicate EUI-64 identifiers assigned by manufacturers.

Application restrictions

Given the minimal probability of consuming all the EUI-64 identifiers, the IEEE/RAC places minimal restrictions on their use within standards. However, if used within the context of an IEEE standard, the documentation shall be reviewed by the IEEE/RAC for correctness and clarity. The IEEE/RAC shall not otherwise restrict the use of EUI-64 identifiers within standards. If the EUI-64 is referenced within non-IEEE standards, there shall not be any reference to IEEE unless approved by the IEEE/RAC.

Distribution restrictions

Given the minimal probability of consuming all the EUI-64 identifiers, the IEEE/RAC places minimal restrictions on their redistribution through third parties, as follows:

Allocation: The EUI-64 values shall be sold within electronically-readable parts; no more than one EUI-64 value shall be contained within each component that is manufactured.

Packaging: A component containing the EUI-64 value shall have a distinguishing characteristic (such as color or shape) to distinguish it from other commonly-used identifier components.

Documentation: Readily available documentation.

Legal indemnification: Any organization producing EUI-64 values shall indemnify the IEEE for damages arising from duplicate number assignments.

The term EUI-64 is trademarked by the IEEE. Companies are allowed to use this term for commercial purposes, but only if their use of this term has been reviewed by the IEEE/RAC and the proposed products using the EUI-64 conform to these restrictions.

Application documentation

As a condition for receiving a company_id assignment, a manufacturer of EUI-64 values accepts the following responsibilities:

- This documentation shall be readily available (at no cost) to any purchaser of EUI-64 values.
- The manufacturer's part specification should include an unambiguous description of how the EUI-64 value is accessed (pin and/or address descriptions).

Manufacturer-assigned identifiers

The manufacturer identifier assignment allows the assignee to generate approximately 1 trillion (10^{12}) unique EUI-64 values, by varying the last 40 bits. The IEEE intends not to assign another OUI/company_id value to a manufacturer of EUI-64 values until the manufacturer has consumed, in product, the preponderance (more than 90%) of this block of potential unique words. It is incumbent upon the manufacturer to ensure that large portions of the unique word block are not left unused in manufacturing.

6.16 Additional Technical Details

As indicated in Appendix B of Chapter 1, the IPv6 protocol apparatus is described by the 100+ RFCs identified in the appendix (some have been obsoleted and/or replaced.) The interested reader, particularly developers, should work himself/herself through that body of information for additional technical details.

CHAPTER **7**

Using IPv6 to Support 3G VoIP

This chapter approaches the discussion of application of IPv6 to VoIP by reviewing some of the information that was introduced in Chapter 6 from a more-applied perspective than done in that chapter, and discuss these issues in the context of a VoIPv6 environment.

7.1 Overview of VoIPv6 Positioning

Figure 7.1 depicts typical Network Elements found in an enterprise network, many of which come into play in VoIP environments. A number of these will need to support IPv6, particularly those that operate at Layer 3 of the protocol model. Figure 7.2 depicts the medium-stage environment IPv6/VoIP under discussion. The idea is that:

- any telephone device shown can freely, easily, and reliably communicate with any other telephone at any point at any time; and,
- whenever possible, an end-to-end IP-based path should be taken by the call, if a path exists (naturally, telephone sets that are hard-wired to the PSTN have no choice but to use the PSTN at least for a portion of the call path).

At the graphical level Figure 7.2 illustrates how a 3G VoIP based on IPv6 will operate. As seen in this figure, many of the VoIP elements discussed in Chapter 1, Chapter 2, and Chapter 5 need to be upgraded to support IPv6 and/or dual-stacks. As we noted in Chapter 3, SIP operates in either environment (also see Figure 5.8). SIP UAs (User Agents) often support different network address types. For example, a UA may have an IPv6 address and an IPv4 address. Such a UA will typically be willing to use any of its addresses to establish a media session with a remote UA. If the remote UA only supports IPv6, for instance, both UAs will use IPv6 to send and receive media [CAM200501]. H.323-based applications require the H.323 elements listed in Chapter 2 (for example, gatekeeper, gateway, and so forth) to be IPv6-ready and/or support a dual-stack.

Up to the present various perceptions may have existed in the user community at large as shown below [ISL200501], but proponents see IPv6 as being the best of all worlds, specifically as it relates to end-to-end security and reachability, QoS, and overall cost-effectiveness.

Frame Relay	Leased line replacement
Multiservice	ATM too broad and complex
MPLS	(Too) complex
IPv6	Will be great once it is deployed

Figure 7.3 depicts a simplified version of the environment pre- and post-IPv6 vis-à-vis security and the NAT-issue. Figure 7.4 depicts the simplicity of IPv6-based VPNs. IPv6 will be able to support 3G VoIP as well as 3G hotspot services (including VoWi-Fi).

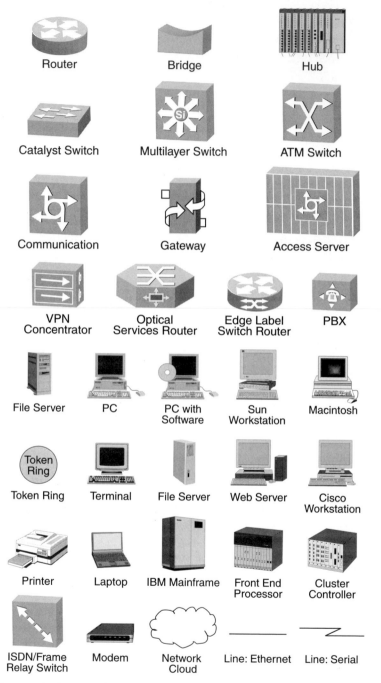

Figure 7.1: Typical network elements found in an enterprise network, some of which come into play in VoIP.

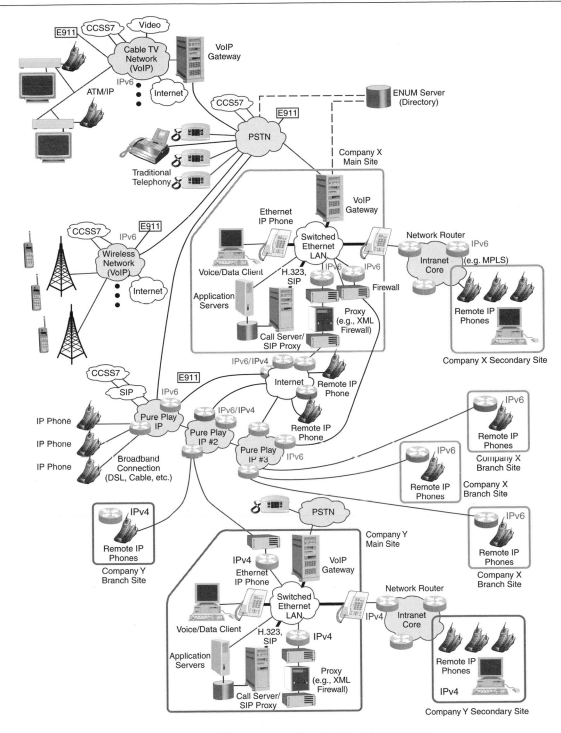

Figure 7.2: Mid-stage state of an IPv6-Enabled 3G VoIP.

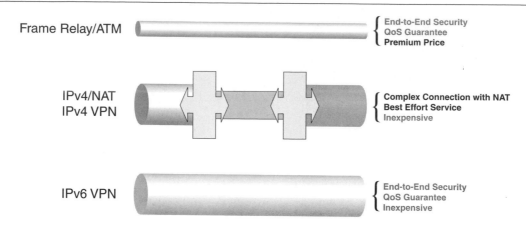

Figure 7.3: Security arrangements for various technologies (100,000-foot view).

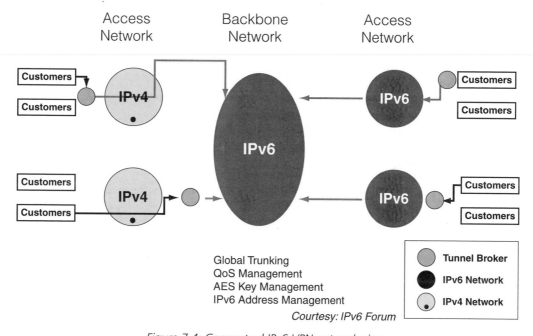

Figure 7.4: Conceptual IPv6 VPN network view.

The discussion that follows highlights important IPv6 concepts that are expected to be of importance in an VoIPv6 context. Some of the information is loosely based on [MSD200401].

7.2 IPv6 Infrastructure

7.2.1 Protocol Mechanisms

As we saw in Chapter 6, an IPv6 Protocol Data Unit (PDU) consists of an IPv6 header and an IPv6 payload, as depicted in Figure 7.5. The IPv6 header consists of two parts: the IPv6 base header, and optional extension headers. The optional extension headers are considered part of the IPv6 payload, as are the TCP/UDP/RTP PDUs (including the voice bit stream). Obviously, IPv4 headers and IPv6 headers are not interoperable; hence, a router operating in a mixed environment must use an implementation of both IPv4 and IPv6 in order to deal both header formats; this is also the case in VoIP arrangements that support both environments as well as transition environments. Figure 7.6 shows the flows of IPv6 PDUs in a VoIPv6 environment.

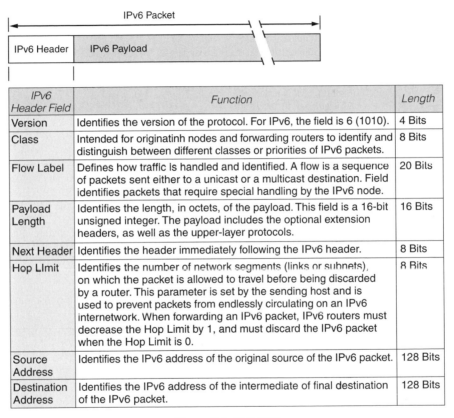

IPv6 Header Field	Function	Length
Version	Identifies the version of the protocol. For IPv6, the field is 6 (1010).	4 Bits
Class	Intended for originatinh nodes and forwarding routers to identify and distinguish between different classes or priorities of IPv6 packets.	8 Bits
Flow Label	Defines how traffic Is handled and identified. A flow is a sequence of packets sent either to a unicast or a multicast destination. Field identifies packets that require special handling by the IPv6 node.	20 Bits
Payload Length	Identifies the length, in octets, of the payload. This field is a 16-bit unsigned integer. The payload includes the optional extension headers, as well as the upper-layer protocols.	16 Bits
Next Header	Identifies the header immediately following the IPv6 header.	8 Bits
Hop LImit	Identifies the number of network segments (links or subnets), on which the packet is allowed to travel before being discarded by a router. This parameter is set by the sending host and is used to prevent packets from endlessly circulating on an IPv6 internetwork. When forwarding an IPv6 packet, IPv6 routers must decrease the Hop Limit by 1, and must discard the IPv6 packet when the Hop Limit is 0.	8 Bits
Source Address	Identifies the IPv6 address of the original source of the IPv6 packet.	128 Bits
Destination Address	Identifies the IPv6 address of the intermediate of final destination of the IPv6 packet.	128 Bits

Figure 7.5: IPv6 PDU.

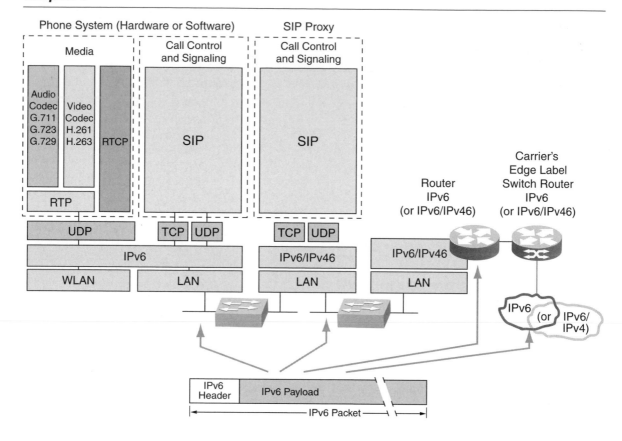

Figure 7.6: Flows of IPv6 packets in a VoIPv6 environment.

IPv6 allows for 2^{128} or $\sim 3.4 \times 10^{38}$ possible addresses. As we noted in the previous chapter, the large size of the IPv6 address allows it to be subdivided into hierarchical routing domains that are supportive the topology of the today's ubiquitous Internet (IPv4-based Internet lacks this flexibility). Conveniently, the use of 128 bits provides multiple levels of hierarchy and flexibility in designing hierarchical addressing and routing. As we have stated, and as we have implied in Figure 7.2, there is interest in end-to-end reliable VoIP and this IPv6 hierarchy supports such goals. It should be kept in mind that today's global voice networking, the global PSTN, is indeed arranged in a hierarchal manner for reasons of administrative oversight, administrative ownership, routing, billing, etc.

7.2.2 Protocol-Support Mechanisms

Two support mechanisms are of interest: (1) a mechanism to deal with communication transmission issues; and (2) a mechanism to support multicast.

Internet Control Message Protocol for IPv6 (ICMPv6) (RFC 2463) is designed to enable hosts and routers that use IPv6 protocols to report errors and forward along other basic status messages. For example, ICMPv6 messages are sent by Network Elements when an IPv6 PDU cannot be forwarded further along to reach its intended destination. ICMPv6 messages are carried as the payload of IPv6 PDUs (see Figure 7.7), hence, there is no guarantee on their delivery. ICMPv6 will find usage and application in 3G VoIP systems.

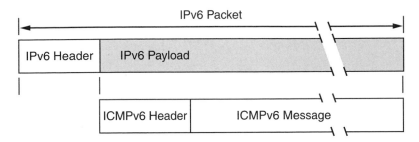

Figure 7.7: ICMPv6 message.

The following list identifies the functionality supported by the basic ICMPv6 mechanisms:

- *Destination Unreachable*: An error message that informs the sending host that a PDU cannot be delivered.
- *Packet Too Big*: An error message that informs the sending host that the PDU is too large to forward.
- *Time Exceeded*: An error message that informs the sending host that the Hop Limit of an IPv6 PDU has expired.
- *Parameter Problem*: An error message that informs the sending host that an error was encountered in processing the IPv6 header or an IPv6 extension header.
- *Echo Request*: An informational message that is used to determine whether an IPv6 node is available on the network.
- *Echo Reply*: An informational message that is used to reply to the ICMPv6 Echo Request message.

(The ping command is basically an ICMPv6 Echo Request messages along with the receipt of ICMPv6 Echo Reply messages; just as is the case in IPv4, one can use pings to detect network or host communication failures and troubleshoot connectivity problems.)

ICMPv6 also supports Multicast Listener Discovery (MLD). MLD (RFC 2710, RFC 3590, RFC 3810) enables one to manage subnet multicast membership for IPv6. MLD is a collection of three ICMPv6 messages that replace the Internet Group Management Protocol (IGMP) version 3 that is employed in IPv4. MLD messages are used to determine group membership on a network segment, also known as a link or subnet. As implied, MLD messages are sent as ICMPv6 messages. They are used in the context of multicast communications (see below):

- *Multicast Listener Query*: Message issued by a multicast router to poll a network segment for group members. Queries can be general, requesting group membership for all groups, or can request group membership for a specific group.
- *Multicast Listener Report*: Message issued by a host when it joins a multicast group, or in response to an MLD Multicast Listener Query sent by a router.
- *Multicast Listener Done*: Message issued by a host when it leaves a host group and is the last member of that group on the network segment.

ICMPv6 also supports Neighbor Discovery (ND). ND (RFC 2461) is a collection of five ICMPv6 messages that manage node-to-node communication on a link. Nodes on the same link are also called *neighboring nodes*. ND replaces Address Resolution Protocol (ARP), ICMPv4 Router Discovery, and the ICMPv4 Redirect message. Table 7.1 identifies key ND processes [MSD200401]. Hosts (e.g., SIP proxies, H.323 gatekeepers, etc.) make use of ND to discover neighboring routers, addresses, address prefixes, and other configuration parameters. Routers make use of ND to advertise their presence, host configuration parameters, and on-link prefixes; also, they use of ND to inform hosts of a better next-hop address to forward PDUs for a specific destination. Nodes make use of ND to resolve the link-layer address of a neighboring node to which an IPv6 PDU is being forwarded. Nodes also use ND to determine when the link-layer address of a neighboring node has changed, and whether IPv6 PDUs can be sent to and received from a neighbor.

Table 7.1: Key ND processes.

Process	Description
Router Discovery	The process by which a host discovers the local routers on an attached link and automatically configures a default router. In IPv4, this is equivalent to using ICMPv4 router discovery to configure a default gateway.
Prefix Discovery	The process by which a host discovers the network prefixes for local destinations.
Parameter Discovery	The process by which a host discovers additional operating parameters, including the link Maximum Transmission Unit (MTU) and the default hop limit for outbound PDUs.
Address Autoconfiguration	The process for configuring IP addresses for interfaces in either the absence of a stateful address configuration server, such as Dynamic Host Configuration Protocol version 6 (DHCPv6).
Address Resolution	The process by which a node resolves a neighboring node's IPv6 address to its link-layer address. The resolved link-layer address becomes an entry in a neighbor cache in the node. The link layer address is equivalent to ARP in IPv4, and the neighbor cache is equivalent to the ARP cache. The neighbor cache displays the interface identifier for the neighbor cache entry, the neighboring node IPv6 address, the corresponding link-layer address, and the state of the neighbor cache entry.
Next-Hop Determination	The process by which a node determines the IPv6 address of the neighbor to which a PDU is being forwarded. The determination is made based on the destination address. The forwarding or next-hop address is either the destination address of the PDU being sent or the address of a neighboring router. The resolved next-hop address for a destination becomes an entry in a node's destination cache, also known as a route cache. The route cache displays the destination address, the interface identifier and next-hop address, the interface identifier and address used as a source address when sending to the destination, and the path MTU for the destination.
Neighbor Unreachability Detection	The process by which a node determines that neighboring hosts or routers are no longer available on the local network segment. After the link-layer address for a neighbor has been determined, the state of the entry in the neighbor cache is tracked. If the neighbor is no longer receiving and sending back PDUs, the neighbor cache entry is eventually removed.
Duplicate Address Detection	The process by which a node determines that an address considered for use is not already in use by a neighboring node. This is equivalent to the use of gratuitous ARP frames in IPv4.
Redirect Function	The process by which a router informs a host of a better first-hop IPv6 address to reach a destination. This is equivalent to the function of the IPv4 ICMP Redirect message.

A useful feature supported in IPv6 is multicasting. Besides a variety of protocol-level functionally supported by multicasting (for example, the just-mentioned MLD and ND), one can also use this mechanism to support VoIP/IPTV functionality (e.g., audioconferencing/bridging and program distribution). (The use of multicasting in IP networks is defined in RFC 1112 which describes addresses and host extensions for the way IP hosts support multicasting—the concepts originally developed for IPv4 also apply to IPv6.) Multicast traffic is promulgated by utilizing a single address in the IPv6 PDU header, but is processed by multiple hosts. Hosts and devices listening on a specific multicast address comprise a multicast group; these devices receive and process traffic sent to the group address. IPv6 multicast addresses have the Format Prefix 1111 1111 (0xFF).

Group membership in multicast lists is dynamic, allowing hosts to join and leave the group at any time. Groups can be from multiple network segments (links or subnets) if the connecting routers support forwarding of multicast traffic and group membership information [MSD200401]. A host (for example, a VoIP SIP proxy or a H.323 gatekeeper) can send traffic to a group address without belonging to the group. In fact, to join a group, a host sends a group membership message. Multicast routers periodically poll membership status. Each multicast group is identified by one IPv6 multicast address. All group members who listen and receive IPv6 messages sent to the group address share the group address. Some of the reserved IPv6 multicast addresses are (RFC 2375) shown in Table 7.2.

Table 7.2: Reserved multicast IPv6 addresses.

IPv6 Multicast Address	Description
FF02::1	The all-nodes address used to reach all nodes on the same link.
FF02::2	The all-routers address used to reach all routers on the same link.
FF02::4	The address used to reach all Distance Vector Multicast Routing Protocol (DVMRP) multicast routers on the same link.
FF02::5	The address used to reach all Open Shortest Path First (OSPF) routers on the same link.
FF02::1:FFXX:XXXX	The solicited-node address used in the address resolution process to resolve the IPv6 address of a link-local node to its link-layer address. The rightmost 24 bits (XX:XXXX) of the solicited-node address are the rightmost 24 bits of an IPv6 unicast address.

7.3 IPv6 Addressing Mechanisms

7.3.1 Conventions

The IPv6 128-bit address is divided along 16-bit boundaries. Each 16-bit block is then converted to a 4-digit hexadecimal number, separated by colons. The resulting representation is called *colon-hexadecimal*. This approach should now be familiar to the reader having covered this in Chapters 1 and 6. This is in contrast to the 32-bit IPv4 address represented in dotted-decimal format, divided along 8-bit boundaries, and then converted to its decimal equivalent, separated by periods.

The following examples show 128-bit IPv6 address in binary form:

```
Address 1:  00100001110110100000000001101001100000000000000000010111100111011
0000001010101010000000001111111111111110001010001001110001011010

Address 2:  00100001110110100000000001101001100000000000000000010111100111011
00000010101010100000000011111111000000000000000001001110001011010

Address 3:  00100001110110100000000001101001100000000000000000001001110001011010
00000010101010100000000011111111000000000000000001001110001011010
```

```
Address 4: 0010000111011010000000000110100110000000000000000001001110001011010
00000010101010100000000001111111100000000000000000010111100111011
```

The following example shows these same addresses divided along 16-bit boundaries:

```
Address 1: 0010000111011010:0000000011010011:0000000000000000:0010111100111011:
0000001010101010:0000000011111111:1111111000101000:1001110001011010:

Address 2: 0010000111011010:0000000011010011:0000000000000000:0010111100111011:
0000001010101010:0000000011111111:0000000000000000:1001110001011010:

Address 3: 0010000111011010:0000000011010011:0000000000000000:1001110001011010:
0000001010101010:0000000011111111:0000000000000000:1001110001011010:

Address 4: 0010000111011010:0000000011010011:0000000000000000:1001110001011010:
0000001010101010:0000000011111111:0000000000000000:0010111100111011:
```

The following shows each 16-bit block in the address converted to hexadecimal and delimited with colons.

```
Address 1: 21DA:00D3:0000:2F3B:02AA:00FF:FE28:9C5A
Address 2: 21DA:00D3:0000:2F3B:02AA:00FF:0000:9C5A
Address 3: 21DA:00D3:0000: 9C5A:02AA:00FF:0000:9C5A
Address 4: 21DA:00D3:0000: 9C5A:02AA:00FF:0000:2F3B
```

IPv6 representation can be further simplified by removing the leading zeros (trailing zeros are not removed) within each 16-bit block. However, each block must have, in the abbreviated nomenclature, at least a single digit. The following example shows the addresses without the *leading* zeros:

```
Address 1: 21DA:D3:0:2F3B:2AA:FF:FE28:9C5A
Address 2: 21DA:D3:0:2F3B:2AA:FF:0:9C5A
Address 3: 21DA:D3:0: 9C5A:2AA:FF:0:9C5A
Address 4: 21DA:D3:0: 9C5A:2AA:FF:0:2F3B
```

Some types of addresses contain long sequences of zeros. In IPv6 addresses, a contiguous sequence of 16-bit blocks set to 0 in the colon-hexadecimal format can be compressed to :: (known as double-colon).

The following list shows examples of compressing zeros:
- The address 21DA:0:0:0:2AA:FF:9C5A:2F3B can be compressed to 21DA::2AA:FF:9C5A:2F3B.
- The multicast address of FF02:0:0:0:0:0:0:2 can be compressed to FF02::2.

Note: Zero-compression can only be used to compress a single contiguous series of 16-bit blocks expressed in colon-hexadecimal notation – one cannot use zero-compression to include part of a 16-bit block; e.g., one cannot abbreviate FF01:30:0:0:0:0:0:8 as FF01:3::8.) Also, zero-compression can be used only once in an address, which enables one to determine the number of 0 bits represented by each instance of a double-colon (::). To determine how many 0 bits are represented by the ::, one can count the number of blocks in the compressed address, subtract this number from 8, and then multiply the result by 16. For example, in the address FF02::2, there are two blocks (the FF02 block and the 2 block); the number of bits expressed by the :: is 96 (= (8 − 2) × 16) [MSD200401].

7.3.2 Addressing issues/reachability

Every IPv6 address has a reachability scope. IPv6 interfaces can have multiple addresses that have different reachability scopes. For example, a node may have a link-local address, a site-local address, and a global address. Table 7.3 shows the address and associated reachability scopes. The reachability of Node-local addresses is "the same node"; the reachability of Link-local addresses is "the local link"; the reachability of Site-local addresses is "the private intranet"; and, the reachability of Global addresses is "the IPv6-enabled Internet."

Table 7.3: Address and associated reachability scopes.

Address scope/ Reachability	Description
Node-local addresses to reach same node	Used to send PDUs to the same node: • Loopback address (PDUs addressed to the loopback address are never sent on a link or forwarded by an IPv6 router—this is equivalent to the IPv4 loopback address) • Node-local multicast address
Link-local addresses to reach local link (*)	Used to communicate between hosts (and/or VoIP devices) on the link; these addresses are always configured automatically: • Unspecified address. It indicates the absence of an address, and is typically used as a source address for PDUs that are attempting to verify the uniqueness of a tentative address (it is equivalent to the IPv4 unspecified address.) The unspecified address is never assigned to an interface or used as a destination address. • Link-local Unicast address • Link-local Multicast address
Site-local addresses to reach the private intranet (internetwork) (*)	Used between nodes that communicate with other (VoIP) nodes in the same site; site-local addresses are configured by router advertisement: • Site-local Unicast address—these addresses are not reachable from other sites, and routers must not forward site-local traffic outside of the site. Site-local addresses can be used in addition to aggregatable global unicast addresses. • Site-local Multicast address
Global addresses to reach the Internet (IPv6-enabled); also known as aggregatable global unicast addresses	Globally routable and reachable addresses on the IPv6 portion of the Internet (they are equivalent to public IPv4 addresses); global addresses are configured by router advertisement: • Global Unicast address • Other scope Multicast address Global addresses are designed to be aggregated or summarized to produce an efficient, hierarchical addressing and routing structure.

(*) When one specifies a link-local or site-local address, one needs to also specify a scope ID, which further defines the reachability scope for these (nonglobal) addresses.

Similarly to the IPv4 address space, the IPv6 address space is partitioned according to the value of high order bits (known as a Format Prefix) in the address. Table 7.4 depicts the IPv6 address space allocation by Format Prefixes. The (current) set of unicast addresses that can be employed by IPv6 nodes consists of aggregatable global unicast addresses, link-local unicast addresses, and site-local unicast addresses (these addresses represent about 12.6% of the entire IPv6 address space, but it is still ~3.4 × 10^{37}). The prefix is the portion of the address that indicates the bits that have fixed values or are the bits of the network identifier. Prefixes for IPv6 routes and subnet identifiers are expressed in the same way as classless interdomain routing notation for IPv4. An IPv6 prefix is written in address/prefix-length notation (IPv4 environments use a dotted decimal representation known as the subnet mask in order to establish the network prefix of a given IP address; the subnet mask approach is not used in IPv6, rather, only the prefix-length notation is used.)

Table 7.4: IPv6 address space allocation.

Address Space Allocation	Format Prefix	Percentage of the Address Space	Hex Notation	Fraction of the Address Space
Reserved	0000 0000	0.391%	0x00	1/256
Reserved for NSAP allocation	0000 001	0.781%	0x0 001	1/128
Aggregatable global unicast addresses	001	12.500%	001	1/8
Link-local unicast addresses	1111 1110 10	0.098%	0xFE 10	1/1024
Site-local unicast addresses	1111 1110 11	0.098%	0xFE 11	1/1024
Multicast addresses	1111 1111	0.391%	0xFF	1/256
The remainder of the IPv6 address	Unassigned	85.742%		

Note: 0xY is the hexadecimal notation for digit "Y".

As noted earlier, the prefix is the part of the address that indicates the bits that have fixed values or are the bits of the network identifier. For example,

21DA:D3::/48 is a 48-bit route prefix

21DA	00D3	0000	16 bits	16 bits	16 bits	16 bits	16 bits
<- route prefix ->							

and,

21DA	00D3	0000	2F3B	16 bits	16 bits	16 bits	16 bits
<- route prefix ->			<- subnet prefix ->	16 bits	16 bits	16 bits	16 bits

7.3.3 Scope/Reachability

The scope ID identifies a specific area within the reachability scope for nonglobal addresses (recall that the reachability scope is related to the address scope as shown in Table7.5.) A node identifies each area of the same scope with a unique scope ID.

Table 7.5: Address scope vs. reachability scope.

Address scope	Reachability Scope
Node-local	Same node
Link-local	Local link (LAN)
Site-local	A private internetwork (intranet)

Figure 7.8 shows an example how the scope ID indicates an interface or site identifier, depending on the scope of the address. In this example the node is connected to three links and two sites. In this example,

- The sites (specifically, site identifiers 1 and 2) are identified by the site-local (intranet) scope ID.
- The links (specifically, interface identifiers 1, 2, and 3) are identified by the link-local (LAN) scope ID.

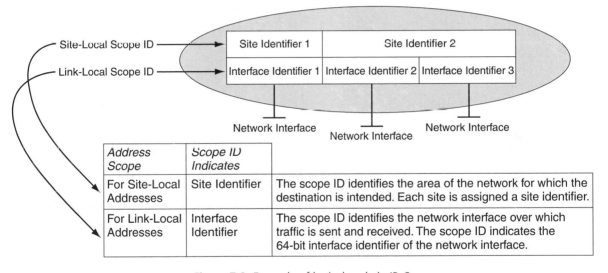

Figure 7.8: Example of logical node in IPv6.

The notation utilized to specify the scope ID with an address is Address%ScopeID. Figure 7.9 depicts an example of how the nodes use the scope ID to identify site scope zones. As one can see in the figure, the interface identifier scope ID is used only by the local node; other nodes may use a different network interface or site identifier for the same link, e.g., the link that has scope ID (interface identifier) 4 for Node A has scope ID 1 for Node B. For example of Figure 7.9, the following describes the link-local address FE70::3 qualified with a scope ID on the link between Node A and Node B:

For Node A, the address is FE70::3%4

For Node B, the address is FE70::3%1

Each attached zone of the same scope must be assigned a different site identifier, but attached zones of different scopes can re-use the same index.

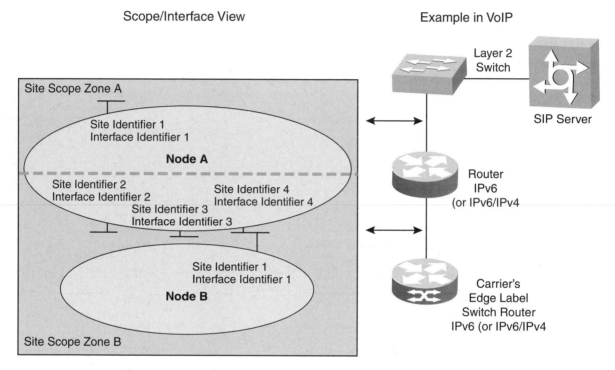

Figure 7.9: How the nodes use the scope ID to identify site scope zones.

Implicit in Figure 7.9 is a topology hierarchy, as follows:

- Public topology: The collection of larger and smaller ISPs that provides access to the IPv6 Internet.
- Site topology: The collection of subnets within an organization's site (namely, this is the intranet, although it need not be strictly contained at a single location—the term *site* here has more an implication related to an organization's domain rather than a single physical site.)
- Interface identifier topology: Identifies a specific interface on a subnet within an organization's site.

7.3.4 Address Types

This section looks at some more detailed information related to address types.

Unicast IPv6 Addresses

A unicast address identifies a single interface within the scope of the unicast address type. This could be a VoIP handset in a VoIPv6 environment. Utilizing an up-to-date unicast routing topology, PDUs addressed to a unicast address are delivered to a single interface. Unicast addresses fall into the following categories:

- Aggregatable global unicast addresses (e.g., used to reach an Internet-connected VoIP phone);
- Link-local addresses (e.g., used to reach an VoIP phone on the same LAN segment);
- Site-local addresses (e.g., used to reach an VoIP phone on corporate intranet);
- Special addresses, including unspecified and loopback addresses; and,
- Compatibility addresses, including 6to4 addresses.
- Aggregatable Global Unicast Addresses

The IPv6-based Internet has been designed to support efficient, hierarchical addressing and routing (this is in contrast IPv4-based Internet which has a mixture of both flat and hierarchical routing.) Aggregatable global unicast addresses are globally-routable and globally-reachable on the IPv6 portion of the (IPv6) Internet. The region of the Internet over which the aggregatable global unicast address is unique (the scope) is the entire IPv6 Internet. As we saw in Table 7.4, aggregatable global unicast addresses (aka global addresses), are identified by the Format Prefix of 001. This type of addressing can be used to reach an Internet-connected VoIP (SIP) phone (say, the author's phone given to him by his company and utilized by him while traveling on business and using the Internet for connectivity), from any origination point, be such origination point on the firm's intranet or on any other-company's intranet, or even at another Internet point. This enables the end-to-connectivity that we have been alluding to throughout this book.

Figure 7.10 shows how the fields within the aggregatable global unicast address create a three-level topological structure with globally-unique addresses. The first 48 bits are comprised of the 3-bit prefix; the Top Level Aggregator (TLA) id comprises the next 13 bits; the next 8 bits are reserved; and, the next 24 bits represent the Next Level Aggregator (NLA) id. This combination gives the first two levels. The next 16 bits represents the site topology, namely, the Site Level Aggregator (SLA) id. The SLA is used by a firm or organization to identify subnets within its site (intranet); the organization can use the 16 bits within its site to create 65,536 subnets or multiple levels of addressing hierarchy, which, can also facilitate the routing process (note that with a 2-octet of address space for subnetting, an aggregatable global unicast prefix assigned to a firm is equivalent to that firm being granted an IPv4 Class A network ID; the structure of the customer's network is not visible to the ISP). Finally, the Interface ID point to the interface of a node on a specific subnet.

Addresses of this type can, by design, be aggregated (summarized) to produce an efficient routing infrastructure.

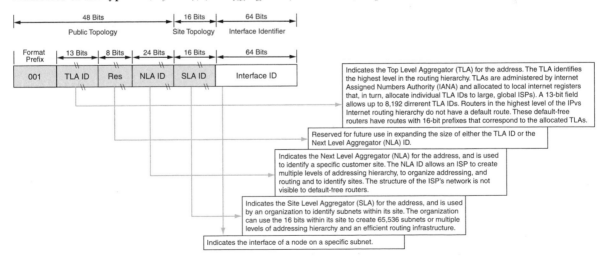

Figure 7.10: Aggregatable global unicast address.

Link-Local (Unicast) Addresses

Link-local addresses are utilized by nodes when communicating with neighboring nodes on the same link. For example, link-local addresses are used to communicate between hosts on the link on a single link IPv6 network without the intervention/utilization of a router (e.g., in a LAN segment, a VLAN, etc.). This type of addressing can be used to reach a company colleague an LAN-connected VoIP phone (say, for colleagues working in the same building—assuming that both are on the same LAN.)

The scope of a link-local address is the local link. An IPv6 router does not forwards link-local traffic beyond the link. A link-local address is required for Neighbor Discovery processes and is always automatically configured, even in the absence of all other unicast addresses. As seen in Table 7.4, link-local addresses are identified by the Format Prefix of 1111 1110 10. The address starts with FE (for example 1111 1110 1000 is 0xFE8; 1111 1110 1001 is 0xFE9; 1111 1110 1010 is 0xFEA; and, 1111 1110 1011 is 0xFEB.) With the 64-bit interface identifier, the prefix for link-local addresses is, by convention, always FE80::/64.

Site-Local (Unicast) Addresses

Site-local addresses are utilized between nodes that communicate with other nodes in the same site (organization). The scope of a site-local address is the site, which is the organization intranet (internetwork.) This type of addressing can be used to reach a company colleague an intranet-connected VoIP phone, say for colleagues working in the same company but perhaps at two company locations in two cities (in this arrangement, however, those VoIP phones would not be directly reachable from anywhere that happens to be on an IP-network—see next paragraph—they may, nonetheless, be reacheable through a gateway.)

As seen in Table 7.4, site-local addresses are identified by the Format Prefix of 1111 1110 11 (they are equivalent to the IPv4 private address space, 10.0.0.0/8, 172.16.0.0/12, and 192.168.0.0/16.) Hence, if there are private intranets that do not have a direct, routed connection to the IPv6 Internet they can use site-local addresses without conflicting with aggregatable global unicast addresses; however, one should keep in mind that increasingly virtually all private intranets have connections to the Internet. Also, if end-to-end anytime anyplace VoIP communication is to be supported, aggregatable global unicast may be indicated for each VoIP device. This is because site-local addresses are not reachable from other sites (namely, other organizations), and routers must not forward site-local traffic outside of the site (organization). Fortunately aggregatable global unicast addresses may be assigned to VoIP devices in addition to site-local addresses. Unlike link-local addresses, site-local addresses are not automatically configured and must be assigned through the stateless address configuration process [MSD200401].

Referring to Figure 7.8, one notes that the first 48-bits are always fixed for site-local addresses, beginning with FEC0::/48. Beyond the 48-fixed bits is a 16-bit subnet identifier (Subnet ID field) with which the network administrator can create subnets for use within the organization (one can create up to 65,536 subnets in a flat subnet structure; or, one can partition the high-order bits of the Subnet ID field to create a hierarchical and aggregatable routing infrastructure.) Beyond the Subnet ID field is a 64-bit Interface ID field that identifies a specific interface on a subnet.

Notice from this discussion that the *aggregatable global unicast address* and the *site-local address* share the same structure beyond the first 48 bits of the address. In aggregatable global unicast addresses, the SLA ID identifies the subnet within an organization and for site-local addresses, the Subnet ID performs the same function. Because of this characteristic, one can assign a specific subnet number to identify a subnet that is used for both site-local and aggregatable global unicast addresses.

Unspecified (Unicast) Address

The unspecified address, 0:0:0:0:0:0:0:0 (that is, ::) indicates the absence of an address, and is typically used as a source address for PDUs that are attempting to verify the uniqueness of a tentative address. It is equivalent to the IPv4 unspecified address of 0.0.0.0. The unspecified address is never assigned to an interface or used as a destination address.

Loopback (Unicast) Address

The loopback address, 0:0:0:0:0:0:0:1 or ::1, identifies a loopback interface, enabling a node to send PDUs to itself. It is equivalent to the IPv4 loopback address of 127.0.0.1. PDUs addressed to the loopback address are never sent on a link or forwarded by an IPv6 router.

Compatibility (Unicast) Addresses

IPv6 provides what are called *6to4 addresses* to facilitate the coexistence of IPv4-to-IPv6 environments and the migration from the IPv4 to the IPv6 environment. The 6to4 address is used for communicating between two nodes operating both IPv4 stacks and IPv6 stacks over an IPv4 routing infrastructure (more on this in Chapter 8). The 6to4 address is formed by combining the prefix 2002::/16 with the 32 bits of the public IPv4 address of the node, forming a 48-bit prefix. For example, for the IPv4 address of 231.207.10.11, the 6to4 address prefix is 2002:836B:1::/48.

Multicast IPv6 Addresses

A multicast address is an addressing mechanism that identifies multiple interfaces; it is used for one-to-many communication. As seen in Table 7.4, IPv6 multicast addresses have the Format Prefix of 1111 1111; namely, the multicast address always begins with 0xFF. With the appropriate multicast routing topology, PDUs addressed to a multicast address are delivered to all interfaces that are identified by the address. Multicast addresses cannot be utilized as source addresses. Multicast addresses flags, scope, and multicast group, as shown if Figure 7.11.

To identify all nodes for the node-local and link-local scopes, the following multicast addresses are defined:

```
FF01::1 (node-local scope all-nodes address)
FF02::1 (link-local scope all-nodes address)
```

To identify all routers for the node-local, link-local, and site-local scopes, the following multicast addresses are defined:

```
FF01::2 (node-local scope all-routers address)
FF02::2 (link-local scope all-routers address)
FF05::2 (site-local scope all-routers address)
```

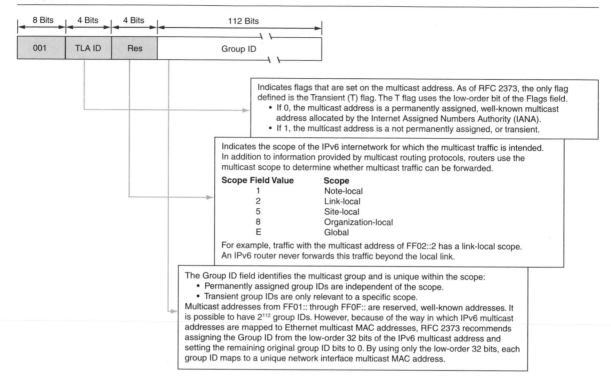

Figure 7.11: Multicast address.

Next, we briefly look at solicited-node addresses. The solicited-node address supports efficient querying of network nodes for the purpose of address resolution. IPv6 uses the Neighbor Solicitation message to perform address resolution. This multicast address consists of the prefix FF02::1:FF00:0/104 along with the last 24-bits of the IPv6 address that is being resolved. In contrast to IPv4 where the ARP Request frame is sent via a MAC-level broadcast, and in doing so imposing on all nodes on the network segment, in IPv6 the solicited-node multicast address is used as the Neighbor Solicitation message destination. This avoids imposing on all IPv6 nodes on the local link by using the local-link scope all-nodes address.

Anycast IPv6 Addresses

An anycast address identifies multiple interfaces, but not an entire broadcast universe. This could be used, for example, to support VoIP Voice Mail group distribution. With the appropriate routing topology, PDUs addressed to an anycast address are delivered to a single interface for further appropriate handling (a PDU addressed to an anycast address is delivered to the nearest interface identified by the address.) To make possible the delivery to the nearest anycast group member, the routing infrastructure must be aware of the interfaces that are assigned anycast addresses and must know their distances in terms of routing metrics. At present, anycast addresses are used only as destination addresses and are assigned only to routers.

7.3.5 Addresses for Hosts and Routers

In contrast to an IPv4 where a host with a single network adapter has a single IPv4 address assigned to that adapter, an IPv6 host (e.g., a SIP proxy) typically has multiple IPv6 addresses (even in the case of a single interface.) (When a computer is configured with more than one IP address, it is referred to as a *multihomed* system.) IPv6 host and router address usage is as follows [MSD200401]:

Host: Typical IPv6 hosts are logically multihomed because they have at least two addresses with which they can receive PDUs. Each host is assigned the following unicast addresses:

- A link-local address for each interface. This address is used for local traffic.
- An address for each interface. This could be a routable site-local address and one or more global addresses.
- The loopback address (::1) for the loopback interface.

Additionally, each host is listening for traffic on the following multicast addresses:

- The node-local scope all-nodes address (FF01::1).
- The link-local scope all-nodes address (FF02::1).
- The solicited-node address for each unicast address on each interface.
- The multicast addresses of joined groups on each interface.

Router: An IPv6 router is assigned the following unicast addresses:

- A link-local address for each interface. This address is used for local traffic.
- An address for each interface. This could be a routable site-local address and one or more global addresses.
- The loopback address (::1) for the loopback interface.

An IPv6 router is assigned the following anycast addresses:

- A subnet-router anycast address for each subnet
- Additional anycast addresses (optional)

Each router is listening for traffic on the following multicast addresses:

- The node-local scope all-nodes address (FF01::1)
- The node-local scope all-routers address (FF01::2)
- The link-local scope all-nodes address (FF02::1)
- The link-local scope all-routers address (FF02::2)
- The site-local scope all-routers address (FF05::2)
- The solicited-node address for each unicast address on each interface
- The addresses of joined groups on each interface

Interface Determination

As noticed in Figure 7.8, the last 64 bits of an IPv6 address are the interface identifier that is unique to the 64-bit prefix of the IPv6 address. There are two ways for interface identifier determination: (1) derived from the Electrical and Electronic Engineers (IEEE) Extended Unique Identifier (EUI)-64 address; and, (2) randomly-generated and randomly-changed over time. IETF RFC 2373 stipulates that unicast addresses that use format prefixes 001 through 111 must use a 64-bit interface identifier that is derived from the EUI-64 address. Related to the second approach, RFC 3041 states that to provide a level of anonymity, the identifier can be randomly generated, and changed over time.

EUI-64 addresses are either assigned to a network adapter or derived from IEEE 802 addresses. LAN Network Interface Cards (NICs) that (at this point in the development of hardware) typically comprise the physical interface (network adapters) of hosts and devices identifiers use the 48-bit IEEE 802 address. This address (also called the *physical*, *hardware*, or *Media Access Control (MAC)* address) consists of two parts: (1) Company ID; and (2) Extension ID. The Company ID is 24-bit ID uniquely assigned to each manufacturer of network adapters; this is also known as the manufacturer ID. The Extension ID (also known as the board ID) is a 24-bit uniquely assigned to each network adapter at the time of assembly. The IEEE 802 address is a globally-unique 48-bit address. The IEEE EUI-64 address is a newly-defined standard for network interface addressing. The company ID is 24-bits in length but the extension ID is 40 bits, supporting a larger address space for a network adapter manufacturer. See Figure 7.12.

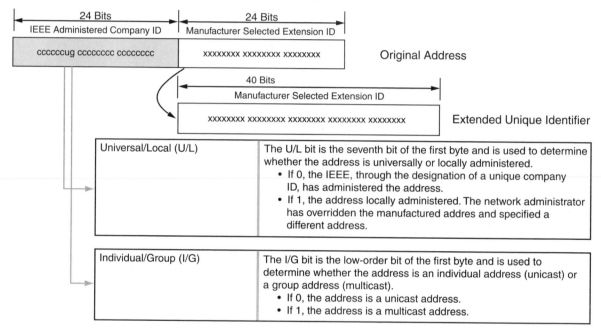

Figure 7.12: IEEE address along with the extended unique identifier.

To generate an EUI-64 address from an IEEE 802 address, 16 bits of 11111111 11111110 (0xFFFE) are inserted into the IEEE 802 address between the company ID and the extension ID. See Figure 7.13.

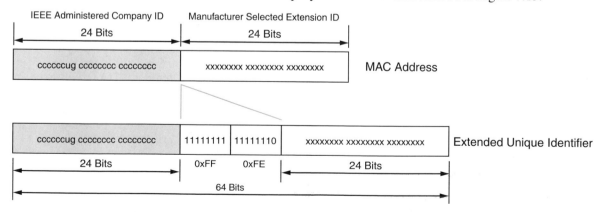

Figure 7.13: Extended unique identifier generated from MAC address.

Mapping EUI-64 addresses to IPv6 interface identifiers

An IPv6 unicast address utilizes a 64-bit interface identifier. To obtain this identifier from a EUI-64 address, the U/L bit in the EUI-64 address is complemented (if it is a 1, it is set to 0; if it is a 0, it is set to 1). The resulting bitstream is used as a universally-administered unicast EUI-64 address.

Mapping IEEE 802 Addresses to IPv6 Interface Identifiers

To obtain an IPv6 interface identifier from an IEEE 802 address, one must first map the IEEE 802 address to an EUI-64 address, as discussed previously; then one must complement (flip) the U/L bit. The resulting bitstream is used as a universally-administered unicast IEEE 802 address.

Randomly-Generated Interface Identifiers

IPv6 address identifiers remain static over time, hence, for security reasons, a capability is needed to generate temporary addresses. (Because of NAT/DHCP, in an IPv4 environment it is difficult to track a user's traffic on the basis of IP address.) (However, it should be noted that many—if not most—hacking techniques do not rely on knowing the IP address of a specific device on a network; instead, such techniques simply look for any available entry point; after that, a deposited Trojan Horse may do the job perpetrating a full infraction.)

In IPv6 after the connection is made through router discovery and stateless address autoconfiguration, the end-user device is assigned a 64-bit prefix. If the interface identifier is based on a EUI-64 address (which, as we saw earlier, is derived from the static IEEE 802 address), the traffic of a specific node can be identified; this opens up the possibility to track a specific user (should that be of interest to an intruder). To address this issue, an alternative IPv6 interface identifier can be randomly generated and changed over time, as described in RFC 3041. For IPv6 systems that have storage capabilities, a history value is stored; when the IPv6 protocol is initialized, a new interface identifier is created through the following process (the IPv6 address based on this random interface identifier is known as a temporary address):

- Retrieve the history value from storage and append the interface identifier based on the EUI-64 address of the adapter.
- Compute the Message Digest-5 (MD5) one-way encryption hash over the quantity in step a.
- Save the last 64 bits of the MD5 hash computed in step b as the history value for the next interface identifier computation.
- Take the first 64 bits of the MD5 hash computed in Step b and set the seventh bit to zero. The seventh bit corresponds to the U/L bit which, when set to 0, indicates a locally administered interface identifier. The result is the interface identifier.

Temporary addresses are generated for public address prefixes that use stateless address autoconfiguration.

7.4 Configuration Methods

As we have seen in previous chapters, the IPv6 protocol can use two address configuration methods: (1) Automatic configuration; and, (2) Manual configuration. Address autoconfiguration for stateless addresses is described in RFC 2462. Autoconfigured addresses exist in one or more of the states depicted in Figure 7.14: tentative, preferred, deprecated, valid (= preferred + deprecated), and, invalid. IPv6 nodes (hosts and routers) automatically create unique link-local addresses for all LAN interfaces that appear to be Ethernet interfaces. IPv6 hosts use received Router Advertisement messages to automatically configure [MSD200401]:

- A default router;
- The default setting for the Hop Limit field in the IPv6 header;
- The timers used in Neighbor Discovery processes;

- The Maximum Transmission Unit (MTU) of the local link;
- The list of network prefixes that are defined for the link. Each network prefix contains both the IPv6 network prefix and its valid and preferred lifetimes. If indicated, a network prefix is combined with the interface identifier to create a stateless IPv6 address configuration for the receiving interface. A network prefix also defines the range of addresses for nodes on the local link;
- 6to4 addresses on a 6to4 tunneling interface for all public IPv4 addresses that are assigned to the computer (some implementations);
- Intrasite Automatic Tunnel Addressing Protocol (ISATAP) addresses on an automatic interface for all IPv4 addresses that are assigned to the computer (some implementations);
- The stack to query for IPv6 ISATAP routers in an IPv4 environment (some implementations); and,
- Routes to off-link prefixes, if the off-link address prefix is advertised by a router (some implementations).

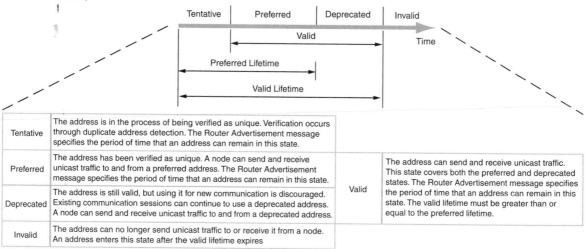

Figure 7.14: Address states.

DHCP is *not* utilized in IPv6 to configure a link-local scope IP address: the link-local scope of an IPv6 addresses is always configured automatically; addresses with other scopes, such as site-local and global, are configured by router advertisements. Specifically, link-local addresses are automatically configured for each interface on each IPv6 node (host or router) with a unique link-local IPv6 address (that is, the IPv6 host configures a link-local address for each interface.) To communicate with IPv6 nodes that are not on attached links, the host must have additional site-local or global unicast addresses. Additional addresses for hosts are obtained from router advertisements; additional addresses for routers must be assigned manually. To communicate with IPv6 nodes on other network segments, IPv6 uses a default router. A default router is automatically assigned based on the receipt of a router advertisement. Alternately, one can add a default route to the IPv6 routing table. Note that one does not need to configure a default router for a network that consists of a single network segment.

The following sequence identifies the address autoconfiguration process for an IPv6 node, such as an IPv6-based VoIP phone:

- A tentative link-local address is derived, based on the link-local prefix of FE80::/64 and the 64-bit interface identifier.

- Duplicate address detection is performed to verify the uniqueness of the tentative link-local address.
- If duplicate address detection fails, one must manually configure the node.

—or—

If duplicate address detection succeeds, the tentative address is assumed to be valid and unique. The link-local address is initialized for the interface. The corresponding solicited-node multicast link-layer address is registered with the network adapter.

For an IPv6 host, such as a SIP server, address autoconfiguration continues as follows:

- The host sends a Router Solicitation message.
- If a Router Advertisement message is received, the configuration information that is included in the message is set on the host.

For each stateless autoconfiguration address prefix that is included, the following processes occurs:

- The address prefix and the appropriate 64-bit interface identifier are used to derive a tentative address.

Duplicate address detection is used to verify the uniqueness of the tentative address. If the tentative address is in use, the address is not initialized for the interface. If the tentative address is not in use, the address is initialized. This includes setting the valid and preferred lifetimes based on information included in the Router Advertisement message.

Other configuration processes are shown in Table 7.6 [MSD200401].

Table 7.6: Configurations of interest.

Configuration	Description
Single Subnet with Link-Local Addresses	This configuration supports the installation of the IPv6 protocol on at least two nodes on the same network segment without intermediate routers.
IPv6 Traffic Between Nodes on Different Subnets of an IPv6 Internetwork	This configuration includes two separate network segments (also known as links or subnets), and an IPv6-capable router that connects the network segments and forwards IPv6 PDUs between the hosts.
IPv6 Traffic Between Nodes on Different Subnets of an IPv4 Internetwork	This configuration supports IPv6 traffic that is carried as the payload of an IPv4 PDU (treating the IPv4 infrastructure as an IPv6 link-layer) without the deployment of IPv6 routers.
IPv6 Traffic Between Nodes in Different Sites Across the Internet (6to4)	This configuration supports the 6to4 tunneling technique. The IPv6 traffic is encapsulated with an IPv4 header before it is sent over an IPv4 internetwork such as the Internet.

7.5 Routing and Route Management

Routing is the process of forwarding PDUs between connected network segments (also known as links or subnets). Routing is a primary function of a network layer protocol, whether it is IP version 4 or version 6. IPv6 routers provide the primary means for joining together two or more IPv6 network segments. Network segments are identified by using an IPv6 network prefix and prefix length. Routers pass IPv6 PDUs from one network segment to another. IPv6 routers are attached to two or more IPv6 network segments and enable hosts on those segments to forward IPv6 PDUs between them. IPv6 PDUs are exchanged and processed on each host by using IPv6 at layer 3 (the Internet layer).

Datagrams with a source and destination IP address identified in the header are handed to the IP protocol engine/layer. Above the IPv6 layer, transport services on the source host pass data in the form of TCP segments or UDP PDUs down to the IPv6 layer. The IPv6 layer creates IPv6 PDUs with source and destination address information that is used to route the data through the network. The IPv6 layer then passes PDUs down to the link layer, where the PDUs are converted into frames for transmission over network-specific media on a physical network. This process occurs in reverse order on the destination host [MSD200401].

IPv6 layer services on each sending host (whether a data server, a multimedia server, or a VoIP server) examine the destination address of each PDU, compare this address to a locally maintained routing table, and then determine what additional forwarding is required.

IPv6 hosts utilize routing tables to maintain information about other IPv6 networks and IPv6 hosts. The routing tables provide important information about how to communicate with remote networks and hosts. Every device that runs IPv6 determines how to forward PDUs based on the contents of the IPv6 routing table. The following list identifies the information contained in the IPv6 routing table:

- An address prefix
- The interface over which PDUs that match the address prefix are sent
- A forwarding or next-hop address
- A preference value used to select between multiple routes with the same prefix
- The lifetime of the route
- The specification of whether the route is published (advertised in a Routing Advertisement)
- The specification of how the route is aged
- The route type

The IPv6 routing table is built automatically, based on the current IPv6 configuration of the router. When forwarding IPv6 PDUs, the router searches the routing table for an entry that is the most specific match to the destination IPv6 address. A route for the link-local prefix (FE80::/64) is not displayed.

Typically, a default router is used by an end device because it is not practical to maintain a routing table for each communication device on an IPv6 network that lists communication information for every other device. The default route (a route with a prefix of ::/0) is typically used to forward an IPv6 PDU to a default router on the local link. Because the router that corresponds to the default router contains information about the network prefixes of the other IPv6 subnets within the larger IPv6 internetwork, it forwards the PDU to other routers until the PDU is eventually delivered to the destination.

The following steps occur during the routing process [MSD200401]:

- Before a communication device sends an IPv6 PDU, it inserts its source IPv6 address and the destination IPv6 address (for the recipient) into the IPv6 header.
- The device then examines the destination IPv6 address, compares it to a locally maintained IPv6 routing table, and takes appropriate action. The device does one of the following:
- It passes the PDU to a protocol layer above IPv6 on the local host.
- It forwards the PDU through one of its attached network interfaces.
- It discards the PDU.

IPv6 searches the routing table for the route that is the closest match to the destination IPv6 address. The most specific to the least specific route is determined in the following order:

1. A route that matches the destination IPv6 address (a host route with a 128-bit prefix length).
2. A route that matches the destination with the longest prefix length.
3. The default route (the network prefix ::/0).

If a matching route is not found, the destination is determined to be an on-link destination.

7.6 Deployment Status

7.6.1 Deployment Approach

Initial[1] network pilot deployments of IPv6 by institutions, research labs, and academia were somewhat open-ended, testing the underlying protocol capabilities of IPv6. Of late, a focus is beginning to emerge from these pilots and trials on how to deploy IPv6. The current focus is network infrastructure deployment, driven by provider, enterprise, consumer, multimedia, and mobility requirements for next generation networks. 3G VoIP is one example of these next-generation mobility networks. Indeed, multimedia is the market driver, according to industry observers: users want to be mobile when using their multimedia. These requirements invariably lead to the need for new network infrastructure components within Provider, Enterprise, and Consumer Networks (PECN). The network pilots now underway will assist stakeholders to prepare for the network infrastructure deployment for PECN, and will help define a set of deployment and transition models that can be used by industry and government.

To support a successful IPv6 deployment the network infrastructure, the applications, the middleware, the security, and the management for the PECN environment and for affiliated end users must first be deployed. The planning and operational analysis to deploy IPv6 pervasively within a network requires planning and testing. Some of this IPv6 planning and testing is still to be done, however, it is not required to have all of this completed in order to begin network infrastructure deployment. Current IPv6 deployments support this pragmatic view.

IPv6 deployment also faces some technology and business challenges: the market benefits from IPv6 assume an end-to-end model; unfortunately this is not the model of most networks today. Thus a technology transformation for the new model is required, in addition to a transition to IPv6. The business strategy to determine the costs and benefits of an approach to deploy IPv6 is an initiative that is now in progress.

The PECN environment has a common foundation: it requires the Service Provider (SP), in the view of stakeholders, to implement a successful deployment of IPv6. The enterprise and consumer deployments will require interoperation with a provider, and each of them can also be a provider to their environment. The SP provides prefixes to an enterprise and the enterprise provides prefixes to their Intranet, or the consumer to their home network devices. IPv6 address assignment is similar across the PECN. This is also true of the deployment models being tested within network pilots and prototype implementations.

Network pilots are testing several deployment models: (1) IPv6 support within the Internet routing core, (2) IPv6 support at the provider and customer edge, and, (3) IPv6 support on the client networks. Then within this model, both sparse and wide-use views exist for IPv6 Intranet nodal and sub-networks deployment. The Internet or provider core is most difficult to test transition to IPv6. The Internet core initially will either tunnel PDUs across the core, encapsulating IPv6 within IPv4 or use the Multiprotocol Label Switching (MPLS) protocol to move IPv6 PDUs across the Internet core transparent to the IPv4 infrastructure. Network pilots exist that can test moving IPv6 PDUs over an Internet core and those network pilots are beginning to connect with each other across multiple geographies, which is good for testing an Internet core paradigm.

The provider and customer edge of network pilots currently are testing native IPv6 and IPv6-in-IPv4 tunnel PDUs to the edge of an Internet core. If not native IPv6 to the Internet core, then various IPv6 transition mechanisms are being used to move IPv6 through an IPv4 infrastructure using a dual-stack method for IPv6 deployment. What the dual-stack method states is that the network and nodes transitioning to IPv6 are capable of supporting both IPv4 and IPv6. This permits the PECN environments to be able to test and verify a deployment model that fits their business requirement to support a sparse or wide-use view for the IPv6 deployment model.

---

[1] This section is based in large measure on reference [BOU200501]. The author thanks the IPv6 Forum for this material.

The provider edge can also use IPv6 with MPLS at the edge to move IPv6 PDUs across an Internet core supporting MPLS, whether that core supports IPv6 or IPv4, and is being tested in several network pilots. The nodal and sub-networks implementations within an Intranet or PECN network pilot currently deploy assuming a dual-stack environment for either sparse or wide-use views for the IPv6 deployment model. The sparse view of deployment is that only nodes or networks that require IPv6 will be upgraded to use IPv6 within the PECN environments. The wide use view of deployment is that IPv6 routing will be dominant (preferred over IPv4) on the Intranets backbone and the sub-networks.

7.6.2 Network Infrastructure Deployment

Current deployment is verifying the network infrastructure to support the installation of IPv6 networks within the PECN environments. Network infrastructure includes the hardware, software, and infrastructure applications for an IPv6 network to begin data communications and support the Internet Protocol Suite implementation on a network and across an Internet core network for end-to-end communications.

Deployment has products participating from IT vendors from multiple geographies, and has demonstrated the network infrastructure can provide IPv6 connectivity and interoperability across multiple implementations. The routing implementation for the IPv6 network infrastructure has been verified. The core network infrastructure applications have been used and tested widely such as node-to-node communications for autoconfiguration, configuration of network parameters for the network and nodes, file transfer, electronic messaging, web access and services. The Application Program Interfaces (APIs) for IPv6 have been verified and tested so application providers can perform the necessary porting of those applications to support IPv6.

Transition mechanisms have been implemented and also tested on currently deployed networks and have demonstrated the ability to support a matrix of combinations of IPv6 and IPv4 interoperation. Sparse and wide-use views have been implemented on several network pilots supporting native IPv6 peering networks such as Moonv6 and 6net. The deployment has verified that PECN users will have a set of options for transition depending on their business and technology view to deploy IPv6 and no single transition mechanism will support all use cases required for transition.

The IPv6 network infrastructure deployment thus far supports the following assertions for PCEN environments:

- IPv6 capable dual-stack products exist on the market and can be purchased.
- IPv6 link or subnet communications between nodes can be supported today.
- IPv6 links and subnets can communicate over an Internet core network.
- IPv6 core applications infrastructure can be supported over an IPv6 network.
- IPv6 transition mechanisms exist to support the interoperation of IPv4 and IPv6 on a network.

Current network pilots have begun to deploy mobility using IPv6 and have started to verify the advantages of IPv6 for Mobile Ad Hoc Networks, and Seamless Mobility. The IPv6 network infrastructure deployed above provides a base for wider IPv6 deployment to support the development of next generation networks within the PECN environments.

7.6.3 Applications, Middleware and Management for IPv6 Deployment

The applications and middleware being used for current deployment are usually freeware software. These systems have permitted the testing of multimedia and web services. The results are that applications can run over an IPv6 network and can perform well.

But, the production applications for streaming media, VoIP, web proxy caches, security applications infrastructure (such as intrusion prevention, or public key infrastructure), database, manufacturing applications, and enterprise resource applications simply have not as of yet been ported to IPv6 as of mid-decade. This is a significant roadblock to the deployment of IPv6, and it is critical for 2006–2007 that applications be available for PECN environments to begin production deployment adoption at that time.

Another functional requirement for IPv6 that has had limited testing with current deployment is the network management of IPv6 and management in the context of interoperation of IPv4 and IPv6. Network views for IPv6 using SNMP have been done for IPv6, but not integrated with IPv4 that will be a requirement for production deployment on most networks. The range of management software for IPv4 networks must be ported to support IPv6.

7.6.4 Security Deployment and Business Challenge for IPv6

As we saw in Chapter 5, today many users who access networks enter the network within a security model where authentication is based on a Firewall or the use of the Authentication, Authorization, and Accounting (AAA) protocol suite implementation. Many users are behind NAT routers that perform translation of the IP header source addresses and keep the state of those addresses for communications with nodes and applications remote from their Intranet network. In addition network access for remote users is often accomplished with Virtual Private Network (VPN) tunnels, where the security is enforced at the edge of the network. Generally speaking, the security model of many users is based on a model where security is at the edge of the network as depicted in Figure 7.15.

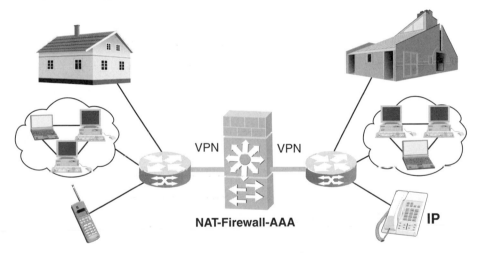

Figure 7.15: Typical security model in IPv4.

Users often today connect to the network trusting a third party usually with NAT on the edge of their network. Approaches such as the IPsec for end-to-end and peer-to-peer applications with encryption cannot be achieved because the IP address is used as a key for secure communications or the IP address is required to be globally routable on an Internet network. The current model prohibits the end-to-end trust model between two nodes, users, or applications whether stationary on a network or mobile. In addition NAT, prevents many applications from operating in a peer-to-peer manner, once they must operate external from an Intranet and across an Internet network and prevents seamless mobility across Internet networks. IPv6 will restore the use of applications using both the models, but that technology evolution will have disruptive ramifications to the security model that the Internet currently assumes operationally, for deployment.

A new model emerging with IPv6 can support the current and a new end-to-end security model, but how that is architected, managed, deployed, and implemented operationally is a question to be discussed. One view is presented in Figure 7.16. The updated model in Figure 7.16 permits the current model but removes NAT to support the evolution of peer-to-peer applications in addition to end-to-end security. The VPN is still available, but the Security Manager permits an end-to-end pass-through trust model for security protocols like IPsec.

The current Firewall model becomes an security management domain for the network edge permitting multiple security models. The Security Manager also will support network Intrusion Detection (IDS), and if there is a breach on the network can shut down the end-to-end communications, and force all communications through the firewall perimeter as an Intranet operation for Internet communications. The security view now takes on a network wide view not a single point of entry view, which begins to support an ambient and a network centric view for network security.

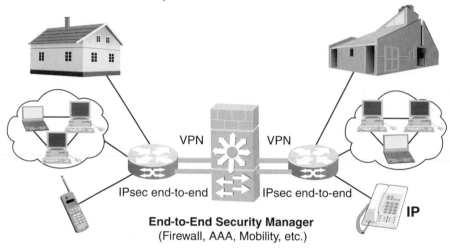

Figure 7.16: End-to-end security model.

This end-to-end model can also support the emerging use of wireless networks with seamless mobility as depicted in Figure 7.17.

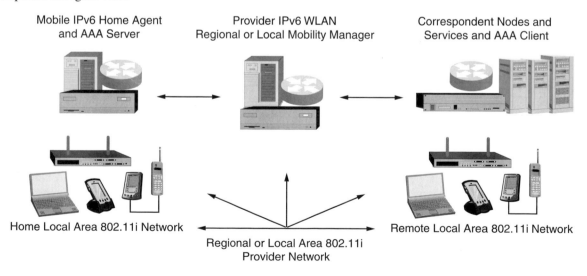

Figure 7.17: Benefits of the security manager.

In Figure 7.17, the benefits of the Security Manager can be used with AAA methods to ensure secure access to wireless networks in addition to the encryption supported by IEEE 802.11i, which supports the encryption of Layer 2 PDUs access to the wireless networks. The End-to-End (E2E) Security Manager with 802.11i will support seamless secure mobility in conjunction with the Mobile IPv6 extensions to the IPv6 architecture for deployment. The emerging IEEE 802.11n work to provide higher throughput will further reinforce 802.11i wireless access security and provide enhanced performance to this emerging security method.

The integral technology to move networks to an end-to-end and peer-to-peer secure model have been defined, but the deployment of this model will be an extremely disruptive technology in the market. The evolution will have an impact to current network operational methods and business practices across an Internet network. An example of the technical challenges are that current firewalls, filters, and IDS assume knowledge below the IP header within the transport data payload, which will not be available to implementations, when the payload is encrypted for example by IPsec or 802.11i entering the wireless network. Only the IP header will be exposed to the edge devices on an end-to-end supported network. From a business practice perspective today deployment and operational models for Internet networks are usually based on encryption from the edge network node view.

An end-to-end security model will be disruptive, but also provides a required new security model that is superior and more efficient for peer-to-peer communications for networks that want to support an end-to-end trust model as an operational requirement. The end-to-end security model also has performance and management advantages operationally, once the infrastructure is created to support an ambient secure model for peer-to-peer applications, which will be driven by the evolution of a seamless mobile communications for applications, and the rise of a mobile society for businesses and people in general.

This new model can also be an economic stimulus for new business, early adopters, and suppliers who provide products and services for the transition to an end-to-end security model, and these early adopters will be the ones potentially who will gain the most profit from this disruptive technology event.

In addition, IPv6 provides many benefits to next generation networks and mobility because of its ability to perform stateless node discovery and network operations. But, on a wireless network the nodes and network infrastructure supporting that stateless environment brings new security concerns that must be addressed for network operations. Current deployment has begun to test IPsec end-to-end, and the above security model is in its design stages in being prepared to be deployed in several network pilots. The security software infrastructure for IPv4 must be ported to IPv6 for the pervasive deployment of IPv6 on production networks.

One approach to deal with the IPSec limitations cited above is to use the newly-proposed Multilayer IP-security (ML-IPsec). ML-IPsec uses a multilayer protection model to replace the single end-to-end model. Unlike IPsec where the scope of encryption and authentication apply to the entire IP datagram payload (sometimes IP header as well), ML-IPsec divides the IP datagram into zones. It applies different protection schemes to different zones. Each zone has its own sets of security associations, its own set of private keys (secrets) that are not shared with other zones, and its own sets of access control rules (defining which nodes in the network have access to the zone) [ZHA200401].

Multilayer IPsec applies separate encryption/authentication with different keys on different parts of an IP datagram. It allows intermediate routers to have limited and controllable access to part of IP datagram (usually headers) but not the user data, for applications such as flow classification, diffserv, transparent proxy, and so on (and those "intelligent routing" that need access to higher-layer protocol headers). The idea is to divide the IP datagram into several parts and apply different forms of protection to different parts. For example, the TCP payload part can be protected between two end points while the TCP/IP header part can be protected but accessible to two end points plus certain routers in the network. It allows TCP PEP to coexist with IPsec, and provides both performance improvement and security protection to wireless networks [ZHA200401].

When ML-IPsec protects a traffic stream from its source to its destination, it will first rearrange the IP datagram into zones and apply cryptographic protections. When the ML-IPsec protected datagram flows through an authorized intermediate gateway, a certain part of the datagram may be decrypted and/or modified and re-encrypted, but the other parts will not be compromised. When the PDU reaches its destination, ML-IPsec will be able to reconstruct the entire datagram. ML-IPsec defines a complex security relationship that involves both the sender and the receiver of a security service, but also selected intermediate nodes along the delivery path [ZHA200401].

Issues Related to Transitioning to IPv6

While it is possible to deploy (some) IPv6 systems as stand-alone islands, this approach does not support the ubiquitous end-to-end carrier-class VoIP services we have advocated in this text. To achieve this end-to-end connectivity significant portions of the IP infrastructure must be upgraded. This evolution starts with the introduction of interconnected mixed-technology networks.

As IPv6 is introduced in the IPv4-based Internet, a plethora of interworking issues will arise, including, but not limited to routing, addressing, and Domain Naming System (DNS). An important key to a successful IPv6 transition is compatibility with the large installed base of IPv4 hosts and routers. Customers at all levels will expect seamless interworking. The interworking issues apply to all types of IP-based networks, including VoIP networks; in particular, there may be a need at some point in the medium-term future to interworking 2G VoIP systems with newer 3G VoIP systems.

Because IPv6 and IPv4 will need to coexist on the intranet and in the Internet for some time to come, network applications, devices, and VoIP elements (proxies, phones, gatekeepers, and so forth) need to be able to communicate transparently with both IPv4 and IPv6 nodes. It follows that there is a need for IPv4 compatibility mechanisms that can be implemented by IPv6 hosts and routers. Such mechanisms, as specified in RFC 2893, include providing complete implementations of both versions of the Internet Protocol (IPv4 and IPv6) (aka dual-stack deployment), and tunneling IPv6 packets over IPv4 routing infrastructures. These mechanisms are designed to allow IPv6 nodes to maintain complete compatibility with IPv4, which is expected to simplify the deployment of IPv6 in the Internet, and facilitate the eventual transition of the entire Internet to IPv6.

The discussion that follows herewith is based on RFC 2893 [GIL200001].

8.1 Introduction

The key to a successful IPv6 transition is compatibility with the large installed base of IPv4 hosts and routers. Maintaining compatibility with IPv4 while deploying IPv6 will streamline the task of transitioning the Internet to IPv6. IETF RFC 2893 defines a set of mechanisms that IPv6 hosts and routers may implement in order to be compatible with IPv4 hosts and routers. The mechanisms described in RFC 2893 are designed to be employed by IPv6 hosts and routers that need to interoperate with IPv4 hosts and utilize IPv4 routing infrastructures. On can expect that most nodes in the Internet will need such compatibility for a long time to come, and perhaps even indefinitely. (However, IPv6 may be used in some environments where interoperability with IPv4 is not required; IPv6 nodes that are designed to be used in such environments need not use or even implement these mechanisms.)

Interworking mechanisms include:

- *Dual-IP layer (also known as dual-stack)*: A technique for providing complete support for both Internet protocols—IPv4 and IPv6–in hosts and routers.

- *Configured tunneling of IPv6 over IPv4*: Point-to-point tunnels made by encapsulating IPv6 packets within IPv4 headers to carry them over IPv4 routing infrastructures.
- *IPv4-compatible IPv6 addresses*: An IPv6 address format that employs embedded IPv4 addresses.
- *Automatic tunneling of IPv6 over IPv4*: A mechanism for using IPv4-compatible addresses to automatically tunnel IPv6 packets over IPv4 networks.

The mechanisms defined here are intended to be part of a "transition toolbox"—a growing collection of techniques that implementations and users may employ to ease the transition. The tools may be used as needed. Implementations and sites decide which techniques are appropriate to their specific needs. RFC 2893 defines the initial core set of transition mechanisms, but these are not expected to be the only tools available; additional transition and compatibility mechanisms are expected to be developed in the future, with new IETF RFCs and documentation being written to specify them.

8.1.1 Terminology

The following terms are used in this discussion:

Types of Nodes:

- *IPv4-only node*: A host or router that implements only IPv4. An IPv4-only node does not understand IPv6. The installed base of IPv4 hosts and routers existing before the transition begins are IPv4-only nodes.
- *IPv6/IPv4 node*: A host or router that implements both IPv4 and IPv6.
- *IPv6-only node*: A host or router that implements IPv6, and does not implement IPv4. The operation of IPv6-only nodes is not addressed here.
- *IPv6 node*: Any host or router that implements IPv6. IPv6/IPv4 and IPv6-only nodes are both IPv6 nodes.
- *IPv4 node*: Any host or router that implements IPv4. IPv6/IPv4 and IPv4-only nodes are both IPv4 nodes.

Types of IPv6 Addresses:

IPv4-compatible IPv6 address: An IPv6 address bearing the high-order 96-bit prefix 0:0:0:0:0:0, and an IPv4 address in the low-order 32-bits. IPv4-compatible addresses are used by IPv6/IPv4 nodes which perform automatic tunneling,

IPv6-native address: The remainder of the IPv6 address space. An IPv6 address that bears a prefix other than 0:0:0:0:0:0.

Techniques Used in the Transition:

- *IPv6-over-IPv4 tunneling*: The technique of encapsulating IPv6 packets within IPv4 so that they can be carried across IPv4 routing infrastructures.
- *Configured tunneling*: IPv6-over-IPv4 tunneling where the IPv4 tunnel endpoint address is determined by configuration information on the encapsulating node. The tunnels can be either unidirectional or bidirectional. Bidirectional configured tunnels behave as virtual point-to-point links.
- *Automatic tunneling*: IPv6-over-IPv4 tunneling where the IPv4 tunnel endpoint address is determined from the IPv4 address embedded in the IPv4-compatible destination address of the IPv6 packet being tunneled.
- *IPv4 multicast tunneling*: IPv6-over-IPv4 tunneling where the IPv4 tunnel endpoint address is determined using Neighbor Discovery. Unlike configured tunneling this does not require any address configuration and unlike automatic tunneling it does not require the use of IPv4-compatible addresses. However, the mechanism assumes that the IPv4 infrastructure supports IPv4 multicast.

Other transition mechanisms include the following:

- IPv6-only operation: An IPv6/IPv4 node with its IPv6 stack enabled and its IPv4 stack disabled.
- IPv4-only operation: An IPv6/IPv4 node with its IPv4 stack enabled and its IPv6 stack disabled.
- IPv6/IPv4 operation: An IPv6/IPv4 node with both stacks enabled.

8.1.2 Approach

The remainder of this section is organized as follows:

- Section 8.2 discusses the operation of nodes with a dual-IP layer, IPv6/IPv4 nodes.
- Section 8.3 discusses the common mechanisms used in both of the IPv6-over-IPv4 tunneling techniques.
- Section 8.4 discusses configured tunneling.
- Section 8.5 discusses automatic tunneling and the IPv4-compatible IPv6 address format.
- Section 8.6 looks at application aspects of transition.

8.2 Dual-IP Layer Operation

The most straightforward way for IPv6 nodes to remain compatible with IPv4-only nodes is by providing a complete IPv4 implementation. IPv6 nodes that provide a complete IPv4 and IPv6 implementations are called *IPv6/IPv4 nodes*. IPv6/IPv4 nodes have the ability to send and receive both IPv4 and IPv6 packets. They can directly interoperate with IPv4 nodes using IPv4 packets, and also directly interoperate with IPv6 nodes using IPv6 packets.

Even though a node may be equipped to support both protocols, one or the other stack may be disabled for operational reasons. Thus IPv6/IPv4 nodes may be operated in one of three modes:

1. With their IPv4 stack enabled and their IPv6 stack disabled;
2. With their IPv6 stack enabled and their IPv4 stack disabled; or
3. With both stacks enabled.

IPv6/IPv4 nodes with their IPv6 stack disabled will operate like IPv4-only nodes. Similarly, IPv6/IPv4 nodes with their IPv4 stacks disabled will operate like IPv6-only nodes. IPv6/IPv4 nodes may provide a configuration switch to disable either their IPv4 or IPv6 stack.

The dual-IP layer technique may or may not be used in conjunction with the IPv6-over-IPv4 tunneling techniques, which are described in Sections 8.3, 8.4 and 8.5. An IPv6/IPv4 node that supports tunneling may support only configured tunneling, or both configured and automatic tunneling. Thus three modes of tunneling support are possible:

1. IPv6/IPv4 node that does not perform tunneling;
2. IPv6/IPv4 node that performs configured tunneling only; and
3. IPv6/IPv4 node that performs configured tunneling and automatic tunneling.

8.2.1 Address Configuration

Because they support both protocols, IPv6/IPv4 nodes may be configured with both IPv4 and IPv6 addresses. IPv6/IPv4 nodes use IPv4 mechanisms (e.g., DHCP) to acquire their IPv4 addresses, and IPv6 protocol mechanisms (e.g., stateless address autoconfiguration) to acquire their IPv6-native addresses. Section 8.5.2 describes a mechanism by which IPv6/IPv4 nodes that support automatic tunneling may use IPv4 protocol mechanisms to acquire their IPv4-compatible IPv6 address.

8.2.2 Domain Naming System

The DNS is used in both IPv4 and IPv6 to map between hostnames and IP addresses. A new resource record type named "A6" has been defined for IPv6 addresses with support for an earlier record named "AAAA". Since IPv6/IPv4 nodes must be able to interoperate directly with both IPv4 and IPv6 nodes, they must provide resolver libraries capable of dealing with IPv4 "A" records as well as IPv6 "A6" and "AAAA" records.

DNS resolver libraries on IPv6/IPv4 nodes must be capable of handling both A6/AAAA and A records. However, when a query locates an A6/AAAA record holding an IPv6 address, and an A record holding an IPv4 address, the resolver library may filter or order the results returned to the application in order to influence the version of IP packets used to communicate with that node. In terms of filtering, the resolver library has three alternatives:

1. Return only the IPv6 address to the application;
2. Return only the IPv4 address to the application; or
3. Return both addresses to the application.

If it returns only the IPv6 address, the application will communicate with the node using IPv6. If it returns only the IPv4 address, the application will communicate with the node using IPv4. If it returns both addresses, the application will have the choice as to which address to use, and thus which IP protocol to employ.

If it returns both, the resolver may elect to order the addresses—IPv6 first, or IPv4 first. Since most applications try the addresses in the order they are returned by the resolver, this can affect the IP version "preference" of applications.

The decision to filter or order DNS results is implementation specific. IPv6/IPv4 nodes may provide policy configuration to control filtering or ordering of addresses returned by the resolver, or leave the decision entirely up to the application.

An implementation must allow the application to control whether or not such filtering takes place.

8.2.3 Advertising Addresses in the DNS

There are some constraints placed on the use of the DNS during transition. Most of these are obvious but are stated here for completeness.

The recommendation is that A6/AAAA records for a node should not be added to the DNS until all of these are true:

1. The address is assigned to the interface on the node.
2. The address is configured on the interface.
3. The interface is on a link which is connected to the IPv6 infrastructure.

If an IPv6 node is isolated from an IPv6 perspective (e.g., it is not connected to the 6bone to take a concrete example) constraint #3 would mean that it should not have an address in the DNS.

This works great when other dual-stack nodes tries to contact the isolated dual-stack node. There is no IPv6 address in the DNS thus the peer does not even try communicating using IPv6 but goes directly to IPv4 (we are assuming both nodes have A records in the DNS).

However, this does not work well when the isolated node is trying to establish communication. Even though it does not have an IPv6 address in the DNS it will find A6/AAAA records in the DNS for the peer. Since the isolated node has IPv6 addresses assigned to at least one interface it will try to communicate using IPv6. If it has no IPv6 route to the 6bone (e.g., because the local router was upgraded to advertise IPv6 addresses using Neighbor Discovery but that router does not have any IPv6 routes) this communication will fail. Typically this means a few minutes of delay as TCP times out. The TCP specification says that ICMP unreachable messages could be due to routing transients thus they should not immediately terminate the TCP connection. This means that the normal TCP timeout of a few minutes apply. Once TCP times out the application will hopefully try the IPv4 addresses based on the A records in the DNS, but this will be painfully slow.

A possible implication of the recommendations above is that, if one enables IPv6 on a node on a link without IPv6 infrastructure, and choose to add A6/AAAA records to the DNS for that node, then external IPv6 nodes that might see these A6/AAAA records will possibly try to reach that node using IPv6 and suffer delays or communication failure due to unreachability. (A delay is incurred if the application correctly falls back to using IPv4 if it cannot establish communication using IPv6 addresses. If this fallback is not done the application would fail to communicate in this case.) Thus it is suggested that either the recommendations be followed, or care be taken to only do so with nodes that will not be impacted by external accessing delays and/or communication failure.

In the future when a site or node removes the support for IPv4 the above recommendations apply to when the A records for the node(s) should be removed from the DNS.

8.3 Common Tunneling Mechanisms

In most deployment scenarios, the IPv6 routing infrastructure will be built up over time. While the IPv6 infrastructure is being deployed, the existing IPv4 routing infrastructure can remain functional, and can be used to carry IPv6 traffic. Tunneling provides a way to utilize an existing IPv4 routing infrastructure to carry IPv6 traffic.

IPv6/IPv4 hosts and routers can tunnel IPv6 datagrams over regions of IPv4 routing topology by encapsulating them within IPv4 packets. Tunneling can be used in a variety of ways:

- *Router-to-Router.* IPv6/IPv4 routers interconnected by an IPv4 infrastructure can tunnel IPv6 packets between themselves. In this case, the tunnel spans one segment of the end-to-end path that the IPv6 packet takes.
- *Host-to-Router.* IPv6/IPv4 hosts can tunnel IPv6 packets to an intermediary IPv6/IPv4 router that is reachable via an IPv4 infrastructure. This type of tunnel spans the first segment of the packet's end-to-end path.
- *Host-to-Host.* IPv6/IPv4 hosts that are interconnected by an IPv4 infrastructure can tunnel IPv6 packets between themselves. In this case, the tunnel spans the entire end-to-end path that the packet takes.
- *Router-to-Host.* IPv6/IPv4 routers can tunnel IPv6 packets to their final destination IPv6/IPv4 host. This tunnel spans only the last segment of the end-to-end path.

Tunneling techniques are usually classified according to the mechanism by which the encapsulating node determines the address of the node at the end of the tunnel. In the first two tunneling methods listed above—router-to-router and host-to-router—the IPv6 packet is being tunneled to a router. The endpoint of this type of tunnel is an intermediary router which must decapsulate the IPv6 packet and forward it on to its final destination. When tunneling to a router, the endpoint of the tunnel is different from the destination of the packet being tunneled. So the addresses in the IPv6 packet being tunneled can not provide the IPv4 address of the tunnel endpoint. Instead, the tunnel endpoint address must be determined from configuration information on the node performing the tunneling. We use the term "configured tunneling" to describe the type of tunneling where the endpoint is explicitly configured.

In the last two tunneling methods—host-to-host and router-to-host—the IPv6 packet is tunneled all the way to its final destination. In this case, the destination address of both the IPv6 packet and the encapsulating IPv4 header identify the same node! This fact can be exploited by encoding information in the

IPv6 destination address that will allow the encapsulating node to determine tunnel endpoint IPv4 address automatically. Automatic tunneling employs this technique, using a special IPv6 address format with an embedded IPv4 address to allow tunneling nodes to automatically derive the tunnel endpoint IPv4 address. This eliminates the need to explicitly configure the tunnel endpoint address, greatly simplifying configuration.

The two tunneling techniques: automatic and configured, differ primarily in how they determine the tunnel endpoint address. Most of the underlying mechanisms are the same:

- The entry node of the tunnel (the encapsulating node) creates an encapsulating IPv4 header and transmits the encapsulated packet.
- The exit node of the tunnel (the decapsulating node) receives the encapsulated packet, reassembles the packet if needed, removes the IPv4 header, updates the IPv6 header, and processes the received IPv6 packet.
- The encapsulating node may need to maintain soft state information for each tunnel recording such parameters as the MTU of the tunnel in order to process IPv6 packets forwarded into the tunnel. Since the number of tunnels that any one host or router may be using may grow to be quite large, this state information can be cached and discarded when not in use.

The remainder of this section discusses the common mechanisms that apply to both types of tunneling. Subsequent sections discuss how the tunnel endpoint address is determined for automatic and configured tunneling.

8.3.1 Encapsulation

The encapsulation of an IPv6 datagram in IPv4 is shown in Figure 8.1.

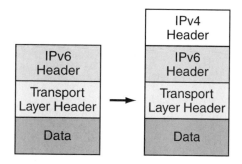

Figure 8.1: Encapsulating IPv6 in IPv4.

In addition to adding an IPv4 header, the encapsulating node also has to handle some more complex issues:

- Determine when to fragment and when to report an ICMP "packet too big" error back to the source.
- How to reflect IPv4 ICMP errors from routers along the tunnel path back to the source as IPv6 ICMP errors.

Those issues are discussed in the following sections.

8.3.2 Tunnel MTU and Fragmentation

The encapsulating node could view encapsulation as IPv6 using IPv4 as a link layer with a very large MTU (65535-20 bytes to be exact; 20 bytes "extra" are needed for the encapsulating IPv4 header). The encapsulating node would need only to report IPv6 ICMP "packet too big" errors back to the source for packets that exceed this MTU. However, such a scheme would be inefficient for two reasons:

1. It would result in more fragmentation than needed. IPv4 layer fragmentation should be avoided due to the performance problems caused by the loss unit being smaller than the retransmission unit.

2. Any IPv4 fragmentation occurring inside the tunnel would have to be reassembled at the tunnel endpoint. For tunnels that terminate at a router, this would require additional memory to reassemble the IPv4 fragments into a complete IPv6 packet before that packet could be forwarded onward.

The fragmentation inside the tunnel can be reduced to a minimum by having the encapsulating node track the IPv4 Path MTU across the tunnel, using the IPv4 Path MTU Discovery Protocol and recording the resulting path MTU. The IPv6 layer in the encapsulating node can then view a tunnel as a link layer with an MTU equal to the IPv4 path MTU, minus the size of the encapsulating IPv4 header.

Note that this does not completely eliminate IPv4 fragmentation in the case when the IPv4 path MTU would result in an IPv6 MTU less than 1,280 bytes. (Any link layer used by IPv6 has to have an MTU of at least 1,280 bytes.) In this case, the IPv6 layer has to "see" a link layer with an MTU of 1,280 bytes and the encapsulating node has to use IPv4 fragmentation in order to forward the 1,280 byte IPv6 packets.

The encapsulating node can employ the following algorithm to determine when to forward an IPv6 packet that is larger than the tunnel's path MTU using IPv4 fragmentation, and when to return an IPv6 ICMP "packet too big" message:

```
if (IPv4 path MTU - 20) is less than or equal to 1280
    if packet is larger than 1280 bytes
            Send IPv6 ICMP "packet too big" with MTU = 1280.
            Drop packet.
    else
            Encapsulate but do not set the Don't Fragment
            flag in the IPv4 header. The resulting IPv4
            packet might be fragmented by the IPv4 layer on
            the encapsulating node or by some router along
            the IPv4 path.
    endif
else
            if packet is larger than (IPv4 path MTU - 20)
                    Send IPv6 ICMP "packet too big" with
                    MTU = (IPv4 path MTU - 20).
                    Drop packet.
            else
                    Encapsulate and set the Don't Fragment flag
                    in the IPv4 header.
            endif
    endif
```

Encapsulating nodes that have a large number of tunnels might not be able to store the IPv4 Path MTU for all tunnels. Such nodes can, at the expense of additional fragmentation in the network, avoid using the IPv4 Path MTU algorithm across the tunnel and instead use the MTU of the link layer (under IPv4) in the above algorithm instead of the IPv4 path MTU. In this case the Don't Fragment bit must not be set in the encapsulating IPv4 header.

8.3.3 Hop Limit

IPv6-over-IPv4 tunnels are modeled as "single-hop." That is, the IPv6 hop limit is decremented by 1 when an IPv6 packet traverses the tunnel. The single-hop model serves to hide the existence of a tunnel. The tunnel is opaque to users of the network, and is not detectable by network diagnostic tools such as traceroute.

The single-hop model is implemented by having the encapsulating and decapsulating nodes process the IPv6 hop limit field as they would if they were forwarding a packet on to any other datalink. That is, they decrement the hop limit by 1 when forwarding an IPv6 packet. (The originating node and final destination do not decrement the hop limit.)

The TTL of the encapsulating IPv4 header is selected in an implementation dependent manner. The current suggested value is published in the "Assigned Numbers RFC. Implementations may provide a mechanism to allow the administrator to configure the IPv4 TTL such as the one specified in the IP Tunnel MIB.

8.3.4 Handling IPv4 ICMP errors

In response to encapsulated packets it has sent into the tunnel, the encapsulating node might receive IPv4 ICMP error messages from IPv4 routers inside the tunnel. These packets are addressed to the encapsulating node because it is the IPv4 source of the encapsulated packet.

The ICMP "packet too big" error messages are handled according to IPv4 Path MTU Discovery and the resulting path MTU is recorded in the IPv4 layer. The recorded path MTU is used by IPv6 to determine if an IPv6 ICMP "packet too big" error has to be generated as described in section 8.3.2. The handling of other types of ICMP error messages depends on how much information is included in the "packet in error" field, which holds the encapsulated packet that caused the error.

Many older IPv4 routers return only 8 bytes of data beyond the IPv4 header of the packet in error, which is not enough to include the address fields of the IPv6 header. More modern IPv4 routers are likely to return enough data beyond the IPv4 header to include the entire IPv6 header and possibly even the data beyond that.

If the offending packet includes enough data, the encapsulating node may extract the encapsulated IPv6 packet and use it to generate an IPv6 ICMP message directed back to the originating IPv6 node, as shown in Figure 8.2.

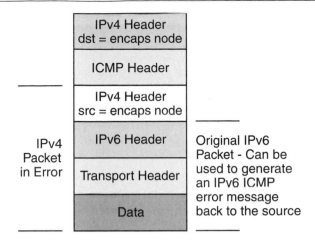

Figure 8.2: IPv4 ICMP error message returned to encapsulating node.

8.3.5 IPv4 Header Construction

When encapsulating an IPv6 packet in an IPv4 datagram, the IPv4 header fields are set as follows:

- Version: 4
- IP Header Length in 32-bit words: 5 (There are no IPv4 options in the encapsulating header.)
- Type of Service: 0. (Note that work underway in the IETF is redefining the Type of Service byte and as a result future RFCs might define a different behavior for the ToS byte when tunneling.)
- Total Length: Payload length from IPv6 header plus length of IPv6 and IPv4 headers (i.e., a constant 60 bytes).
- Identification: Generated uniquely as for any IPv4 packet transmitted by the system.
- Flags:
  Set the Don't Fragment (DF) flag as specified in Section 3.2.
  Set the More Fragments (MF) bit as necessary if fragmenting.

Fragment Offset:
- Set as necessary if fragmenting.

Time to Live:
- Set in implementation-specific manner.

Protocol:
- 41 (Assigned payload type number for IPv6).

Header Checksum:
- Calculate the checksum of the IPv4 header.

Source Address:
- IPv4 address of outgoing interface of the encapsulating node.

Destination Address:
- IPv4 address of tunnel endpoint.

Any IPv6 options are preserved in the packet (after the IPv6 header).

8.3.6 Decapsulation

When an IPv6/IPv4 host or a router receives an IPv4 datagram that is addressed to one of its own IPv4 address, and the value of the protocol field is 41, it reassembles if the packet if it is fragmented at the IPv4 level, then it removes the IPv4 header and submits the IPv6 datagram to its IPv6 layer code. The decapsulating node must be capable of reassembling an IPv4 packet that is 1300 bytes (1280 bytes plus IPv4 header). The decapsulation is shown in Figure 8.3.

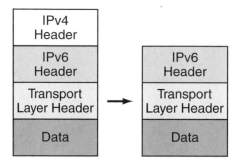

Figure 8.3: Decapsulating IPv6 from IPv4.

When decapsulating the packet, the IPv6 header is not modified. [Note that work underway in the IETF is redefining the Type of Service byte and as a result future RFCs might define a different behavior for the ToS byte when decapsulating a tunneled packet.] If the packet is subsequently forwarded, its hop limit is decremented by one. As part of the decapsulation the node should silently discard a packet with an invalid IPv4 source address such as a multicast address, a broadcast address, 0.0.0.0, and 127.0.0.1. In general it should apply the rules for martian filtering in described in RFC 1812 and ingress filtering described in RFC 2267 on the IPv4 source address.

The encapsulating IPv4 header is discarded.

After the decapsulation the node should silently discard a packet with an invalid IPv6 source address. This includes IPv6 multicast addresses, the unspecified address, and the loopback address but also IPv4-compatible IPv6 source addresses where the IPv4 part of the address is an (IPv4) multicast address, broadcast address, 0.0.0.0, or 127.0.0.1. In general it should apply the rules for martian filtering described in RFC 1812 and ingress filtering described in RFC 2267 on the IPv4-compatible source address. The decapsulating node performs IPv4 reassembly before decapsulating the IPv6 packet. All IPv6 options are preserved even if the encapsulating IPv4 packet is fragmented.

After the IPv6 packet is decapsulated, it is processed almost the same as any received IPv6 packet. The only difference being that a decapsulated packet must not be forwarded unless the node has been explicitly configured to forward such packets for the given IPv4 source address. This configuration can be implicit in e.g., having a configured tunnel which matches the IPv4 source address. This restriction is needed to prevent tunneling to be used as a tool to circumvent ingress filtering described in RFC 2267.

8.3.7 Link-Local Addresses

Both the configured and automatic tunnels are IPv6 interfaces (over the IPv4 "link layer") thus must have link-local addresses. The link-local addresses are used by routing protocols operating over the tunnels.

The Interface Identifier described in RFC 2373 (now updated by RFC 3513) for such an Interface should be the 32-bit IPv4 address of that interface, with the bytes in the same order in which they would appear in the header of an IPv4 packet, padded at the left with zeros to a total of 64 bits. Note that the "Universal/Local" bit is zero, indicating that the Interface Identifier is not globally unique. When the host has more than one IPv4 address in use on the physical interface concerned, an administrative choice of one of these IPv4 addresses is made.

The IPv6 Link-local address for an IPv4 virtual interface is formed by appending the Interface Identifier, as defined above, to the prefix FE80::/64.

```
+-------+-------+-------+-------+-------+-------+------+------+
|  FE      80      00      00      00      00      00     00  |
+-------+-------+-------+-------+-------+-------+------+------+
|  00      00   |  00   |  00   |  IPv4 Address               |
+-------+-------+-------+-------+-------+-------+------+------+
```

8.3.8 Neighbor Discovery over Tunnels

Automatic tunnels and unidirectional configured tunnels are considered to be unidirectional. Thus the only aspects of Neighbor Discovery and Stateless Address Autoconfiguration that apply to these tunnels is the formation of the link-local address.

If an implementation provides bidirectional configured tunnels it must at least accept and respond to the probe packets used by Neighbor Unreachability Detection. Such implementations should also send NUD probe packets to detect when the configured tunnel fails at which point the implementation can use an alternate path to reach the destination. Note that Neighbor Discovery allows that the sending of NUD probes be omitted for router to router links if the routing protocol tracks bidirectional reachability.

For the purposes of Neighbor Discovery the automatic and configured tunnels specified in this RFC as assumed to not have a link- layer address, even though the link-layer (IPv4) does have address. This means that a sender of Neighbor Discovery packets:

*   should not include Source Link Layer Address (SLLA) options or Target Link Layer Address (TLLA) options on the tunnel link.
*   must silently ignore any received SLLA or TLLA options on the tunnel link.

8.4 Configured Tunneling

In configured tunneling, the tunnel endpoint address is determined from configuration information in the encapsulating node. For each tunnel, the encapsulating node must store the tunnel endpoint address. When an IPv6 packet is transmitted over a tunnel, the tunnel endpoint address configured for that tunnel is used as the destination address for the encapsulating IPv4 header.

The determination of which packets to tunnel is usually made by routing information on the encapsulating node. This is usually done via a routing table, which directs packets based on their destination address using the prefix mask and match technique.

8.4.1 Default Configured Tunnel

IPv6/IPv4 hosts that are connected to datalinks with no IPv6 routers may use a configured tunnel to reach an IPv6 router. This tunnel allows the host to communicate with the rest of the IPv6 Internet (i.e., nodes with IPv6-native addresses). If the IPv4 address of an IPv6/IPv4 router bordering the IPv6 backbone is known, this can be used as the tunnel endpoint address. This tunnel can be configured into the routing table as an IPv6 "default route." That is, all IPv6 destination addresses will match the route and could potentially traverse the tunnel. Since the "mask length" of such a default route is zero, it will be used only if there are no other routes with a longer mask that match the destination. The default configured tunnel can be used in conjunction with automatic tunneling, as described in Section 8.5.4.

8.4.2 Default Configured Tunnel Using IPv4 "Anycast Address"

The tunnel endpoint address of such a default tunnel could be the IPv4 address of one IPv6/IPv4 router at the border of the IPv6 backbone. Alternatively, the tunnel endpoint could be an IPv4 "anycast address." With this approach, multiple IPv6/IPv4 routers at the border advertise IPv4 reachability to the same IPv4 address. All of these routers accept packets to this address as their own, and will decapsulate IPv6 packets tunneled to this address. When an IPv6/IPv4 node sends an encapsulated packet to this address, it will be delivered to only one of the border routers, but the sending node will not know which one. The IPv4 routing system will generally carry the traffic to the closest router.

Using a default tunnel to an IPv4 "anycast address" provides a high degree of robustness since multiple border router can be provided, and, using the normal fallback mechanisms of IPv4 routing, traffic will automatically switch to another router when one goes down. However, care must be taking when using such a default tunnel to prevent different IPv4 fragments from arriving at different routers for reassembly. This can be prevented by either avoiding fragmentation of the encapsulated packets (by ensuring an IPv4 MTU of at least 1,300 bytes) or by preventing frequent changes to IPv4 routing.

8.4.3 Ingress Filtering

The decapsulating node must verify that the tunnel source address is acceptable before forwarding decapsulated packets to avoid circumventing ingress filtering. Note that packets which are delivered to transport protocols on the decapsulating node should not be subject to these checks. For bidirectional configured tunnels this is done by verifying that the source address is the IPv4 address of the other end of the tunnel. For unidirectional configured tunnels the decapsulating node must be configured with a list of source IPv4 address prefixes that are acceptable. Such a list must default to not having any entries i.e., the node has to be explicitly configured to forward decapsulated packets received over unidirectional configured tunnels.

8.5 Automatic Tunneling

In automatic tunneling, the tunnel endpoint address is determined by the IPv4-compatible destination address of the IPv6 packet being tunneled. Automatic tunneling allows IPv6/IPv4 nodes to communicate over IPv4 routing infrastructures without pre-configuring tunnels.

8.5.1 IPv4-Compatible Address Format

IPv6/IPv4 nodes that perform automatic tunneling are assigned IPv4- compatible address. An IPv4-compatible address is identified by an all-zeros 96-bit prefix, and holds an IPv4 address in the low-order 32-bits. IPv4-compatible addresses are structured as shown in Figure 8.4.

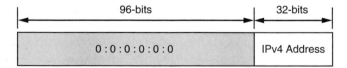

Figure 8.4: IPv4-compatible IPv6 address format.

IPv4-compatible addresses are assigned exclusively to nodes that support automatic tunneling. A node should be configured with an IPv4-compatible address only if it is prepared to accept IPv6 packets destined to that address encapsulated in IPv4 packets destined to the embedded IPv4 address.

An IPv4-compatible address is globally unique as long as the IPv4 address is not from the private IPv4 address space. An implementation should behave as if its IPv4-compatible address(es) are assigned to the node's automatic tunneling interface, even if the implementation does not implement automatic tunneling using a concept of interfaces. Thus, the IPv4-compatible address should not be viewed as being attached to, for example, an Ethernet interface, that is, implications should not use the Neighbor Discovery mechanisms such as NUD (RFC 2461) at the Ethernet. Any such interactions should be done using the encapsulated packets i.e., over the automatic tunneling (conceptual) interface.

8.5.2 IPv4-Compatible Address Configuration

An IPv6/IPv4 node with an IPv4-compatible address uses that address as one of its IPv6 addresses, while the IPv4 address embedded in the low-order 32-bits serves as the IPv4 address for one of its interfaces.

An IPv6/IPv4 node may acquire its IPv4-compatible IPv6 addresses via IPv4 address configuration protocols. It may use any IPv4 address configuration mechanism to acquire its IPv4 address, then "map" that address into an IPv4-compatible IPv6 address by pre-pending it with the 96-bit prefix 0:0:0:0:0:0. This mode of configuration allows IPv6/IPv4 nodes to "leverage" the installed base of IPv4 address configuration servers.

The specific algorithm for acquiring an IPv4-compatible address using IPv4-based address configuration protocols is as follows:

1. The IPv6/IPv4 node uses standard IPv4 mechanisms or protocols to acquire the IPv4 address for one of its interfaces. These include:
   • The Dynamic Host Configuration Protocol (DHCP)
   • The Bootstrap Protocol (BOOTP)
   • The Reverse Address Resolution Protocol (RARP)
   • Manual configuration
   • Any other mechanism which accurately yields the node's own IPv4 address
2. The node uses this address as the IPv4 address for this interface.
3. The node prepends the 96-bit prefix 0:0:0:0:0:0 to the 32-bit IPv4 address that it acquired in step (1). The result is an IPv4-compatible IPv6 address with one of the node's IPv4-addresses embedded in the low-order 32-bits. The node uses this address as one of its IPv6 addresses.

8.5.3 Automatic Tunneling Operation

In automatic tunneling, the tunnel endpoint address is determined from the packet being tunneled. If the destination IPv6 address is IPv4-compatible, then the packet can be sent via automatic tunneling. If the destination is IPv6-native, the packet cannot be sent via automatic tunneling.

A routing table entry can be used to direct automatic tunneling. An implementation can have a special static routing table entry for the prefix 0:0:0:0:0:0/96. (That is, a route to the all-zeros prefix with a 96-bit mask.) Packets that match this prefix are sent to a pseudo-interface driver which performs automatic tunneling. Since all IPv4-compatible IPv6 addresses will match this prefix, all packets to those destinations will be auto-tunneled.

Once it is delivered to the automatic tunneling module, the IPv6 packet is encapsulated within an IPv4 header according to the rules described in Section 8.3. The source and destination addresses of the encapsulating IPv4 header are assigned as follows:

Destination IPv4 address:

> Low-order 32-bits of IPv6 destination address

Source IPv4 address:

> IPv4 address of interface the packet is sent via

The automatic tunneling module always sends packets in this encapsulated form, even if the destination is on an attached datalink.

The automatic tunneling module must not send to IPv4 broadcast or multicast destinations. It must drop all IPv6 packets destined to IPv4-compatible destinations when the embedded IPv4 address is broadcast, multicast, the unspecified (0.0.0.0) address, or the loopback address (127.0.0.1). Note that the sender can only tell if an address is a network or subnet broadcast for broadcast addresses assigned to directly attached links.

8.5.4 Use With Default Configured Tunnels

Automatic tunneling is often used in conjunction with the default configured tunnel technique. "Isolated" IPv6/IPv4 hosts—those with no on-link IPv6 routers—are configured to use automatic tunneling and IPv4-compatible IPv6 addresses, and have at least one default configured tunnel to an IPv6 router. That IPv6 router is configured to perform automatic tunneling as well. These isolated hosts send packets to IPv4-compatible destinations via automatic tunneling and packets for IPv6-native destinations via the default configured tunnel. IPv4-compatible destinations will match the 96- bit all-zeros prefix route discussed in the previous section, while IPv6-native destinations will match the default route via the configured tunnel. Reply packets from IPv6-native destinations are routed back to an IPv6/IPv4 router which delivers them to the original host via automatic tunneling.

8.5.5 Source Address Selection

When an IPv6/IPv4 node originates an IPv6 packet, it must select the source IPv6 address to use. IPv6/IPv4 nodes that are configured to perform automatic tunneling may be configured with global IPv6-native addresses as well as IPv4-compatible addresses. The selection of which source address to use will determine what form the return traffic is sent via. If the IPv4-compatible address is used, the return traffic will have to be delivered via automatic tunneling, but if the IPv6-native address is used, the return traffic will not be automatic-tunneled. In order to make traffic as symmetric as possible, the following source address selection preference is recommended:

- *Destination is IPv4-compatible*: Use IPv4-compatible source address associated with IPv4 address of outgoing interface.
- *Destination is IPv6-native*: Use IPv6-native address of outgoing interface.

If an IPv6/IPv4 node does not have a global IPv6-native address, but is originating a packet to an IPv6-native destination, it may use its IPv4-compatible address as its source address.

8.5.6 Ingress Filtering

The decapsulating node must verify that the encapsulated packets are acceptable before forwarding decapsulated packets to avoid circumventing ingress filtering. Note that packets which are delivered to transport protocols on the decapsulating node should not be subject to these checks. Since automatic tunnels always encapsulate to the destination (i.e., the IPv4 destination will be the destination) any packet received over an automatic tunnel should not be forwarded.

8.6 Application Aspects of IPv6 Transition

8.6.1 Transition Issues

The transition mechanisms discussed previously do not consider whether the applications support IPv6. Two interrelated topics in this arena need consideration:

1. How different network transition techniques affect applications, and strategies for applications to support IPv6 and IPv4, and,
2. How to develop IPv6-capable or protocol-independent applications ("application porting guidelines") using standard APIs (e.g., RFC 3493, RFC 3542).

Applications will have to be modified to support IPv6 (and IPv4) by using one of a number of techniques described herewith. In what follows is a quick summary of some of the issues, approaches, and techniques discussed in IETF RFC 4038 [SHI200501], on which this section is based.

8.6.2 Overview of IPv6 Application Transition

The transition of an application can be classified by using four different cases (excluding the first case when there is no IPv6 support in either the application or the operating system) as seen in Figure 8.5.

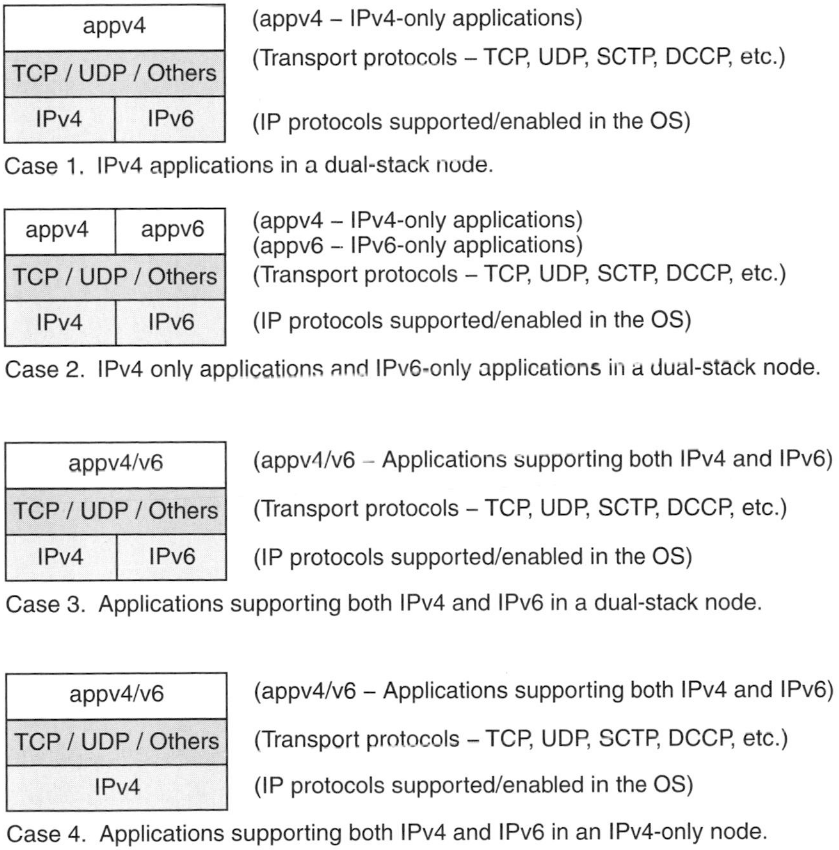

appv4		(appv4 – IPv4-only applications)
TCP / UDP / Others		(Transport protocols – TCP, UDP, SCTP, DCCP, etc.)
IPv4	IPv6	(IP protocols supported/enabled in the OS)

Case 1. IPv4 applications in a dual-stack node.

appv4	appv6	(appv4 – IPv4-only applications) (appv6 – IPv6-only applications)
TCP / UDP / Others		(Transport protocols – TCP, UDP, SCTP, DCCP, etc.)
IPv4	IPv6	(IP protocols supported/enabled in the OS)

Case 2. IPv4 only applications and IPv6-only applications in a dual-stack node.

appv4/v6		(appv4/v6 – Applications supporting both IPv4 and IPv6)
TCP / UDP / Others		(Transport protocols – TCP, UDP, SCTP, DCCP, etc.)
IPv4	IPv6	(IP protocols supported/enabled in the OS)

Case 3. Applications supporting both IPv4 and IPv6 in a dual-stack node.

appv4/v6	(appv4/v6 – Applications supporting both IPv4 and IPv6)
TCP / UDP / Others	(Transport protocols – TCP, UDP, SCTP, DCCP, etc.)
IPv4	(IP protocols supported/enabled in the OS)

Case 4. Applications supporting both IPv4 and IPv6 in an IPv4-only node.

Figure 8.5: Overview of application transition.

Case 1: IPv4-only applications in a dual-stack node. IPv6 protocol is introduced in a node, but applications are not yet ported to support IPv6.

Case 2: IPv4-only applications and IPv6-only applications in a dual-stack node. Applications are ported for IPv6-only. Therefore there are two similar applications, one for each protocol version (e.g., ping and ping6).

Case 3: Applications supporting both IPv4 and IPv6 in a dual-stack node. Applications are ported for both IPv4 and IPv6 support. Therefore, the existing IPv4 applications can be removed.

Case 4: Applications supporting both IPv4 and IPv6 in an IPv4-only node. Applications are ported for both IPv4 and IPv6 support, but the same applications may also have to work when IPv6 is not being used (e.g., disabled from the OS).

The first two cases are not interesting in the longer term; only few applications are inherently IPv4- or IPv6-specific, and should work with both protocols without having to care about which one is being used.

8.6.3 Problems with IPv6 Application Transition

There are several reasons why the transition period between IPv4 and IPv6 applications may not be straightforward. These issues are described in this section.

8.6.3.1 IPv6 Support in the OS and Applications Are Unrelated

Considering the cases described in the previous section, IPv4 and IPv6 protocol stacks are likely to co-exist in a node for a long time. Similarly, most applications are expected to be able to handle both IPv4 and IPv6 during another long period. A dual-stack operating system is not intended to have both IPv4 and IPv6 applications. Therefore, IPv6-capable application transition may be independent of protocol stacks in a node. Applications capable of both IPv4 and IPv6 will probably have to work properly in IPv4-only nodes (whether the IPv6 protocol is completely disabled or there is no IPv6 connectivity at all).

8.6.3.2 DNS Does Not Indicate Which IP Version Is Used

In a node, the DNS name resolver gathers the list of destination addresses. DNS queries and responses are sent by using either IPv4 or IPv6 to carry the queries, regardless of the protocol version of the data records. The DNS name resolution issue related to application transition is that by only doing a DNS name lookup a client application can not be certain of the version of the peer application. For example, if a server application does not support IPv6 yet but runs on a dual-stack machine for other IPv6 services, and this host is listed with an AAAA record in the DNS, the client application will fail to connect to the server application. This is caused by a mismatch between the DNS query result (i.e., IPv6 addresses) and a server application version (i.e., IPv4).

Using SRV records would avoid these problems. Unfortunately, they are not used widely enough to be applicable in most cases. Hence an operational solution is to use "service names" in the DNS. If a node offers multiple services, but only some of them over IPv6, a DNS name may be added for each of these services or group of services (with the associated A/AAAA records), not just a single name for the physical machine, also including the AAAA records. However, the applications cannot depend on this operational practice.

The application should request all IP addresses without address family constraints and try all the records returned from the DNS, in some order, until a working address is found. In particular, the application has to be able to handle all IP versions returned from the DNS.

8.6.3.3 Supporting Many Versions of an Application is Difficult

During the application transition period, system administrators may have various versions of the same application (an IPv4-only application, an IPv6-only application, or an application supporting both IPv4 and IPv6). Typically one cannot know which IP versions must be supported prior to doing a DNS lookup *and* trying (see section 8.6.3.2) the addresses returned. Therefore if multiple versions of the same application are available, the local users have difficulty selecting the right version supporting the exact IP version required. To avoid problems with one application not supporting the specified protocol version, it is desirable to have hybrid applications supporting both.

An alternative approach for local client applications could be to have a "wrapper application" that performs certain tasks (such as figuring out which protocol version will be used) and calls the IPv4/IPv6-only applications as necessary. This application would perform connection establishment (or similar tasks) and pass the opened socket to another application. However, as applications such as this would have to do more than just perform a DNS lookup or determine the literal IP address given, they will become complex -- likely much more so than a hybrid application. Furthermore, writing "wrapping" applications that perform complex operations with IP addresses (such as FTP clients) might be even more challenging or even impossible. In short, wrapper applications do not look like a robust approach for application transition.

8.6.4 Description of Transition Scenarios and Guidelines

Once the IPv6 network is deployed, applications supporting IPv6 can use IPv6 network services to establish IPv6 connections. However, upgrading every node to IPv6 at the same time is not feasible, and transition from IPv4 to IPv6 will be a gradual process. Dual-stack nodes provide one solution to maintaining IPv4 compatibility in unicast communications. In this section we will analyze different application transition scenarios (as introduced in section 8.6.2) and guidelines for maintaining interoperability between applications running in different types of nodes.

Note that the first two cases, IPv4-only and IPv6-only applications, are not interesting in the longer term; only few applications are inherently IPv4- or IPv6-specific, and should work with both protocols without having to care about which one is being used.

8.6.4.1 IPv4 Applications in a Dual-Stack Node

In this scenario, the IPv6 protocol is added in a node, but IPv6- capable applications are not yet available or installed. Although the node implements the dual-stack, IPv4 applications can only manage IPv4 communications and accept/establish connections from/to nodes that implement an IPv4 stack. To allow an application to communicate with other nodes using IPv6, the first priority is to port applications to IPv6.

In some cases (e.g., when no source code is available), existing IPv4 applications can work if the Bump-in-the-Stack (BIS) or Bump-in-the- API (BIA) mechanism is installed in the node (BIS is defined in RFC 2767, BIA is defined in RFC 3338). It is strongly recommended that application developers not use these mechanisms when application source code is available. Also, they should not be used as an excuse not to port software or to delay porting.

8.6.4.2 IPv6 Applications in a Dual-Stack Node

As it was seen in the previous section, applications should be ported to IPv6. The easiest way to port an IPv4 application is to substitute the old IPv4 API references with the new IPv6 APIs with one-to-one mapping. This way the application will be IPv6-only. This IPv6-only source code cannot work in IPv4-only nodes, so the old IPv4 application should be maintained in these nodes. This necessitates having two similar applications working with different protocol versions, depending on the node they are running (for example, telnet and telnet6). This case is undesirable, as maintaining two versions of the same source code per application could be difficult. This approach would also cause problems for users having to select which version of the application to use, as described in section 8.6.3.3.

Most implementations of dual-stack allow IPv6-only applications to interoperate with both IPv4 and IPv6 nodes. IPv4 packets going to IPv6 applications on a dual-stack node reach their destination because their addresses are mapped by using IPv4-mapped IPv6 addresses: the IPv6 address ::FFFF:x.y.z.w represents the IPv4 address x.y.z.w (see Figure 8.6.)

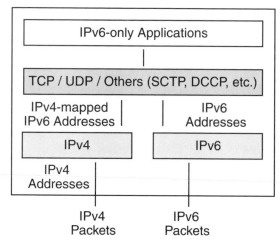

Figure 8.6: Address mapping.

One can analyze the behavior of IPv6-applications that exchange IPv4 packets with IPv4 applications by using the client/server model. We consider the default case to be when the IPV6_V6ONLY socket option has not been set. In these dual-stack nodes, this default behavior allows a limited amount of IPv4 communication using the IPv4-mapped IPv6 addresses.

IPv6-only server: When an IPv4 client application sends data to an IPv6-only server application running on a dual-stack node by using the wildcard address, the IPv4 client address is interpreted as the IPv4-mapped IPv6 address in the dual-stack node. This allows the IPv6 application to manage the communication. The IPv6 server will use this mapped address as if it were a regular IPv6 address, and a usual IPv6 connection. However, IPv4 packets will be exchanged between the nodes. Kernels with dual-stack properly interpret IPv4-mapped IPv6 addresses as IPv4 ones, and vice versa.

IPv6-only client: IPv6-only client applications in a dual-stack node will not receive IPv4-mapped addresses from the hostname resolution API functions unless a special hint, AI_V4MAPPED, is given. If it is, the IPv6 client will use the returned mapped address as if it were a regular IPv6 address, and a usual IPv6 connection. However, IPv4 packets will be exchanged between applications.

Respectively, with IPV6_V6ONLY set, an IPv6-only server application will only communicate with IPv6 nodes, and an IPv6-only client only with IPv6 servers, as the mapped addresses have been disabled. This option could be useful if applications use new IPv6 features such as Flow Label. If communication with IPv4 is needed, either IPV6_V6ONLY must not be used, or dual-stack applications must be used, as described in section 8.6.4.3.

Some implementations of dual-stack do not allow IPv4-mapped IPv6 addresses to be used for interoperability between IPv4 and IPv6 applications. In these cases, there are two ways to handle the problem:

1. Deploy two different versions of the application (possibly attached with '6' in the name).

2. Deploy just one application supporting both protocol versions as described in the next section.

The first method is not recommended because of a significant number of problems associated with selecting the right applications. These problems are described in sections 8.6.3.2 and 8.6.3.3. Therefore, there are two distinct cases to consider when writing one application to support both protocols:

Whether the application can (or should) support both IPv4 and IPv6 through IPv4-mapped IPv6 addresses or the applications should support both explicitly, and,

Whether the systems in which the applications are used support IPv6.

8.6.4.3 IPv4/IPv6 Applications in a Dual-Stack Node

Applications should be ported to support both IPv4 and IPv6. Over time, the existing IPv4-only applications could be removed. As we have only one version of each application, the source code will typically be easy to maintain and to modify, and there are no problems managing which application to select for which communication. This transition case is the most advisable. During the IPv6 transition period, applications supporting both IPv4 and IPv6 should be able to communicate with other applications, irrespective of the version of the protocol stack or the application in the node. Dual-applications allow more interoperability between heterogeneous applications and nodes.

If the source code is written in a protocol-independent way, without dependencies on either IPv4 or IPv6, applications will be able to communicate with any combination of applications and types of nodes.

Implementations typically prefer IPv6 by default if the remote node and application support it. However, if IPv6 connections fail, version-independent applications will automatically try IPv4 ones. The resolver returns a list of valid addresses for the remote node, and applications can iterate through all of them until connection succeeds.

Application writers should be aware of this protocol ordering, which is typically the default, but the applications themselves usually need not be (RFC 3484).

If the source code is written in a protocol-dependent way, the application will support IPv4 and IPv6 explicitly by using two separate sockets. Note that there are some differences in bind() implementation - that is, in whether one can first bind to IPv6 wildcard addresses, and then to those for IPv4. Writing applications that cope with this can be a pain. Implementing IPV6_V6ONLY simplifies this. The IPv4 wildcard bind fails on some systems because the IPv4 address space is embedded into IPv6 address space when IPv4-mapped IPv6 addresses are used.

8.6.4.4 IPv4/IPv6 Applications in an IPv4-Only Node

As the transition is likely to take place over a longer time frame, applications already ported to support both IPv4 and IPv6 may be run on IPv4-only nodes. This would typically be done to avoid supporting two application versions for older and newer operating systems, or to support a case in which the user wants to disable IPv6 for some reason.

The most important case is the application support on systems where IPv6 support can be dynamically enabled or disabled by the users. Applications on such a system should be able to handle a situation IPv6 would not be enabled. Another scenario is when an application is deployed on older systems that do not support IPv6 at all (even the basic APIs such as getaddrinfo). In this case, the application designer has to make a case-by-case judgment call as to whether it makes sense to have compile-time toggle between an older and a newer API (having to support both in the code), or whether to provide getaddrinfo and so on function support on older platforms as part of the application libraries.

8.6.5 Application Porting Considerations

The minimum changes for IPv4 applications to work with IPv6 are based on the different size and format of IPv4 and IPv6 addresses. Applications have been developed with IPv4 network protocol in mind. This assumption has resulted in many IP dependencies through source code. The following list summarizes the more common IP version dependencies in applications:

1. *Presentation format for an IP address*: An ASCII string that represents the IP address, a dotted-decimal string for IPv4, and a hexadecimal string for IPv6.
2. *Transport layer API*: Functions to establish communications and to exchange information.
3. *Name and address resolution*: Conversion functions between hostnames and IP addresses.
4. *Specific IP dependencies*: More specific IP version dependencies, such as IP address selection, application framing, and storage of IP addresses.
5. *Multicast applications*: One must find the IPv6 equivalents to the IPv4 multicast addresses and use the right socket configuration options.

Refer to RFC 4038 for more details on these issues.

References

[3GP200501] 3GPP Web page: http://www.3GPP.org.

[3GP200502] 3GPP2 Web page: http://www.3GPP2.org.

[3GS200501] 3G staff, "Wireless Voice Over IP: Technical and Commercial Prospects," 3G Online Magazine, March 22, 2005. http://www.3g.co.uk/PR/March2005/1232.htm.

[AND200501] L. Andersson, T. Madsen, "Provider Provisioned Virtual Private Network (VPN) Terminology," RFC 4026, March 2005.

[ASP200401] *Promotional Material, Polycom Installed Voice Business Group*, 9040 Roswell Road, Suite 450, Atlanta, GA 30350. http://www.aspi.com/products.

[ATL200401] *Promotional Material, Atlanta Signal Processors, Inc.*, Atlanta, Georgia.

[BOS200101] L. Bos and S. Leroy, "Toward an All-IP-Based UMTS System Architecture," IEEE Network, January 2001.

[BOU200501] J. Bound, "IPv6 Deployment State 2005," Special Issue entitled IPv6 More than a Protocol, Upgrade, The European Journal for the Informatics Professional, Vol. VI, No. 2, April 2005.

[BRA200501] T. Bradley, Glossary, http://netsecurity.about.com/library/glossary/bldef-appg.htm.

[CAB200501] A. Cabellos-Aparicio and J. Domingo-Pascual, "Internet Protocol version 6 Overview," Special Issue entitled IPv6 More than a Protocol, Upgrade, The European Journal for the Informatics Professional, Vol. VI, No. 2, April 2005. http://www www.upgrade-cepis.org.

[CAL200301] R. Callon and M. Suzuki, "A Framework for Layer 3 Provider Provisioned Virtual Private Networks," Work in Progress, July 2003.

[CAM200501] G. Camarillo, J. Rosenberg, "Usage of the Session Description Protocol (SDP) Alternative Network Address Types (ANAT) Semantics in the Session Initiation Protocol (SIP)," Request for Comments: 4092, June 2005.

[CHA200101] A. E. Cha, "Showdown at the Digital Corral – Internet-based Single Number Plan Starts a Tug of War Over Control," Washington Post Article on ENUM, April 22, 2001.

[CIS200501] Cisco Systems Promotional Material, 2005.

[CSO200501] http://www.csoonline.com/glossary/category.cfm?ID=13.

[DAV200201] J. Davies, Understanding IPv6, Microsoft Press, 2002.

[DEE199801] S. Deering, R. Hinden, "Internet Protocol, Version 6 (IPv6) Specification," RFC 2460, December 1998. Copyright © The Internet Society (1998). All Rights Reserved. This document and translations of it may be copied and furnished to others, and derivative works that comment on or otherwise explain it or assist in its implementation may be prepared, copied, published and distributed, in whole or in part, without restriction of any kind, provided that the above copyright notice and this paragraph are included on all such copies and derivative works.

[DEM200301] Desmeules, Cisco Self-Study: Implementing IPv6 Networks (IPV6), Pearson Education, May 2003.

[DOM200501] J. Domingo-Pascual, A. García-Martínez, and M. Ford, "IPv6: A New Network Paradigm," Special Issue entitled IPv6 More than a Protocol, Upgrade, The European Journal for the Informatics Professional, Vol. VI, No. 2, April 2005.

[ENU200301] Frequently Asked Questions, ENUM Organization, http://www.enum.org/information/faq.cfm.

[GEI200501] R. Geib, E. Azañón-Teruel, S. Donaire-Arroyo, A. Ferrándiz-Cancio, C. Ralli-Ucendo, and F. Romero-Bueno " Service Deployment Experience in Pre-Commercial IPv6 Networks," Special Issue entitled IPv6 More than a Protocol, Upgrade, The European Journal for the Informatics Professional, Vol. VI, No. 2, April 2005. http://www www.upgrade-cepis.org.

[AGG200001] M. Day, S. Aggarwal, G. Mohr, J. Vincent' Instant Messaging / Presence Protocol Requirements, RFC 2779, (February 2000).

[GIL200001] R. Gilligan, E. Nordmark, Transition Mechanisms for IPv6 Hosts and Routers, RFC 2893, August 2000. Copyright © The Internet Society (2000). All Rights Reserved. This document and translations of it may be copied and furnished to others, and derivative works that comment on or otherwise explain it or assist in its implementation may be prepared, copied, published and distributed, in whole or in part, without restriction of any kind, provided that the above copyright notice and this paragraph are included on all such copies and derivative works.

[GON199801] M. Goncalves, K. Niles, IPv6 Networks, McGraw-Hill Osborne, 1998.

[GOS200301] S. Goswami, Internet Protocols: Advances, Technologies, and Applications, May 2003, Kluwer Academic Publishers

[GRA200001] B. Graham, TCP/IP Addressing: Designing and Optimizing your IP Addressing Scheme (2nd edition). Morgan Kaufmann, 2000.

[HAG200201] S. Hagen, IPv6 Essentials, O'Reilly, 2002.

[HAN199901] M. Handley, H. Schulzrinne, E. Schooler, J. Rosenberg, SIP: Session Initiation Protocol, Request for Comments: 2543, March 1999.

[HIN200301] R. Hinden, S. Deering, "Internet Protocol Version 6 (IPv6) Addressing Architecture" RFC 3513, April 2003. Copyright © The Internet Society (2003). All Rights Reserved. This document and translations of it may be copied and furnished to others, and derivative works that comment on or otherwise explain it or assist in its implementation may be prepared, copied, published and distributed, in whole or in part, without restriction of any kind, provided that the above copyright notice and this paragraph are included on all such copies and derivative works.

[HOW200401] Andrew Wilson (Wil) Howitt, "Acoustics, and Signal Processing," 2004, Massachusetts Institute of Technology, 77 Massachusetts Avenue, Cambridge, MA 02139.

[HUI199701] C. Huitema, IPv6 the New Internet Protocol (2nd edition), Prentice Hall, 1997.

[IMT200001] ITU, "The IMT-2000 initiative," ITU-R Draft Rec. M, "Detailed Specifications of the Radio Interfaces of MT-2000," Doc 8/126. http://www.itu.int/imt.

[INF200401] Information Week, August 2, 2004, pages 43 ff.

[INF200501] http://www.infosec.uga.edu/glossary.html

[INS200401] Insight Research, Press Release: "New Telecom Research Study Quantifies VoIP Equipment Penetration into the Enterprise Marketplace," Boonton, NJ. June 17, 2004.

[INT200401] Integrated Research, "Avoiding the Pitfalls of VoIP," July 1, 2004, Integrated Research White Paper, Denver, CO.

[IPV200401] IPv6Forum, "IPv6 Vendors Test Voice, Wireless and Firewalls on Moonv6," November 15, 2004.

[IPV200501] IPv6 Portal, http://www.ipv6tf.org.

[ISL200501] J. Islam, "IPv6 Applications," v6 Application Initiative, 2005, IPv6 Forum.

[ITO200401] J. Itojun Hagino, IPv6 Network Programming, Butterworth-Heinemann, 2004.

[KUI200501] L. Kuiper, "Security for the Enterprise Network Engineer," Cisco Networkers Conference, 2005.

[LEE200501] H. K. Lee, Understanding IPv6, Springer-Verlag New, York, 2005.

[LIG200501] Lightreading Online Magazine, http://www.lightreading.com

[LIG200502] Staff, "VoIP Session Border Controllers: A Heavy Reading Competitive Analysis," Heavy Reading Magazine, July 2005.

[LOR200501] M. De Lorenzi and A. Brunner, SIP.edu Deployment, http://www.ethworld.ethz.ch/technologies/sipeth/notes

[LOS200301] P. Loshin, IPv6: Theory, Protocol, and Practice (2nd edition). Elsevier Science & Technology Books, 2003.

[MAL200501] T. Maltagliati, "Reports: Agencies Must Prepare for New Internet," FederalTimes.com online magazine, June 01, 2005.

[MAV200201] N. Mavrakis, "3G Wireless Systems, A Comprehensive Study," December 15, 2002, Stevens Institute of Technology White Paper. Hoboken, NJ.

[MIC200501] Microsoft Promotional Material.

[MIL199701] M. A. Miller, *Implementing IPv6: Migrating to the Next Generation Internet Protocol*, December 1997, Wiley, John & Sons, Incorporated.

[MIL200001] M. Miller, P. E. Miller, *Implementing IPV6: Supporting the Next Generation Internet Protocols (2nd Edition)*. Hungry Minds, 2000.

[MIN197901] D. Minoli, "Optimal Packet Length For Packet Voice Communication," IEEE Trans. on Comm., Concise paper, March 1979, Vol COMM-27, pp. 607–611.

[MIN197902] D. Minoli, "Packetized Speech Network, Part 3: Delay Behavior And Performance Characteristics, Australian Electronics Engineer, Aug. 1979, pp. 59–68.

[MIN197903] D. Minoli, "Packetized Speech Networks, Part 2: Queeing Model," Australian Electronics Engineer, July 1979, pp. 68–76.

[MIN197904] D. Minoli, "Packetized Speech Networks, Part 1: Overview," Australian Electronics Engineer, April 1979, pp. 38–52.

[MIN197905] D. Minoli, "Satellite On-Board Processing Of Packetized Voice," ICC 1979 Conference Record, pp. 58.4.1–58.4.5.

[MIN197906] D. Minoli, "Issues In Packet Voice Communication," Proceedings of IEE, Aug. 1979, Vol. 126, No. 8, pp. 729–740.

[MIN197907] D. Minoli, "Some Design Parameters For PCM Based Packet Voice Communication," 1979 International Electrical/Electronics Conference Record.

[MIN197908] D. Minoli, "Digital Voice Communication Over Digital Radio Links," SIGCOMM Computer Communications Review, Vol 9, No. 4, Oct 1979, pp. 6–22.

[MIN199401] Daniel Minoli, George Dobrowski, Signaling Principles for Frame Relay and Cell Relay Services, Artech House, 1994.

[MIN199801] D. Minoli, E. Minoli, *Delivering Voice over IP*, Wiley, April 1, 1998, New York.

[MIN199802] D. Minoli, E. Minoli, *Delivering Voice over Frame Relay and ATM*, April 1, 1998, New York.

[MIN200201] D. Minoli, E. Minoli, *Delivering Voice over IP, 2nd Edition*, Wiley, August 2002, New York.

[MIN200202] D. Minoli, *Voice over MPLS*, McGraw, 2002, New York.

[MIN200203] D. Minoli, Hotspot Networks: Wi-Fi for Public Access Locations, McGraw-Hill, 2002.

[MIN200601] D. Minoli and J. Cordovana, Authoritative Computer and Network Security Dictionary, Wiley, 2006. New York.

[MOB200501] *Mobile Pipeline Staff, Mobile, VoIP Replacing Traditional Voice Service, Researcher Says*, June 22, 2005, CMP Publications.

[MOB200502] Mobile Pipeline Staff, "Verizon Looking At VoIP Over 3G," Mobile Pipeline, April 01, 2005. http://www.commweb.com/showArticle.jhtml?articleID=160401527.

[MSD200401] Microsoft Corporation, MSDN Library, Internet Protocol, 2004, http://msdn.microsoft.com.

[MUR200501] N. R. Murphy, D. Malone, *IPv6 Network Administration*, O'Reilly & Associates, 2005.

[NOR200501] Nortel Promotional Material, 2005.

[OUE200501] A. Ouellette, "Acme Packet Defines Role of Session Border Controllers in Converged Fixed-Mobile IMS Architecture," CCNMatthews, June 03, 2005, Acme Packet, +1 781-328-4436.

[PAT200001] G. Patel, S. Dennett, "The 3GPP and 3GPP2 Movements Toward an All-IP Mobile Network," IEEE Personal Communications, August 2000.

[PER200201] Perkins, C., "IP Mobility Support for IPv4", IETF RFC 3344, August 2002.

[POL200401] R. Pollock "Secure Networks," Communications News, April 2004.

[RAD200401] Promotional Materials, RadiSys Corporation, 5445 NE Dawson Creek Drive, Hillsboro, OR 97124 USA.

[REA200401] Marguerite Reardon, "Wi-Fi and VoIP: Is Sum Greater than Parts?" March 1, 2004, Staff Writer, CNET News.com.

[REN200301] Jim Rendon, "VoIP Being Talked up in Contact Centers," SearchNetworking.com News Writer, 23 May 2003, SearchNetworking.com.

[ROS200001] M. Day, J. Rosenberg, H. Sugano, A Model for Presence and Instant Messaging, RFC 2778, February 2000. Copyright © The Internet Society (2000). All Rights Reserved. This document and translations of it may be copied and furnished to others, and derivative works that comment on or otherwise explain it or assist in its implementation may be prepared, copied, published and distributed, in whole or in part, without restriction of any kind, provided that the above copyright notice and this paragraph are included on all such copies and derivative works.

[ROS200301] J. Rosenberg, J. Weinberger, C. Huitema, R. Mahy, STUN - Simple Traversal of User Datagram Protocol (UDP) Through Network Address Translators (NATs), RFC 3489, March 2003. Copyright © The Internet Society (2003). All Rights Reserved. This document and translations of it may be copied and furnished to others, and derivative works that comment on or otherwise explain it or assist in its implementation may be prepared, copied, published and distributed, in whole or in part, without restriction of any kind, provided that the above copyright notice and this paragraph are included on all such copies and derivative works.

[ROS200401] J. Rosenberg, "A Presence Event Package for the Session Initiation Protocol (SIP)," RFC 3856, August 2004.

[SCH200201] J. Rosenberg, H. Schulzrinne, G. Camarillo, A., Peterson, R. Sparks, M. Handley, E. Schooler, "SIP: Session Initiation Protocol," RFC 3261, June 2002. Copyright © The Internet Society (2002). All Rights Reserved. This document document and translations of it may be copied and furnished to others, and derivative works that comment on or otherwise explain it or assist in its implementation may be prepared, copied, published and distributed, in whole or in part, without restriction of any kind, provided that the above copyright notice and this paragraph are included on all such copies and derivative works.

[SHI200501] M-K. Shin, Ed., Y-G. Hong, J. Hagino, P. Savola, E. M. Castro, "Application Aspects of IPv6 Transition," RFC 4038, March 2005.

[SOL200401] H. S. Soliman, Mobile IPv6, Pearson Education, 2004.

[SRI200101] Srisuresh, P. and K. Egevang, "Traditional IP Network Address Translator (Traditional NAT)," RFC 3022, January 2001.

[SRI200201] P. Srisuresh, J. Kuthan, J. Rosenberg, A. Molitor, A. Rayhan, Middlebox communication architecture and framework, RFC 3303, August 2002. Copyright © The Internet Society (2002). This document and translations of it may be copied and furnished to others, and derivative works that comment on or otherwise explain it or assist in its implementation may be prepared, copied, published and distributed, in whole or in part, without restriction of any kind, provided that the above copyright notice and this paragraph are included on all such copies and derivative works.

[SYS200501] SysMaster Corporation, Promotional Materials, 5801 Christie Ave., Ste 400, Emeryville, CA 94608, http://www.sysmaster.com/p_vp_gk.htm, 2005.

[TDD200501] UMTS TDD Alliance, http://www.umtstdd.org.

[TEA200401] D. Teare, C. Paquet, "CCNP Self-Study: Advanced IP Addressing," Cisco Press, Jun 11, 2004.

[TSI200001] G. Tsirtsis and P. Srisuresh, "Network Address Translation—Protocol Translation (NAT-PT)," RFC 2766, February 2000.

[TUR200501] B. Turner, "The Impact of VoIP, Wi-Fi and 3G Data on Wireless Telecom: How Fixed-Mobile Convergence Will Reshape the Wireless Industry," International Wireless Telecom—Carriers Global Edition, World Media Online Limited, pp. 145–157, St John Street, London, EC1V 4PY, United Kingdom, Tel: +44 870 755 7107, Email: info@wirtel.co.uk

[USP199401] Low Complexity CELP Speech Coder, US Patent 5,371,853 Dec. 6, 1994. .

[WEG199901] J. D. Wegner, *IP Addressing and Subnetting*, Including IPv6, September 1999, Elsevier Science & Technology Books.

[WIM200501] WiMAX Forum Web Page, http://www.wimaxforum.org/home.

[WIW200501] Staff, "Broadband Wireless Threatens 3G Voice Ambitions," Wireless Watch, October 26, 2004.

[WRE200401] G. Wrenn, "Securing Web services: A Job for The XML Firewall," 08 Mar 2004, http://searchsecurity.techtarget.com/tip/1,289483,sid14_gci954170,00.html?Offer=SEcpwslg25

[ZHA200401] Yongguang Zhang, "A Multilayer IP Security Protocol for TCP Performance Enhancement in Wireless Networks," IEEE Journal On Selected Areas In Communications, Vol. 22, NO. 4, May 2004.

About the Author

Mr. Minoli has many years of telecom, networking and IT experience for end users, carriers, academia, and Venture Capitalists, including work at ARPA think tanks, Bell Telephone Laboratories, ITT, Bell Communications Research (Bellcore/Telcordia), AT&T, AIG, Prudential Securities, Capital One Financial, New York University, Rutgers University, Stevens Institute of Technology, and Societe' General de Financiament de Quebec (1975-2004). Recently he also played a founding role in the launching of two networking companies through the high-tech incubator Leading Edge Networks Inc., which he ran in the early 2000s: Global Wireless Services, a provider of broadband hotspot mobile Internet and hotspot Wi-Fi VoIP services to high-end marinas; and, InfoPort Communications Group, an optical and Gigabit Ethernet metropolitan carrier supporting Data Center/SAN/channel extension and Grid Computing network access services (2001-2003).

An author of a number of technical references on Information Technology, telecommunications, and data communications he has also written columns for ComputerWorld, NetworkWorld, and Network Computing (1985-present). Mr. Minoli co-authored the first-ever book on VoIP, Delivering Voice over IP (Wiley, April 1, 1998, now in second edition); the first-ever book on VoATM and VoFR, Delivering Voice over Frame Relay and ATM (Wiley, April 1, 1998); and, the first-ever book on VoMPLS, Voice over MPLS (McGraw, 2002). This first-to-the market book on VoIPv6 completes the author's tetrateuch on the topic of voice over packet.

Mr. Minoli has taught at New York University (Information Technology Institute), Rutgers University, and Stevens Institute of Technology (1984-2003). Also, he was a Technology Analyst At-Large, for Gartner/DataPro (1985-2001); based on extensive hand-on work at financial firms and carriers, he tracked technologies and wrote around 50 distinct CTO/CIO-level technical/architectural scans in the area of telephony and data communications systems, including topics on security, disaster recovery, IT outsourcing, network management, LANs, WANs (ATM and MPLS), wireless (LAN and public hotspot Wi-Fi), VoIP, network design/economics, carrier networks (such as metro Ethernet and CWDM/DWDM), and e-commerce. Over the years he has advised Venture Capitalists for investments of $150M in a dozen high-tech companies, and has acted as Expert Witness in a (won) $11B lawsuit regarding an early wireless Air-to-Ground communication system that made use of voice-over-packet technologies in the airplane cabin.

Index

Symbols

1G commercial VoIP 19
1xEV-DO 103
3G cellular systems 28
3G operators 102
3GPP 98
3G wireless network 14
6over4 32
6over4 address 32
6to4 32
6to4 address 32
6to4 machine 32
6to4 router 32

A

adaptive differential pulse code modulation (ADPCM) 22, 55–56
adaptive postfilter 75
adaptive quantization 61
adaptive subband coding 55
address 32
address-of-record 115
address autoconfiguration 33
addressing capabilities 3
address maximum valid time 33
address resolution 33
ADPCM (adaptive differential PCM) 22, 55–56
aggregatable unicast global address 33
ALG 207, 216, 235, 237
algorithm complexity 73
  measures of complexity 73
    quality 73
algorithmic delay 72
always-on integrated communications 89
analog-to-digital (A/D) 60
anycast address 33
application layer gateways (ALGs) 216
application level gateways (ALGs) 235
AS 33
ATM 18, 50, 95
attempt address 33
audio encoder 64

B

back-to-back user agent 115
basic header 14
basic network address translation (basic NAT) 2
bibliography 41
  IPv6 RFC bibliography 41
  NAT RFC bibliography 44
  SIP RFC bibliography 45
bit rate 71
bootstrap protocol (BOOTP) 335
bridging delay 72

C

cable TV environments 54
  multiple system operators (MSOs) 54
call admission (RAS) 84
call control
  H.245 (Q.931) 82
call setup through gatekeeper 84
call state control function (CSCF) 98, 100
CANCEL 134, 135
CDMA EVDO 103
CELP synthesis model 71
CELP vocoder 68
channel vocoder 64
circuit-switched 18
classless interdomain routing (CIDR) 7
code excited linear prediction (CELP) 67
colon hexadecimal notation 33
comfort noise generation (CNG) 72
common open policy service (COPS) 51
compact disc (CD) 56
compatibility addresses 33
compressing zeros 34

authentication, authorization, and accounting (AAA) 319
authorization, authentication, and accounting (AAA) 101
autoconfiguration 3–5, 12–13, 32–33, 36
automatic call distribution (ACD) 1
automatic IPv6 tunnel 33
automatic tunnel 33
autonomous system (AS) 33

computer telephony integration (CTI) 49
conjugate-structure algebraic code excited linear prediction (CS-ACELP) 70
connectionless 10, 20
contact center 49
controlled zone (CZ) 195
converged intranet 203
COPS 51
correspondent node 34
coverage 96
CS-ACELP (conjugate-structure algebraic-code-excited linear-prediction) 76
    ITU G.729 76

D

data rate 96
default path 34
default routers list 34
delay 72
demilitarized zone (DMZ) 198, 252
destination cache 34
DHCP 204
digital signal processing (DSP) 1
directory services 91
distance vector 34
DMZ 198, 252
domain names system 34
double colon 34
DSP 73
DTMF tones 90
dual-IP layer operation 325
    address configuration 325
    advertising addresses in the DNS 326
    domain naming system 326
        DNS resolver libraries 326
    IPv6/IPv4 nodes 325
    IPv6 stack 325
dual-stack transition 17
dual-stack architecture 34
dynamic host configuration protocol (DHCP) 34, 335

E

E.164 92
E.164 ISDN-era addressing scheme 107
E911 services 19
encapsulating security payload 34
end-to-end security 3, 15, 30
enterprise deployment strategies 29
enterprise network
    typical network elements found in an enterprise network 294
enterprise network VoIP security 201
ENUM 91, 93
error weighting filter 75

EUI 34
EUI-64 address 34
EV-DV (evolution-data and voice) 103
excitation energy 65
extended unique identifier (EUI) 34
extensibility 3
extension headers 34
externally-controlled zone (ECZ) 195

F

Federal Communication Commission (FCC) 93
feedback adaptive PCM 62
feed forward adaptive PCM 62
fiber distributed data interface II (FDDI II) 18
filter 198
firewall 198
firewall/DMZ architecture 51
firewall issues for VoIP 204
fixed-mobile convergence (FMC) operators 104
    NTT DoCoMo 104
    PBX 104
    WLANs 104
flow 35
formant vocoder 64
format prefix 35
fragment 35
fragmentation 35
fragmentation header 35
frequency division duplex (FDD) 103
fundamental SIP functionality 108

G

G.711 57
G.711 speech coder 58
G.712 57
G.720 58
G.722 58
G.722.1 58
G.722.2 58
G.722 speech coder 58
G.723.1 58, 71
G.723.1 dual rate speech coder with annex A 58
G.724 58
G.725 58
G.726 58
G.726 ADPCM waveform coder 59
G.727 58
G.728 58
G.729 58, 70
G.729A 71
G.729 with annex-B CS-ACELP voice coder 59
gatekeeper 83
GDOI 51
generic audio encoder 64

global address 35
globally-unique IP address 3
global system for mobile communications (GSM) 96
group domain of interpretation (GDOI) group keying 52
group identifier 35
GSM-FR speech coder 59, 77

H
H.225 78
H.225.0 81
H.225.0 layer 82
H.245 78, 81
H.245 call control 84
H.245 control 82
H.323 79, 90
  call admission (RAS) 84
  call proceeding 83
  call setup through gatekeeper 84
  H.245 call control 84
  registering with gatekeeper (RAS) 84
  setup 83
H.323 entities 81, 82
  gatekeeper 82
  gateways 82
  multipoint control unit (MCU) 83
H.323 signaling protocol 80
H.323 standards 80
H.323 terminal 82
higher level checksum 35
higher level protocol 35
home agent (HA) 101
home subscriber server (HSS) 98
hop-by-hop option header 35
host-to-host tunnel 35
host-to-router tunnel 35
hosts file 35
hotspot 95
hybrid multipulse linear predictive coding 55
hybrid TDM and IP systems 25

I
ICMP messages 210
ICMPV6 35
IEEE 802.11e QoS-support 94
IEEE 802.11g 94
IEEE 802.11i 15, 321
IEEE 802.16 95
IEEE EUI-64 address 312
IM 89
in-path SIP proxy 245
inline authentication 194
integrated services digital network (ISDN) 18
integrated voice/data LAN 18
interactive voice response (IVR) 53

interexchange carrier (IXC) 52
interface 35
interface identifier 12, 13, 33, 35
Internet control message protocol for IPv6 (ICMPv6) 10, 35, 298
  configuration methods 313
    address states 314
      deprecated 314
      invalid 314
      preferred 314
      preferred lifetime 314
      tentative 314
      valid 314
      valid lifetime 314
  deployment status 317
    authentication, authorization, and accounting (AAA) 319
    middleware 318
    network infrastructure deployment 318
    security for IPv6 319
      end-to-end security model 320
      IEEE 802.11i 321
      multilayer IP-security (ML-IPsec) 321
  multicast listener done 299
  multicast listener query 299
  multicast listener report 299
  neighbor discovery 299
    address autoconfiguration 300
    address resolution 300
    duplicate address detection 300
    neighbor unreachability detection 300
    next-hop determination 300
    parameter discovery 300
    prefix discovery 300
    redirect function 300
    router discovery 300
  routing and route management 315
Internet VoIP 105
  Wi-Fi-enabled mobile phones 105
internetwork 5
intra-site automatic tunneling addressing protocol (ISA-TAP) 35
invisible computing or ubiquitous computing 89
INVITE 111–114
IP6.Int 35
IP address 3, 5–7, 9, 30, 35, 37
  IPv6 and IPv4 addresses simultaneously 13
IPng, Internet next generation 5
IPSec 4, 8
IPsec encapsulating security payload (ESP) 243
IPsec IP authentication header (AH) 243
IP trunking 25
IPv4 2–14, 18, 21, 29–39, 42–44, 46
  allocation 14
  compatibility mechanisms 323

IP address 6
  IPv6 and IPv4 addresses simultaneously 13
  IPv4 hosts 323
  IPv6 traffic over IPv4 networks 13
  network address translation issues 6
    IP address 6
      "legal" addresses 7
      address class A 6
      address class B 6
      address class C 7
      address class D 7
      classless interdomain routing (CIDR) 7
IPv4-only node 324
IPv4 compatible address 35
IPv4 fragmentation 329
IPv4 ICMP errors 328
IPv4 node 35, 324
IPv6 4, 259
  64-Bit global identifier (EUI-64) registration authority 291
  address 260
  allocation 14
  application transition 337
    address mapping 340
    application porting 342
    applications in a dual-stack node 338
      IPv6-only client 340
      IPv6-only server 340
    applications supporting both IPv4 and IPv6 in a dual-stack node 338
    applications supporting both IPv4 and IPv6 in an IPv4-only node 338
    DNS queries 338
    IPv4-only applications and IPv6-only applications in a dual-stack node 338
    IPv4/IPv6 applications in a dual-stack node 341
    IPv4/IPv6 applications in an IPv4-only node 341
    IPv4 applications in a dual-stack node 339
    IPv6 applications in a dual-stack node 339
    multiple versions 339
  automatic tunneling 334
    automatic tunneling operation 335
    default configured tunnels 336
      isolated IPv6/IPv4 hosts 336
    ingress filtering 336
    IPv4-compatible address configuration 335
      bootstrap protocol (BOOTP) 335
      dynamic host configuration protocol (DHCP) 335
      reverse address resolution protocol (RARP) 335
    IPv4-compatible address format 334
      neighbor discovery mechanisms 335
    source address selection 336
basic header 14
basic protocol constructs 10
common tunneling mechanisms 327
  automatic 328

configured 328
decapsulation 332
fragmentation 328
hop limit 330
host-to-host tunnel 327
host-to-router tunnel 327
IPv4 header construction 331
  fragment offset 331
  header checksum 331
  time to live 331
IPv4 ICMP errors 330
  IPv6 ICMP message 330
  packet too big 330
link-local addresses 332
neighbor discovery over tunnels 333
router-to-host tunnel 327
router-to-router tunnel 327
tunnel MTU 328
compatibility mechanisms 323
configured tunneling 333
  default configured tunnel 333
  default configured tunnel using IPv4 (anycast address) 334
  ingress filtering 334
connectionless 10
deployment of IPv6 323
dual-stack transition 17
encapsulating IPv6 in IPv4 328
expanded addressing capabilities 259
flow labeling capability 259
host 260
interface 260
interworking in IPv4-based Internet 323
interworking mechanisms 323
  automatic tunneling of IPv6 over IPv4 324
  configured tunneling of IPv6 over IPv4 324
  dual-IP layer 323
  IPv4-compatible IPv6 addresses 324
IPng, Internet next generation 5
IPv6 address 9
  abbreviated format 9
  auto-return or loopback virtual address 10
  deprecated state 13
  groups of IP addresses 9
  interface identifier 12, 13, 33, 35
  IPv4 over IPv6 addresses automatic representation 10
  IPv6 and IPv4 addresses simultaneously 13
  IPv6 over IPv4 dynamic/automatic tunnels 10
  lifetime 13
  not specified address 10
  preferred state 13
IPv6 address space 9
IPv6 autoconfiguration
  serverless autoconfiguration 12
  stateful configuration 13

stateless autoconfiguration 12
IPv6 datagram 10
IPv6 hosts 323
IPv6 packet 10
    IPv6 header 10
        IPv6 base header 11
    IPv6 payload 10
IPv6 traffic over IPv4 networks 10, 13
key IPv6 protocols 10
    Internet control message protocol for IPv6 (ICMPv6) 10
    multicast listener discovery (MLD) 10
    neighbor discovery (ND) 10
link 260
link MTU 260
mobile IPv6 15
modified EUI-64 format 290
neighbors 260
node 259
packet 260
path MTU 260
router 260
transition approaches 17
transition mechanisms 325
transition toolbox 324
upper layer 260
IPv6, techniques
    automatic tunneling 324
    configured tunneling 324
    IPv4 multicast tunneling 324
    IPv6-over-IPv4 tunneling 324
IPv6-enabled 3G VoIP 295
IPv6-only node 324
IPv6/IPv4 node 36, 324, 325
IPv6 address 9
    abbreviated format 9
    groups of IP addresses 9
    interface identifier 12, 13, 33
IPv6 addresses, types of 324
    IPv4-compatible IPv6 address 324
    IPv6-native address 324
IPv6 addressing 279
    address type identification 282
    anycast addresses 285
        required anycast address 286
    multicast addresses 286
        pre-defined multicast addresses 288
    text representation of addresses 280
    text representation of address prefixes 281
    unicast addresses 282
        global unicast addresses 284
        interface identifiers 283
        IPv6 addresses with embedded IPv4 addresses 284
        local-use IPv6 unicast addresses 285
        the loopback address 284
        unspecified address 284

IPv6 addressing mechanisms 301
    addresses for hosts and routers 310
        IEEE address along with the extended unique identifier 312
        IEEE EUI-64 address 312
        interface determination 311
        mapping EUI-64 addresses to IPv6 interface identifiers 313
        mapping IEEE 802 addresses to IPv6 interface identifiers 313
        media access control (MAC) 312
        randomly-generated interface identifiers 313
    address reachability scope 303
    address scope 305
    address types 306
        aggregatable global unicast address 307, 308
        anycast IPv6 address 310
        compatibility (unicast) address 309
        link-local (unicast) address 308
        loopback (unicast) address 309
        multicast IPv6 address 309
        site-local (unicast) address 308
        site-local address 308
        unspecified (unicast) address 309
    IPv6 128-bit address conventions 301
    link-local address 305
    logical node in IPv6 305
    next level aggregator (NLA) 307
    site-local address 305
    site level aggregator (SLA) 307
    top level aggregator (TLA) 307
IPv6 address space 9
IPv6 base header 11
IPv6 benefits 5
    extensibility 6
    mobility 6
    plug-and-play 6
    scalability 5
    security 6
IPv6 datagram 10, 35
IPv6 extension header 261
    destination options header 272
    extension header order 262
    fragment header 268
        fragment packet 270
        reassembly 271
    hop-by-hop options 269
    hop-by-hop options header 264
        next header 264
    IPv6 options 263
    routing header 265
        routing type 265
IPv6 flow label field 276
IPv6 flow label 273
IPv6 header 10

IPv6 base header 11
  destination address 11
  flow label 11
  hop limit 11
  Next header 11
  source address 11
IPv6 header format 260
IPv6 infrastructure 297
  IPv6 header field 297
  IPv6 packet 297
    flows of IPv6 packets in a VoIPv6 environment 298
  protocol Mmechanisms 297
IPv6 in IPv4 35
IPv6 MTU 35
IPv6 network 14
IPv6 node 35, 324
IPv6 over IPv4 tunnel 36
IPv6 packet 10–11, 14, 36–38
  IPv6 header 10
  IPv6 payload 10
IPv6 packet size issues 273
IPv6 prefixes 36
IPv6 RFC bibliography 41
IPv6 routing table 36
IPv6 stack 325
IPv6 terminology 32
  6over4 32
  6over4 address 32
  6to4 32
  6to4 address 32
  6to4 machine 32
  6to4 router 32
  address 32
  address autoconfiguration 33
  address maximum valid time 33
  address resolution 33
  aggregatable unicast global address 33
  anycast address 33
  AS 33
  attempt address 33
  automatic IPv6 tunnel 33
  automatic tunnel 33
  autonomous system (AS) 33
  colon hexadecimal notation 33
  compatibility address 33
  compressing zeros 34
  correspondent node 34
  default path 34
  default routers list 34
  destination cache 34
  distance vector 34
  domain names system 34
  double colon 34
  dual stack architecture 34
  dynamic host configuration protocol (DHCP) 34

encapsulating security payload 34
EUI 34
EUI-64 address 34
extended unique identifier (EUI) 34
extension headers 34
flow 35
format prefix 35
fragment 35
fragmentation 35
fragmentation header 35
global address 35
group identifier 35
higher level checksum 35
higher level protocol 35
hop-by-hop option header 35
host-to-host tunnel 35
host-to-router tunnel 35
hosts file 35
ICMPV6 35
interface 35
interface identifier 35
Internet control message protocol for IPv6 (ICMPV6) 35
intra-site automatic tunneling addressing protocol (ISA-TAP) 35
IP6.Int 35
IPv4 compatible address 35
IPv4 node 35
IPv6/IPv4 node 36
IPv6 in IPv4 35
IPv6 MTU 35
IPv6 node 35
IPv6 over IPv4 tunnel 36
IPv6 prefixes 36
IPv6 routing table 36
ISATAP 36
ISATAP address 36
ISATAP nachine 36
ISATAP name 36
ISATAP router 36
jumbogram 36
jumbo payload option 36
lifetime in preferred state 36
link 36
link maximum transmission Unit (MTU) 36
link state 36
local address 36
local area network segment 36
local interface 36
local loop address 36
local site address 37
MAC address 37
machine (host) 37
mapping IPv4 address 37
maximum-level aggregation identifier 37
maximum transfer unit (MTU) 37

medium access control 37
MTU 37
multicast address 37
multicast group 37
multicast IPv4 tunnel 37
name resolution 37
ND 37
neighbor 37
neighbors cache 37
neighbors discovery (ND) 37
neighbors discovery options 37
network addresses translator 37
network prefix 37
network segment 37
next-level aggregation identifier (NLA ID) 38
next hop obtaining 37
NLA ID 38
no-broadcast multiple access link 38
not specified address 38
own link 38
packet 38
parameters discovery 38
path's MTU discovery 38
path determination system 38
path MTU 38
path vector 38
PDU 38
point-to-point protocol 38
prefix-length notation 38
prefixes list 38
protocol data unit (PDU) 38
pseudo-header 38
pseudo-periodic 38
reassembing 38
redirect 38
relay router 6to4 38
router 39
router's cache 39
router advertisement 39
routers discovery 39
routing loop 39
scope 39
scope ID 39
site-level aggregation identifier (SLA ID) 39
site prefix 39
solicited-node address 39
static routing 39
subnet anycast router address 39
subnetwork 39
subnetwork associated path 39
suitable path selection 39
TLA ID 39
top-level aggregation identifier (TLA ID) 39
transition 39
tunnel 39

unicast address 40
IPv6 traffic classes 274
IPv6 upper-layer protocol 274
    maximum packet lifetime 275
    maximum upper-layer payload size 275
IP version 6 (IPv6) 259
ISATAP 36
ISATAP address 36
ISATAP machine 36
ISATAP name 36
ISATAP router 36
ISDN 79
ITU-G.729 71
ITU-T standards related to voice coding 57
    G.711 57
    G.712 57
    G.720 58
    G.722 58
    G.722.1 58
    G.722.2 58
    G.723.1 58
    G.724 58
    G.725 58
    G.726 58
    G.727 58
    G.728 58
    G.729 58
ITU G.711 60

J
jitter 54
jumbogram 36
jumbo payload option 36

K
kerberos 8

L
layer 2 198
layer 2 VPN (L2VPN) 198
layer 3 198
layer 3 security mechanisms 198
layer 3 VPN (L3VPN) 199
layered security apparatus 196
lifetime In preferred state 36
linear prediction 62
linear prediction analysis-by-synthesis (LPAS) 74
linear predictive coding (LPC) 56, 65
linear predictor 66
link 36
link maximum transmission unit (MTU) 36
link state 36
local address 36
local area network segment 36

local interface 36
local loop address 36
local site address 37
location server 87
logarithmic quantization 61
look-ahead 72
lossless 55
low-delay code-excited linear prediction vocoder standard 76
LPC 66
LPC vocoders 64

M

MAC address 37
machine (host) 37
mapping IPv4 address 37
maximum-level aggregation identifier 37
maximum transfer unit (MTU) 37
mean opinion score (MOS) 54
media access control (MAC) 3, 95, 312
media gateway control protocol (MGCP) 52
medium access control 37
MEGACO 90, 108
  DTMF tones 90
MG at the PSTN side 98
MG at the universal terrestrial access network (UTRAN) side 98
MGCF 98, 100
MGCP 52
MGCP/MEGACO 79
Microsoft's defense-in-depth model 196
MIDCOM 216, 235
  ALG 237
  architectural framework for middleboxes 240
  filter 239
  firewall 237
  MIDCOM agent 238, 241, 246
    end-hosts as in-path MIDCOM agents 242
    MIDCOM confidentiality 243
  MIDCOM agent registration 239
  MIDCOM framework 244
    in-path SIP proxy 245
    middlebox implementing firewall service 245
    NAPT and firewall 249
    NAPT service 247
    SIP phone 244
  MIDCOM PDP functions 243
    IPsec encapsulating security payload (ESP) 243
    IPsec IP authentication header (AH) 243
    MIDCOM authentication 243
    registration and deregistration of MIDCOM agents 244
  MIDCOM policy decision point (PDP) 238
  MIDCOM protocol 240
  MIDCOM session 239

middlebox 236
middlebox communication (MIDCOM) protocol 239
multiple MIDCOM sessions 251
NAT 237
policy action (or) action 239
policy rule(s) 239
proxy 237
signaling 251
timers on middlebox 251
MIDCOM policy decision point (PDP) 238
middlebox 199, 236
middlebox communication (MIDCOM) 193
middlebox communication (MIDCOM) protocol 239
millions of instructions per second (MIPS) 70
MIP foreign agent (FA) 101
mixed-excitation linear predictive (MELP) 77
mobile IP (MIP) 94
mobile IPv6 15
mobile switching center (MSC) server 98
MOS 58, 59, 76
MPLS 18, 50–51
MTU 37
multicast 8, 10, 37, 39, 41–43
multicast address 37
multicast group 37
multicast IPv4 tunnel 37
multicast listener discovery (MLD) 10
multilayer IP-security (ML-IPsec) 321
multimedia applications 8
  real-time control protocol (RTCP) 8
  real-time transport protocol (RTP) 8
multiple system operators (MSOs) 54
multiplexing 52
multipoint control unit (MCU) 72
multipulse excitation vocoder 75
multistage extraction 69

N

name resolution 37
NAPT 209
  small office/home office (SOHO) 209
NAT 197, 202, 204, 237, 252, 319
  address binding 211
  address lookup and translation 211
  address unbinding 211
  arrangement 8
    IPSec 8
    kerberos 8
    multicast 8
    multimedia 8
  basic NAT 208
  definition 206
  limitations 214
    ARP responses to NAT mapped global addresses 215

privacy and security 214
  translation of outbound TCP/UDP fragmented packets in
    NAPT setup 215
NAPT 209
  small office/home office (SOHO) 209
packet translations 211
  checksum 212
  DNS support 213
  FTP support 213
  ICMP error packet modifications 213
  partitioning of local and global addresses 213
  private address space recommendation 214
problems 193
NAT RFC bibliography 44
routing across NAT 214
switch-over from basic NAT to NAPT 214
traditional NAT 207
NAT traversal 89
National Cable & Telecommunications Association
    (NCTA) 54
ND 37
neighbor 37
neighbor discovery (ND) 10, 37, 299, 326
  address autoconfiguration 300
  address resolution 300
  duplicate address detection 300
  mechanisms 335
  neighbor unreachability detection 300
  next-hop determination 300
  options 37
  parameter discovery 300
  prefix discovery 300
  redirect function 300
  router discovery 300
neighbors cache 37
network addresses translator 37
network address port translation (NAPT) 2
network address port translators (NAPT) 205
network address translation (NAT) 2
network prefix 37
network segment 37
next-level aggregation identifier (NLA ID) 38
next hop obtaining 37
next level aggregator (NLA) 307
NLA ID 38
no-broadcast multiple access link 38
node identifier (address) 5
noiseless 55
nonline-of-sight (NLOS) 95
not specified address 38
NTT DoCoMo 104
numbering 91
  E.164 92
  ENUM 91, 93
  telephone numbers 92

O
OAM&P 53
one-way system delay 72
operations, administration, maintenance, and provisioning
    (OAM&P) 50
own link 38

P
packet 38
packet loss 54
packet over synchronous optical network (POS) 18
parameters discovery 38
parametric vocoders 64
path's MTU discovery 38
path determination system 38
path MTU 38
path vector 38
PBXs 21, 24, 104
  hybrid TDM and IP systems 25
  IP trunking 25
  TDM trunking 25
  traditional PBXs with IP adjunct extensions 24–25
PCM (pulse code modulation) 55
  compact disc (CD) 56
PDU 38
physical (PHY) layer 95
pitch 65
plug-and-play 3–6, 12
point-to-point protocol 38
prefix-length notation 38
prefixes list 38
presence 167
  abstract model for a presence and instant messaging 168
    ACCESS RULES 174
    ADMINISTRATOR 177
    CLOSED 174
    COMMUNICATION ADDRESS 174
    COMMUNICATION MEANS 174
    CONTACT ADDRESS 174
    DELIVERY RULES 174
    DOMAIN 177
    ENTITY 177
    FETCHER 174
    FIREWALL 177
    IDENTIFIER 177
    INBOX USER AGENT 174
    INSTANT INBOX 174
    INSTANT INBOX ADDRESS 174
    INSTANT MESSAGE 174
    INSTANT MESSAGE PROTOCOL 174
    INSTANT MESSAGE SERVICE 170, 174
    INSTANT MESSAGING SYSTEM 173
    INTENDED RECIPIENT 177
    NAMESPACE 177

NOTIFICATION 175
OPEN 175
OTHER PRESENCE MARKUP 175
OUT OF CONTACT 177
POLLER 175
PRESENCE INFORMATION 175
PRESENCE PROTOCOL 171, 175
  presence tuple 171
  PRESENCE TUPLES 172
PRESENCE SERVICE 168, 169, 175
  FETCHERS 169
  NOTIFICATION 170
  SUBSCRIBERS 169
  varieties of WATCHER 169
  WATCHERS 169
PRESENCE TUPLE 175
PRESENCE USER AGENT 175
PRESENTITY 175
PRINCIPAL 176
PRINCIPALS 172
PROXY 176
SENDER 176
SENDER USER AGENT 176
SERVER 176
SPAM 176
SPOOFING 176
STALKING 176
STATUS 176
SUBSCRIBER 176
SUBSCRIPTION 176
SUCCESSFUL DELIVERY 177
VISIBILITY RULES 176
WATCHER 176
WATCHER INFORMATION 176
WATCHER USER AGENT 176
INSTANT MESSAGES 180
  common message format 180
  reliability 181
protocol requirements 177
  access control 178
  namespace 178
  network topology 179
    authentication 179
    message encryption 179
  presence caching and replication 180
  presence lookup and notification 180
  scalability 178
SIP applications 181
  edge presence server 182
  learning presence state
    co-location 189
    presence documents 190
    REGISTER 190
  presence agent (PA) 182
  presence event package 185

aggregation 188
authentication 186
authorization 186
DIAMETER 186
forked requests 188
migration 189
NOTIFY bodies 185
NOTIFY requests 187
S/MIME 187
state agents 188
SUBSCRIBE bodies 185
SUBSCRIBE requests 186
subscription duration 185
presence server 182
presence URIs 184
presence user agent (PUA) 182
SUBSCRIBE 183
private branch exchanges (PBX) 4
protocol data unit (PDU) 38
proxy 199
proxy firewall 199
proxying 200
proxy servers 87, 199
proxy services 200
pseudo-header 38
pseudo-periodic 38
PSTN 56, 91, 93
  numbering 91
PSTN-to-IP gateways 88
public switched telephone network (PSTN) 18, 108
pulse code modulation (PCM) 56
pure-play VoIP carriers 17

Q
Q.931 78
QoS 96
quality of service (QoS) 1, 51
quantization error 60
quantization noise 60

R
RADIUS 52
RADIUS (remote access dial-in user service) 194
RAS 78
RAS control (gatekeeper) 82
real-time control protocol (RTCP) 8
real-time streaming protocol (RTSP) 52
real-time transport protocol (RTP) 8, 52, 108, 218
reassembing 38
redirect 38
redirect server 87
regional Internet registries (RIRs) 14
REGISTER 113, 124
registering with gatekeeper (RAS) 84

registrar server 87
relay router 6to4 38
remote authentication dial-in user service (RADIUS) 52
request header fields 122
residential voice service expenditures 4
resilient packet ring (RPR) 18
response header fields 122
restricted zone (RZ) 195
reverse address resolution protocol (RARP) 335
RFC 2893 323
roaming 94
router 39
router's cache 39
router advertisement 39
routers discovery 39
routing loop 39
RTCP 78
RTP 52, 78
RTSP 52, 78

S
S/MIME 187
scope 39
scope ID 39
second-generation VoIP networks 20
  evolution paths for 2G deployments 27
secure(d) zone (SZ) 195
secure real-time transport protocol (SRTP) 52
secure sockets layer (SSL) 88
serverless autoconfiguration 3, 12
service data units (SDU) 95
service provider (SP) 15
services over IP (SoIP) 18
session border controller (SBC) 193
session initiation protocol (SIP) 4, 52, 85, 107
  address-of-record 115
  back-to-back user agent 115
  call 115
  call leg 115
  call stateful 115
  capabilities 140
    OPTIONS 140
    processing of OPTIONS request 141
  client 115
  conference 115
  core 115
  dialogs 116, 142
    call-ID 142
    requests within a dialog 144
      generating the request 144
      processing the responses 145
      UAC behavior 144
      UAS behavior 146
    UAC behavior 143
    UAS behavior 143

downstream 116
E.164 ISDN-era addressing scheme 107
final response 116
framing SIP messages 123
fundamental SIP functionality 108
header 116
header fields 116
  call-ID 110
  contact 110
  CSeq (command sequence) 110
  from 110
  max-forwards 111
  to 110
  via 110
header field value 116
home domain 116
informational response 116
initiating a session 147
  INVITE 147–148
  session description protocol (SDP) 148
  UAC processing 147
    1xx responses 149
    2xx responses 149
    3xx responses 149
    4xx, 5xx and 6xx responses 149
  UAS processing 150
    INVITE 150
    INVITE accepted 151
    INVITE redirected 151
    INVITE rejected 151
initiator, calling party, caller 116
invitation 116
INVITE 111–114
invitee, invited user, called party, callee 116
location server 87
location service 116
loop 116
loose routing 116
MEGACO 108
message 116
method 116
modifying an existing session 152
  UAC behavior 152
  UAS behavior 153
NAT traversal 89
outbound proxy 117
parallel search 117
provisional response 117
proxy, proxy server 117
proxy behavior 155
  proxy route processing 156
    rewriting record-route header field values 159
    SIP trapezoid 157
    traversing a strict-routing proxy 158
  stateful proxy 156

proxy server 87
public switched telephone network (PSTN) 108
real-time transport protocol (RTP) 108
recursion 117
redirect server 87, 117
REGISTER 113
registrar 117
registrar server 87
registrations 136
  REGISTER request 137
    processing REGISTER requests 138
    REGISTER 139
    request-URI 137
    request-URI 136
regular transaction 117
request 117
response 117
ringback 117
route set 117
sequential search 117
server 117
session 117
SIP elements 114
  UAC 114
  UAS 114
SIP environment 86
  gateway 86
  location server 86
  redirect server 86
  registrar server 86
SIP header fields 120
SIP message body length 123
SIP message body type 123
SIP messages 87, 119
  SIP requests 119
SIP responses 120
SIP session setup 109
SIP transaction 117
spiral 118
stateful proxy 118
stateless proxy 118
strict routing 118
structure of the SIP protocol 114
  syntax and encoding 114
  transaction layer 114
  transaction user (TU) 114
  transport layer 114
STUN (simple traversal of UDP through network address
    translators) 89
target refresh request 118
terminating a session 153
  terminating a session with a BYE request 154
transactions 160
  client transaction 161
    sending server responses 164

server transaction 161
transaction user (TU) 118
transport in SIP 162
  error handling 165
  framing 165
  receiving responses 163
  receiving server requests 164
  sending requests 162
UAC core 118
UAS core 118
upstream 118
URL-encoded 118
user agent (UA) 118
user agent behavior 123
user agent client (UAC) 86, 118
user agent server (UAS) 86, 118
signaling 54, 78
  signaling protocols 78
    call control and signaling
      H.225 78
      Q.931 78
      RAS 78
      SIP 78
    H.225 78
    H.245 78
    media
      RTCP 78
      RTP 78
      RTSP 78
    Q.931 78
    RAS 78
    RTCP 78
    RTP 78
    RTSP 78
    signaling and gateway control
      UDP 78
    SIP 78
    UDP 78
signaling gateway (SG) 98
silence compression algorithms 72
simple traversal of user datagram protocol through network
    address translators (STUN) 193
SIP 52, 78, 79, 88
SIP border gateways 253
  access border gateway function (A-BGF) 255
  gateway/session border controller example 256
  interconnect border control function (IBCF) 254
  interconnect border gateway function (I-BGF) 255
  interworking function (IWF) 255
  manageability 254
  proxy-call session control function (P-CSCF) 255
  scalability 254
  security 254
SIP elements 114
SIP environment 86

SIP header fields 120
  header field classification 122
    request header fields 122
    response header fields 122
  header field rows 121
SIP message body length 123
SIP message body type 123
SIP messages 87, 119
  SIP requests 119
    request-URI 119
    SIP-version 119
SIP phone 244
SIP proxy 216
SIP responses 120
SIP RFC bibliography 45
SIP session setup 109
site-level aggregation identifier (SLA ID) 39
site level aggregator (SLA) 307
site prefix 39
small office/home office (SOHO) 209
solicited-node address 39
spectrum 96
speech analysis 64
speech digitization methods 55
  PSTN 56
  pulse code modulation (PCM) 56
  vocoding 55
    adaptive subband coding 55
    hybrid multipulse linear predictive coding 55
    linear predictive coding (LPC) 56
    mathematical analysis 56
    pulse code modulation (PCM) 56
    stochatically-excited linear predictive coding (LPC) 55
  waveform coders 55
    ADPCM (adaptive differential PCM) 55
    PCM (pulse code modulation) 55
speech encoding methods 55
  adaptive differential pulse code modulation (ADPCM) 56
  G.711 speech coder 58
  G.722 speech coder 58
  G.723.1 dual rate speech coder with annex A 58
  G.726 ADPCM waveform coder 59
  G.729 with annex-B CS-ACELP voice coder 59
  GSM-FR speech coder 59
  ITU-T standards related to voice coding 57
    G.711 57
    G.712 57
    G.720 58
    G.722 58
    G.722.1 58
    G.722.2 58
    G.723.1 58
    G.724 58
    G.725 58
    G.726 58

G.727 58
G.728 58
G.729 58
linear predictive coding (LPC) 56
lossless 55
noiseless 55
PSTN 56
pulse code modulation (PCM) 56
quality of coding schemes 74
waveform coding 60
  adaptive quantization 61
  analog-to-digital (A/D) 60
  logarithmic quantization 61
  uniform quantization 61
speech processing delay 72
speech synthesis 64
SRTP 52
stateful configuration 13
stateful inspection 194
stateless autoconfiguration 12
static routing 39
statistical TDM (STDM) 50
stochatically-excited linear predictive coding (LPC) 55
structure of the SIP protocol 114
  syntax and encoding 114
  transaction layer 114
  transport layer 114
STUN (simple traversal of UDP through network address translators) 89, 215
  binding acquisition 229
    UDP bindings 229
  binding lifetime discovery 227
  binding requests 220
    MESSAGE-INTEGRITY 222
  binding response 221
    MESSAGE-INTEGRITY 222
  discovery 223
  discovery process 226
  formulating the binding request 225
  full cone 217
  obtaining a shared secret 224
  port restricted cone 217
  processing binding responses 225
  restricted cone 217
  shared secret requests 222
    PASSWORD attribute 223
  STUN client 217–218
  STUN message 230
  STUN message attributes 231
  STUN message header 230
  STUN message overview 219
  STUN server 217
  symmetric 217
  use cases 226
subnet anycast router address 39

subnetwork 39
subnetwork associated path 39
suitable path selection 39

T
TCP/UDP (transmission control protocol/user datagram
    protocol) 206, 210
TCP ports 200
TDD 103
TDM 1–2, 4, 18, 20, 22, 25, 31, 102
TDM networks 18
TDM switching 88
TDM trunking 25
telephone numbers 92
telephony routing over IP (TRIP) 52
third-generation 3G VoIP networks 28
  "presence" features 29
  3G cellular systems 28
  converged system 28
  enterprise deployment strategies 29
  wireless LAN (WLAN) 28
Third-Generation Partnership Project 2 (3GPP2) 97, 101
  authorization, authentication, and accounting (AAA) 101
  home agent (HA) 101
  MIP foreign agent (FA) 101
third-generation Wi-Fi 16
Third Generation Partnership Project (3GPP) 97, 100
  all-IP 3G cellular service 99
  call state control function (CSCF) 98, 100
  home subscriber server (HSS) 98
  MG at the PSTN side 98
  MG at the universal terrestrial access network (UTRAN)
      side 98
  MGCF 98, 100
  mobile switching center (MSC) server 98
  signaling gateway (SG) 98
time division multiplexing (TDM) 1
TLA ID 39
top-level aggregation identifier (TLA ID) 39
top level aggregator (TLA) 307
traditional interexchange carrier (IXC) 52
transaction user (TU) 114
transition 39
transport in SIP 162
transport layer security (TLS) 88, 186
TRIP 52
triple-play applications 1
triple-play carriers 17
triple play 50
tunnel 39

U
U.S. Dept. of Defense (DoD) 4
U.S. Government Accountability Office (GAO) 4

UDP 78
UMTS release 5 basic architecture 101
uncontrolled zone (UZ) 195
unicast address 40
unified messaging 167
uniform quantization 61
Universal Mobile Telecommunication System (UMTS) 97,
    100, 102
  UMTS release 5 basic architecture 101
URL filtering 194
user agent behavior 123
  canceling a request 134
    CANCEL 134–135
    server behavior 135
  redirect servers 133
  user agent client (UAC) 123
    UAC behavior 124
  user agent server (UAS) 123
user agent client (UAC) 123
  processing responses 128
    processing 3xx responses 128
    processing 4xx responses 129
    transaction layer errors 128
    unrecognized responses 128
  sending the request 127
  UAC behavior 124, 143
    call-ID 125
    CSeq 126
    max-forwards 126
    request-URI 124
      REGISTER 124
    via 126
user agent clients (UAC) 86
user agent server (UAS) 86, 123
  UAS behavior 130, 143
    applying extensions 132
    content processing 132
    generating the response 132
      sending a provisional response 132
    header inspection 130
      merged requests 131
      require 131
    processing the request 132
    stateless UAS behavior 133

V
video-on-demand (VOD) 1
virtual local area network (VLAN) 15
VLAN 202
  VLAN segments 203
vocoder 64
  adaptive postfilter 75
  CELP vocoder 68, 75
    CELP synthesis model 71

G.729 70
  low-complexity CELP 69
channel vocoder 64
CS-ACELP (conjugate-structure algebraic-code-excited
    linear-prediction) 76
  ITU G.729 76
error weighting filter 75
formant vocoder 64
G.723.1 71
G.729A 71
GSM-FR speech coder 77
ITU-G.729 71
linear prediction analysis-by-synthesis (LPAS) 74
linear prediction analysis by synthesis coding 74
low-delay code-excited linear prediction vocoder
    standard 76
  ITU G.728 76
LPC vocoders 64
mixed-excitation linear predictive (MELP) 77
multipulse excitation vocoder 75
parametric vocoder 64
vocoder attributes 71
  bit rate 71
vocoder attributes 71
  bit rate 71
    silence compression algorithms 72
      comfort noise generation (CNG) 72
      voice activity detector (VAD) 72
  delay 72
    algorithmic delay 72
    bridging delay 72
    frame delay 72
    look-ahead 72
    one-way system delay 72
    speech processing delay 72
vocoding 55
  code excited linear prediction (CELP) 67
  excitation energy 65
  generic audio encoder 64
  linear predictive coding (LPC) 65
  millions of instructions per second (MIPS) 70
  multistage extraction 69
  pitch 65
  speech analysis 64
  speech synthesis 64
  standards for low bit rate vocoding 69
  VoWi-Fi 63
vocoding frequency domain 63
voice activity detector (VAD) 72
voice digitization and encoding 54
  jitter 54
  mean opinion score (MOS) 54
  packet loss 54
  signaling 54
voice over asynchronous transfer mode (VoATM) 49

voice over frame relay (VoFR) 49
voice over Internet protocol (VoIP) 1
  1G commercial VoIP 19
  ATM 18
  circuit-switched 18
  E911 services 19
  enterprise VoIP deployment approaches 21
  environment 19
  hybrid environments 26
  integrated services digital network (ISDN) 18
  integrated voice/data LAN 18
  MPLS 18
  packet over synchronous optical network (POS) 18
  PBXs 21, 24
    hybrid TDM and IP systems 25
    traditional PBXs with IP adjunct extensions 24–25
  penetration trends (market information) 20
  platforms 21
  public switched telephone network (PSTN) 18
  pure-play VoIP carriers 17
  resilient packet ring (RPR) 18
  second-generation VoIP networks 20
    evolution paths for 2G deployments 27
    hybrid TDM and IP systems 22
    pure IP server-based telephony systems 22, 26
    traditional PBXs with IP adjunct extensions 22
    VoIP trunking only 22
  services over IP (SoIP) 18
  TDM networks 18
    TDM trunking 25
  third-deneration 3G VoIP networks 28
    3G cellular systems 28
    converged system 28
    enterprise deployment strategies 29
    presence features 29
    SIP 28
    wireless LAN (WLAN) 28
  triple-play carriers 17
  VoIP trunking 23
  VoWi-Fi 21
voice over IP (VoIP) 49
  cable TV environments 54
    multiple system operators (MSOs) 54
  common open policy service (COPS) 51
  COPS 51
  GDOI 51
  group domain of interpretation (GDOI) group keying 52
  media gateway control protocol (MGCP) 52
  MGCP 52
  RADIUS 52
  real-time streaming protocol (RTSP) 52
  real-time transport protocol (RTP) 52
  remote authentication dial-in user service (RADIUS) 52
  RTP 52
  RTSP 52

secure real-time transport protocol (SRTP) 52
session initiation protocol (SIP) 52
SIP 52
SRTP 52
telephony routing over IP (TRIP) 52
TRIP 52
voice over multiprotocol label switching (VoMPLS) 49
voice over packet (VoP) 1, 49
coders in a VoP application 55
MPLS 50
operations, administration, maintenance, and provisioning
(OAM&P) 50
statistical TDM (STDM) 50
triple play 50
voice revenues 50
voice over X.25 (VoX25) 49
VoIP 4
corporate VoIP arrangement 201
firewall issues for VoIP 204
VoIP protocol stack 204
VoIP and wireless networks 94
VoIP environment 19
VoIP in 3G cellular networks 96
VoIP platforms 21
VoIP protocol stack 204
VoIP trunking 23
VoIPv6 positioning 293
VoP/VoIP 51
ATM 50
firewall/DMZ architecture 51
MPLS 50, 51
operations, administration, maintenance, and provisioning
(OAM&P) 50
quality of service (QoS) 51
speech digitization methods 55
speech encoding methods 55
lossless 55
noiseless 55
statistical TDM (STDM) 50
voice revenues 50
VPNs 51

VoWi-Fi 21, 63, 94
VoWi-Fi environment 204
VPN (virtual private network) 51, 235

W
waveform coding 60
adaptive quantization 61
analog-to-digital (A/D) 60
feedback adaptive PCM) 62
feed forward adaptive PCM 62
linear prediction 62
uniform quantization 61
web-based services 105
Wi-Fi 96, 102
Wi-Fi-enabled mobile phones 105
Wi-Fi services 15
third-generation Wi-Fi 16
WiMAX 95–96, 102–103
wireless LAN (WLAN) 28, 94, 104
wireless networks 94
wireless VoIP service offering dynamics 102
3G operators 102
1xEV-DO 103
CDMA EVDO 103
EV-DV (evolution-data and voice) 103
WiMAX 103
Internet VoIP 105
Wi-Fi-enabled mobile phones 105
WLAN 94, 104

X
XML firewall 200